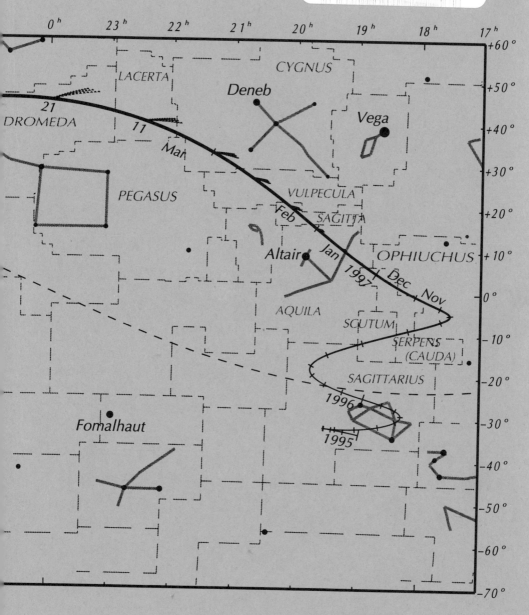

0ʰ   23ʰ   22ʰ   21ʰ   20ʰ   19ʰ   18ʰ   17ʰ

+60°
+50°
+40°
+30°
+20°
+10°
0°
−10°
−20°
−30°
−40°
−50°
−60°
−70°

CYGNUS

Deneb

Vega

LACERTA

21

DROMEDA   11   Mar

PEGASUS

VULPECULA

Feb   SAGITTA

Altair   Jan   1997   OPHIUCHUS

AQUILA   Dec   Nov

SCUTUM

SERPENS
(CAUDA)

SAGITTARIUS

1996

Fomalhaut

1995

COMET
OF THE
CENTURY

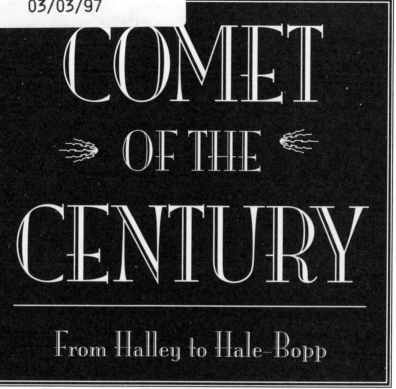

COMET
OF THE
CENTURY

From Halley to Hale-Bopp

Fred Schaaf

With Illustrations by Guy Ottewell

COPERNICUS
AN IMPRINT OF SPRINGER-VERLAG

© 1997 Springer-Verlag New York, Inc.

Published in the United States by Copernicus, an imprint of Springer-Verlag New York, Inc.

Copernicus
Springer-Verlag New York, Inc.
175 Fifth Avenue
New York, NY 10010

Library of Congress Cataloging in Publication Data

Schaaf, Fred.
Comet of the century : from Halley to Hale-Bopp / Fred Schaaf.
p.   cm.
Includes bibliographical references.
ISBN 0-387-94793-0 (hardcover : alk. paper)
1. Comets.   I. Title.
QB721.S365   1996
523.6—dc20   96-26567

Printed in the United States of America.
Printed on acid-free paper.

9 8 7 6 5 4 3 2 1

ISBN 0-387-94793-0   SPIN 10539221

To the late Dennis Milon, co-discoverer of 1975's
Comet Kobayashi-Berger-Milon and one of the truly great
photographers of the heavens. May his work live on
and be seen by whole new generations of sky-lovers.

CONTENTS

THE LATEST GREATS

PREFACE

This book may be loaded with a richness of facts, but what it's about, through and through, is simple and blazing: the way comets, throughout history, and up to the present, have stirred the human spirit.

What does a natural phenomenon need to have in order to stir the spirit? One answer might be as follows: mystery, beauty, and danger. And if that answer is correct, then comets are preeminently equipped to stir the spirit.

Comets are the most mysterious objects in our solar system. As recently as 50 years ago, astronomers were not even close to knowing what the heart of a comet really is or where the seemingly endless supply of new comets was coming from. Mystery: Comets are the astronomical objects most capable of surprises and most likely to contain secrets of the solar system's birth. There's nothing more mysterious than birth and death, and comets may be the greatest purveyors of both that life on Earth has known. Scientists believe that much of Earth's water and even many of the building blocks of life—the organic molecules—may have been brought to Earth by comets. Thus comets may have played an essential role in the birth of life on our planet.

As for the danger of comets—that's where the death comes in. Some of the greatest mass extinctions in Earth's history must have been caused by collisions with comets. The destruction of the dinosaurs (and, along with them, about three-quarters of all species on Earth) was accomplished, we are now almost certain, by the aftereffects of an impact from either an asteroid or (perhaps more likely) a comet. Comets themselves flirt with the danger of their own deaths, especially when they do what no other objects in the universe can: They scrape the very surface of the Sun and, sometimes, escape. But it's our own deaths with which we are naturally more concerned. Thus you'll be interested to learn, in Chapter 7, what has been called "the single most dangerous object known to humankind." (You'll also learn how backyard skywatchers saw this object just a few years ago and how you can see the spectacular debris from it on a few August nights every year.)

Do comets have beauty? If you happened to get a good look—far from city lights on a clear night—at Comet Hyakutake in the spring of 1996, you don't have to ask. Comets can shine in gold and blue (with touches of red, green, and even orange). They can appear as puffs of bright-hearted glow with ever-changing tails cast across whole constellations, even across enormous sections of the sky. Their subtleties (which is all the faint comets have) are beautiful too, at least to the knowledgeable observer. Let me tell you about beauty: that's the word you use to describe something golden-headed and fountain-tailed that moves night by night majestically out of the orange dawn and into the starry heavens (in this case, the beautiful also has a name: Comet West, which you'll meet through my eyes in the first chapter of this book).

This book has been a lot like a comet. Its core lay dormant for a long time but then was energized by some news that you too may have heard: There is a potentially very great comet coming to the skies of the world in 1996, to achieve its peak of grandeur in the spring of 1997. It is Comet Hale-Bopp, named after the two amateur astronomers who discovered it far beyond the orbit of Jupiter in the summer of 1995.

A few astronomers and science writers have already thrown caution to the wind and written that they expect Hale-Bopp to be the "comet of the century." As you'll see, the position that I (and the vast majority of astronomers) take is that Hale-Bopp merely has the potential. In fact, the title of this book is meant to have a touch of irony. Several times in recent decades, objects that were predicted to be the "comet of the century" proved to be tremendous disappointments. The impulse to wish for a spectacle is strong, but there is never an absolute guarantee that an object as capricious as a comet will comply with our wishes. Of course, I don't want to be negative either. After you've read this book, you'll know that even if we can't be sure how a particular comet will perform, some comets deliver a visual majesty that can satisfy even the most ardent longing.

If you're interested in learning about our potential "comet of the century," Hale-Bopp, you'll find a wealth of information about it here, especially in some of the final chapters.

But this book is really about many candidates for that title in many centuries. The superlatives stirred by these comets gush out from their descriptions like their own millions of miles of spouted, glowing, and structured tails.

This book is also about some other comets that, though less visually entertaining, are exciting because of what they do.

What can comets do? They are the ultimate mavericks of the solar system. They can outshine the Full Moon, stretch their tails across most of the sky, become larger than the Sun in appearance and reality, crash into Jupiter with the biggest blasts ever witnessed by human eyes in our solar system, suddenly brighten by a thousand times, eject overnight a tail millions of miles long and grow a new one back just as quickly, almost completely reverse direction in hours while traveling at speeds in excess of a million miles an hour. Comets are the very essence of prodigious size and brightness, unexpectedness, headlong flight, freakish behavior, and mysterious significance. And now is the time to meet all those things full-force, in the great comets of the past and in the prospect of an impending "comet of the century."

Acknowledgments

My first debt of gratitude for this book is owed to my friend of now exactly 20 years, Guy Ottewell. He and I wrote the Halley book *Mankind's Comet* in 1985, and parts of this current book are based closely on the research and writing we did for that earlier work. Guy has been generous in letting me use several diagrams and several quotes from *Mankind's Comet*. Finally, of course, I wish to thank him for all the new, unique, and creative diagrams he has contributed to *Comet of the Century*. These help bring back to life the spectacular comets of the past, and both figuratively and literally add a whole additional dimension to the book.

After the discovery of Comet Hale-Bopp in the summer of 1995, the idea for this book took on new prominence in my mind. But the book would never have been launched were it not for the encouragement and support of my editor, Bill Frucht. It's always been a pleasure to talk with Bill but working with him has demonstrated to me even more than I knew before his wide range of interests and his commitment to intellectual and scholarly endeavor. His belief in this challenging project and in my abilities got the book rolling and kept it on track. He promised a lot—and then proceeded to deliver on every promise.

A book like *Comet of the Century* has a lot of parts, both verbal and visual, and orchestrating them—on a very tight schedule!—has required a skilled and knowledgeable production editor. In this, I could not have done better than to have gotten Steven Pisano. I am tremendously appreciative of Steve's diligent work and his expertise. He has demonstrated an unflagging commitment to this book and to excellence which is admirable—and for which I am extremely grateful.

My thanks also go to those great scientists and gentlemen Donald Yeomans and William Liller; to the Library of Congress's priceless Ruth Freitag; and to my friend Steve Albers for advice and a very special appendix. I also wish to thank all the photo contributors for their masterful and stirring work. A special thanks goes to photo contributors Ray Maher and Paul Ostwald, and to other members of the South Jersey Astronomy Club, who have in recent years helped make my own personal quest for comets more inspiring and enjoyable.

OVERVIEW

New concepts and special terms in this book are explained or defined when first mentioned in the text, and terms are defined again in the glossary. Nevertheless, you may want to read the following overview.

Not So Long, Long Ago, in a Galaxy Not So Far, Far Away (You're in It) . . .

. . . Comets mystified human beings.

As they observed them and pondered what they had seen, however, humans eventually came to understand that there was a solid, more lasting part of comets that they called the *nucleus*.

Humans figured out that this nucleus of ice was the center and source of the big glowing cloud of gas and dust that, together with the nucleus, formed the *head* of a comet.

They noticed that the head grew as a comet began to get closer to the Sun, and they speculated that some kind of ice in the nucleus must be turning into the vapor of the cloud under the influence of the Sun's heat.

They discovered that the majority of comets spend most of their time far out among the outer planets, or even farther away from the Sun, and then hurtle inward to closest approach to the Sun, a position that was called *perihelion*.

Humans did something else, last but not least: They marveled at the magnificent *tail* that was often seen to sprout from a comet's head, a tail that always pointed away from the Sun.

How could we humans keep track of comets, measure them, and learn more about them? Astronomers, the people who studied comets and the rest of the universe, needed to note positions and sizes of things in the sky. To make this task easier, they invented *declination*, which is a projection of Earth's lines of latititude into the sky (thus a "north celestrial pole" is right over Earth's geographic north pole) and *right ascension* (R.A.), which is a projection of Earth's lines of longitude into the sky.

Comets can appear at any declination and at any right ascension, anywhere in the heavens.

Astronomers measured the length of the things like the Big Dipper or a comet's tail as angles that were fractions of the full circle we get when we draw a line from one horizon to overhead, to the opposite horizon, and then all the way under Earth to the original horizon. There are 360 *degrees* (°) in the full circle around the heavens above and below our feet (the Big Dipper is almost 30 degrees long and could therefore stretch almost one-third the way from horizon to overhead). There are 60 *minutes of arc* (') in each degree (the Sun and Moon both appear about 30' wide). There are 60 *seconds of arc* (") in each minute of arc (in a telescope, the planet Jupiter always appears more than 30" wide, which is more than 1/60 the apparent diameter of the Sun or Moon).

Comets have central concentrations of light a few seconds of arc across or smaller and tails 90 degrees long (halfway across the sky!) and longer.

Finally, astronomers needed a system for rating the brightness of astronomical objects in the sky. They divided all the stars that were visible to the naked eye (that is, the unaided eye) into six classes of brightness, each class of brightness being called a *magnitude*. A "first-magnitude" star was in the first class of brightness (brighter than the other classes). A "sixth-magnitude" star was in the sixth and dimmest class of brightness. In modern times, astronomers had to get more precise and say that a star of magnitude 1.0 is exactly 100 times brighter than a star of magnitude 6.0. Objects faint enough to require viewing through a telescope are seventh magnitude, eighth magnitude, and so on—the fainter the object, the higher the magnitude number. The very brightest stars and planets can be 0 in magnitude or even have negative-number magnitudes such as −1.4 (the brightness of Sirius, the brightest star) or −4 (the brightness of

Venus, the brightest planet). The full Moon shines at about –12.7. The Sun shines at a literally blinding magnitude –27.

And comets? They vary in brightness: from dimmer than magnitude 30 (about the faint limit of the Hubble Space Telescope) to brighter than magnitude –15 (many times brighter than the full Moon).

Some of this sounds pretty incredible. But it's all true. And now you know the basic terms and concepts that astronomers need to try to corral comets. Comets are mavericks that are not so easily broken, though. As you'll see in the chapters that follow, they do in fact have a tendency to take us on some breakneck journeys of wonder.

Better hang on for the ride.

About the Diagrams in This Book

ORBITAL DIAGRAMS   In all the orbital diagrams in this book, the "ecliptical plane" is the plane of Earth's orbit. When a comet's orbit is above (north of) that plane, the line for it is boldface; when the orbit is below (south of) that plane the line for it is of standard lightness. The comet's position is indicated on certain dates (usually the start of a month) and is found at the top of a vertical boldface line (a "stalk") to indicate how far above (north of) the ecliptic plane the comet is, or bottom of a standard-lightness line to show how far below (south of) the ecliptic the comet is. The ecliptic plane is divided up into a gridwork of squares 1"a.u."—1 standard Sun-Earth distance (about 93 million miles)—on a side. A generalized idea of the length and orientation of the comet's dust tail is given by the tail representation shown for various dates (usually the start of each month).

SKY-PATH DIAGRAMS   In all the sky-path diagrams in this book, a generalized idea of the apparent length and orientation of the comet's dust tail is given by the tail representation shown for various dates (usually the start of each month).

COMETS IN PERSON

Chapter 1
A COMET PASSES

What is it like to be passed, in the cold and the still of the night, by a ghost 10,000 times the size of the Earth?

That is a question which can be answered only by someone who has looked upon a great comet. I can think of no better way to introduce the majesty and wonder of comets than by describing my own experience of such sightings. I will supply more facts and scientific observations and lore about comets later in this book. But this first chapter is a tale of my recollections of the hopes I had, the efforts I made, and some of the thrilling moments I experienced in pursuing the remarkable comets of the past 30 years. The tale will also serve as an overview of some of the things comets are or can be.

My Missed Sungrazer

If someone had held me up as an infant to see the two fine comets of 1957 and made a special attempt to help me remember them—as many parents did with Halley's Comet in 1910, and as the father of the famed astronomer Kepler did with the Great Comet of 1577—I might have those eerie and beautiful forms in the sky as a hazy but potent memory. That did not happen, however: I do not remember the comets Arend-Roland and Mrkos. And although my science-fiction-reading mother and sister, and the advent of the space program, probably did influence me developing an early interest in astronomy and space, it was to be a long time before I saw my first comet.

I came close in 1965. Arend-Roland and Mrkos in 1957, and several comets of the early 1960s, had been very good comets. But in the autumn of 1965, the newspapers were bristling with headlines about the coming glory of a great comet. It had been discovered by two Japanese astronomers, Ikeya and Seki. It was due to outshine greatly the brightest stars, pass incredibly close to the Sun, and thus be stirred to produce a mighty tail. Later in life I was to learn that Comet Ikeya-Seki was indeed a member of that class of comets which are the most thrilling (if not necessarily the most rewarding) of all—it was a sungrazer.

Unfortunately, what I didn't know was that the trajectory of sungrazers generally favors viewers in the Southern Hemisphere. I had long since taught myself the constellations, had viewed several eclipses and meteor showers, and had even read rather complex books which included topics like the internal structure of stars—but I was still only 10 years old. Ikeya-Seki was viewed by expert observers even in broad daylight when it was virtually right beside the Sun—*Life* magazine ran a daytime photograph, taken in Hawaii, showing the comet's tail curving out from behind a rock which the photographer had used to block out the Sun. Ikeya-Seki was a magnificent sight for weeks in the Southern Hemisphere, where its brilliant tail stretched across a significant fraction of the sky. But in the northern United States, where I lived, the only hope of an easy view of a bright Ikeya-Seki was to catch it on one of just a few mornings, low in the east, just before dawn. Poor weather and lack of a view unobstructed by forest prevented this 10-year-old from glimpsing the awesome visitor.

Missing Ikeya-Seki was a big disappointment for me. I often wondered whether a really great view would have been possible at all from my latitude. I eventually learned from the director of the American Meteor Society, David Meisel, that he had glimpsed the tail of Ikeya-Seki sticking far up into the eastern sky on one of those October mornings in 1965, when he was driving down the wide-open and sparsely populated Delmarva peninsula—hardly any farther south than my New Jersey home. Other observers got good views from much farther north.

At least the next year I managed to see, just before a November dawn, the opening act of another rumored sky spectacle. That was the Leonid meteor shower of 1966, produced by the usually quite dim Comet Tempel-Tuttle. The astonishing Leonid display I saw, which included a brief meteor that rivaled a half-moon in brightness, was splendid in itself and never to be forgotten. But I sometimes have thought that seeing it was a sort of consolation for my having missed the great Ikeya-Seki.

The First Comet

My knowledge of astronomy continued to grow, but not until I was 13 did I obtain my first really useful telescope. The vast majority of comets never become bright enough to see without a telescope, and in the next several years there were only a few cometary targets I could have found with a little telescope like mine. Actually, however, I would have seen a comet with my $4^1/_4$-inch reflector (a telescope with a primary mirror $4^1/_4$-inches wide) in 1968 or 1969 if it hadn't been for another problem: I had no source of information about when and where to look for any old or new comets. Ikeya-Seki had been that rarity, a comet which got lots of publicity in the popular press.

Luckily, in the summer of 1969—the summer of the first manned Moon landing—I at last got a look at what was then the only major magazine for amateur astronomers, *Sky & Telescope*. Country rustic that I was, it took a trip to see a sister in Michigan and a visit to the Abrams Planetarium in East Lansing for me to find a copy of this magazine, which until then I had only read about in books. I was sold on it in a minute. My subscription order went off as soon as I got home.

Without *Sky & Telescope* I probably wouldn't have learned about the one good comet and the one great comet which graced the skies of early 1970.

My first comet was another named for discoverers who were Japanese amateur astronomers. This time they numbered three. (Three is the highest permissible number of independent discoverers whose names can be attached to a comet. If a fourth person makes an independent discovery of a comet before it is publicly announced, he or she is out of luck.)

I remember well going out in the fading January dusk to look for Comet Tago-Sato-Kosaka. I set up my telescope just a bit down the country road from our driveway and looked to the southwest to find the bright star called Fomalhaut. I scanned along from the star with my naked eye. Could I see anything unusual? After an expectant few minutes, as the sky further darkened, I suddenly caught sight of a glimmer that seemed larger and hazier than a star's pinpoint of light.

My pulse raced with an excitement that can perhaps be only partly explained. After almost a decade as a youthful skywatcher, what I was about to see was not just a new sight, but for me an entirely new *type* of sight in the heavens. Yet it was more than that. All comets are individuals, and changing, and temporary in their displays. I knew that in a short few nights, Comet Tago-Sato-Kosaka could be faded beyond the grasp of my telescope—and perhaps in a few weeks from now never seen again from Earth, unless human history continues for tens of thousands or more years (actually about 454,000 years in this comet's case!). After years of reading about the stupendously strange and beautiful—sometimes even terrifying—great comets of the past, I was finally glimpsing one of the lesser of that spectacular company's kin. Or was I? Only the telescope would confirm what this object was.

It took only a moment's look through the little wide-field "finderscope" on my main telescope to find out: The naked-eye blur of luminosity was a comet all right, a comet with a tail! I had found Tago-Sato-Kosaka (a name whose every syllable I now relished).

I cannot completely explain why my heart leaped up when I saw this sight and felt the satisfaction of detecting, all by myself, a comet. There was something which could not—and cannot—be explained about the thrill and sense of connection I felt in seeing this mysterious visitant light.

The Beauty of Comet Bennett

Perhaps only a devoted amateur astronomer seeing his or her first comet would get as excited as I did about the meaningful but only modestly bright flicker of Tago-Sato-Kosaka. But this object turned out to be a warm-up for a comet that I think would amaze and delight anybody who has ever taken even a moment's pleasure in looking at the night sky. The heading of the news note in *Sky & Telescope* that was intended to prepare readers for the coming object was deliciously restrained: "Bright Comet Bennett."

As it turned out, Comet Bennett put on a better show than anyone had expected. It had been discovered by another amateur astronomer, this time one from South Africa. And unlike Ikeya-Seki, this comet would be readily visible in the Northern Hemisphere. The United States, Europe, Soviet Union, China, and Japan would get Comet Bennett fairly

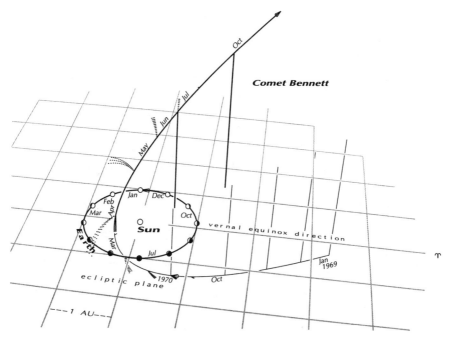

Orbit of Comet Bennett.

high in the sky in the weeks after it passed the Sun and might display its best brightness and tail development.

There is little doubt that Comet Bennett was a great comet. According to the criteria for greatness suggested by comet scientist Donald K. Yeomans (we'll discuss them later in this book), Comet Bennett was the first great comet easily and extensively visible from the United States since the legendary close return of Halley's Comet in 1910.

My first sight of Comet Bennett came as the comet was pulling away from the Sun. It was due to be low in the east in strong morning twilight: Would it really be bright enough to detect yet? The sky grew brighter and brighter as sunrise approached. I was about ready to give up when finally something nudged above my treeline. What I was seeing looked as bright as one of the brightest stars (Altair)—but it was sporting, even in the bright sky, a short yet brilliant spike of tail. I could only begin to imagine what Comet Bennett would look like when it had lofted up into a fully dark sky in the weeks ahead.

Comet Bennett may have been the most beautiful comet of the twentieth century. Not the brightest or the longest. But night after night it hung in ever-higher splendor with a tail that made it look like a white sword or an angel. The tail I saw was perhaps less than half the length of the Big Dipper—not very long for a great comet. But it was dense and detailed and radiant. Even my small telescope revealed silky white strands through which sparkled a multitude of background stars. One of the great beauties of this comet was its path in the heavens: It hung breathlessly like some vast shining swan's feather by the graceful constellation of Cygnus the Swan; it was a snowy sword laid in peace (but power!) near one of the brightest parts of the Milky Way's stream; it floated into the north circumpolar regions of Cassiopeia the Queen to spend ever more of the night aloft.

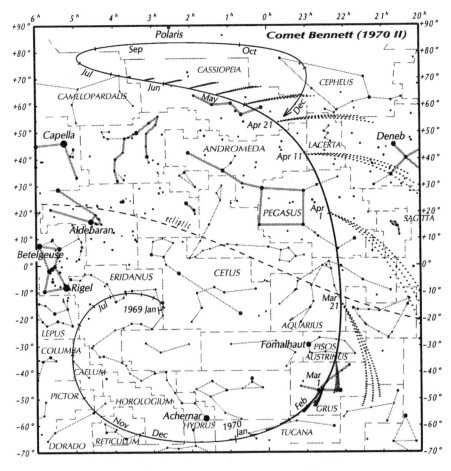

Comet Bennett's path among the constellations.

If Comet Bennett had been visible at any time other than those loneliest hours just before dawn, it would have been the talk of the world. I think that even in the already high-powered and light-polluted society of 1970 it would have made its mark. It would have been talked about in offices and portrayed in artworks like the great comets of past centuries. But Comet Bennett received little publicity. Only astronomers, and a handful of other people whose job or habits found them outside at 4 A.M., were privy to this dreamlike apparition.

To See Kohoutek

The next major experience with a comet in my life came in 1973, when the world thought it had reason to get as excited about a comet as it had for Ikeya-Seki. This time, there were many months of advance warning to plan cruises, mint commemorative medallions, and worry about the comet destroying life as we know it. The object of these preparations was a comet whose very name went on to become a synonym for *flop* and *fiasco*: Kohoutek.

Kohoutek was expected to attain a brightness exhibited by few comets in history that had not been sungrazers and to continue shining brightly until the week it would start appearing high enough after sunset to be seen in a fully dark sky. We will examine later in this book what "went wrong" with Kohoutek and other some other comets that were initially predicted to be great, even to be "the comet of the century." Suffice it to say that as astronomers' doubts grew about Kohoutek's forecasted brilliance, there seemed to be no way to convince the public to temper its expectations. The comet was accorded its own box on the cover of *Time* magazine, and a huge article within, as 1973 drew to a close. But the "Christmas comet" was never seen spectacularly bright, except by the Skylab astronauts. After this burst of brightness when passing near the Sun, Kohoutek faded incredibly quickly. I saw it on several nights in January 1974 as a naked-eye object of moderate brightness with a fairly long but very dim tail. I even became the subject of a (corny and inaccurate) local newspaper article in which I was proclaimed to be the teen who did what virtually no one else in the area had been able to do—see Kohoutek.

I felt as cheated as anyone by Kohoutek's disappointing performance. (Actually it was for amateur astronomers, and especially for professionals, an interesting comet—but our expectations had been too high.) Maybe things evened out, though: I was more disappointed than the public because I knew what the experience of a great comet could really be, but I also felt less cheated because I knew that the wonders of which great comets are capable are no fiction. All that was required was to be patient and wait for the next great comet. As it turned out, the wait was not very long.

The Grandeur of Comet West

In late 1975 or early 1976, the first word I got about Comet West sounded like an echo of the news and predictions there had been about Bennett. Like Bennett, Comet West had been discovered while visible from the Southern Hemisphere (this time by a professional astronomer) and should not be visible from the United States until it passed the Sun and began to emerge, quickly higher each day, in the dawn sky. But the bad experience of overselling Kohoutek before it had proven itself was still fresh in the minds of astronomers. Unfortunately, this probably led to there being far too little advance publicity for Comet West.

Would Comet West turn out, like Bennett, to be even better than expected? I personally got some news to suggest that it would. The news was an exciting report from my friend Steve Albers, a skilled amateur astronomer, who had tried to see if he could spot Comet West within a few days of its closest passage to the Sun in both space and the sky. That ought to have been the time when West was brightest, but the nearness of the comet to our line of sight with the Sun meant that it would have to be terrifically bright to be glimpsed at all in the solar glare. But my friend succeeded: He was one of the few enterprising skywatchers who observed the comet just a few minutes after sunset in late February. We later learned of an expert observer, John Bortle, who even glimpsed West with the naked eye a few minutes before sunset—one of the rare instances of a daylight comet.

At the time these sightings were made, West's brightness may have rivaled or even surpassed that of the second-brightest planet, Jupiter. But would Comet West fade incredibly fast as it moved away from the Sun, as Kohoutek had?

Astronomers in the Northeast and Middle Atlantic states, such as myself, found our first dawn view of Comet West cruelly delayed. The culprit was a long spell of cloudiness. There was no Internet or recorded as-

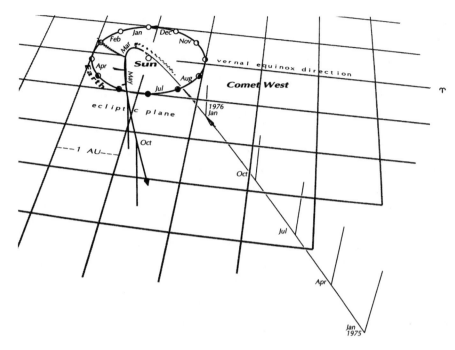

Orbit of Comet West.

tronomy telephone hotlines in those days, and unless you were a social enough amateur astronomer to have contacts around the country (I didn't), there was little to do but wait for the skies to clear.

After over a week of waiting, I knew that the morning of March 7, 1976 (6 years to the day after I saw my first total eclipse of the Sun, I noted) was going to be clear. I was ready. In fact, I was over-ready. Just as a lark, I decided to step out into my backyard long before the comet's head would get above the treeline to the east. Maybe an immensely long tail is already sticking up, I jokingly thought. A few more steps took me to where I could see a low point of the treeline, and then . . . there was a beam of light jutting up as plain as could be from the forest. Surely there was some mistake. Surely it was some kind of humanmade radiance. No, it was no mistake. It was the farther reaches of the mighty tail of Comet West.

I found a place in the edge of the woods where I could get a peek at what was rising, but it took quite a while before I could see the head and most of the tail in one nearly uninterrupted view. Morning twilight was brightening strongly down low, but the lowest part of the comet was so luminous that it didn't matter. The head and the first section of Comet

West's tail were a mass of glow, vividly golden to the naked eye. The naked eye saw in the midst of the head a point of light which itself was as bright as a rather bright star. In my 6-inch telescope and its finderscope, the views of the head and near-head region of the tail were astounding. The "false nucleus," which looked like a star to the naked eye, appeared with magnification as a ball, like some strange new planet. This was the densest cloud of gas and dust hiding the true icy core, or nucleus of the comet. The dramatic parabolic outline characteristic of a bright, active comet's head was present, and there was a quite noticeable shadow running back into the tail from behind the denser central region of the head. I was struck, on this and other mornings, by a greenish hue which I think was in a region of the comet's head where the gold light from sunlit dust and the blue light from fluorescing gas were combined in equal measure. It was mesmerizing to watch the colors changing over the next few weeks, one arm of the parabolic outline gaining in prominence, a thin line of light called a "spine" becoming visible back along the midline of the tail where it departed from the head.

Indeed, although the gold of the head at last dissolved into the coming dawn, and the comet's total brightness was never again as great for me as on that morning, some aspects of Comet West's show were even better on clear mornings in the next week or so. The prime example was the comet's two tails.

Not every comet is active enough or comes close enough to the Sun to produce a tail visible from Earth. But among those that do, there can be two fundamentally different types of tail: a straight, narrow, bluish one, composed of gas, and a curved, broad, yellowish (or even slightly reddish) one, composed of dust. Most bright comets produce both a gas tail and a dust tail, but we usually view the comet from an angle at which the two are more or less superimposed on each other and therefore hard to differentiate. Comet West was one of the rare cases in which the tails were fairly broadside to us and thus were displayed separately.

At mid-northern latitudes the awesome dust tail was seen as an immensely wide fan to the upper left and the gas tail as a long, almost vertical shaft of light. A first look at an excellent photograph of Comet West's dust tail and you're gasping: There are several different kinds of bands visible in its huge fan or fountain of light. We will discuss these features later, but for now let's just say that these features are approximately arrangements of the tail into dust particles sorting themselves by size and date of release from the head. I want to stress that the visual impression of Comet West's dust tail was nowhere near as detailed as these photographs. But the unaided eye could detect several of the branches fairly easily. And of course to behold something so large—tens of millions of miles wide!—spread out and reaching up toward the Milky Way band in real life contains an element of thrill no photographic image could match.

I will have more to say about Comet West's tail in this book. Rest assured that even though it was not so concentrated, or so bright in its entirety as Bennett's, this broader, longer, more complex structure held me enthralled on the clear mornings before moonlight interfered in March 1976. I was impressed too with the gas tail, which I, unlike many observers, was able to trace out for a length almost as great as that of the dust tail—which stretched more than 20 degrees of arc (almost a quarter of the way from the horizon to the zenith).

The reason for Comet West's extra brightness and its complex and huge dust tail was an actual splitting of the comet's icy nucleus into at least four

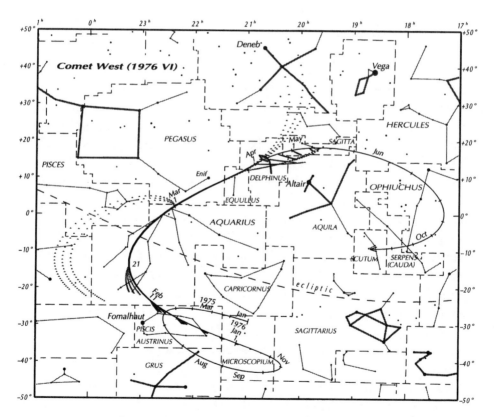

Comet West's path among the constellations.

parts. The four clumps of gas and dust surrounding each piece of the nucleus were seen separately by some expert observers with large telescopes.

When the bright Moon of late March left the pre-dawn sky at last, the show of Comet West was still not over. In April the comet, which had always aimed its gas tail more or less toward the compact diamond shape of Delphinus the Dolphin, now approached this star pattern and extended what was left of its two tails into and, like a radiant veil, over the little diamond.

It was a wistful farewell I gave to this comet when I last glimpsed its finally dim form in my telescope that June. But a lasting consequence of Comet West was its role in launching my writing career. I like to say that I rode the tails of Comet West into my first writing job, as the weekly columnist on astronomy for the Atlantic City newspaper *The Press*. After 20 years—well over a thousand weeks—of doing the column, I have still never failed to submit my copy for publication.

Waiting for Greatness

All of the comets I have mentioned so far are ones that had never previously been seen in historical times (except possibly Comet Bennett, and the progenitor of Ikeya-Seki). We know this because calculations of the orbits of these comets show that the orbits are enormously elongated ellipses whose far ends lie many times farther out than Pluto's. Kohoutek had never visited the vicinity of the Earth and Sun before, and the other comets had not visited in thousands or hundreds of thousands of years. But in the summer of 1976, I saw—with telescope and just barely with naked eye—an unusually close approach of one of the smaller, less active comets whose smaller orbits return them to visibility near Sun and Earth a number of times in an average human lifespan. This particular "short-period comet," D'Arrest, comes back into view every 6 to 7 years. But even in 1976, I was casting my thoughts 10 years ahead to the predicted return of the brightest and most famous of these mostly dim short-period comets—Halley's Comet.

The average orbital period of Comet Halley is about 75 years, so the number of its returns viewed by most people in a lifetime is just one. If Halley happens to come when you are a young child, and you do get a look at it, you may live to see it at a second return.

I knew I'd have to be satisfied with one visit (probably!) and I was already anticipating it in 1976. But seeing the great comets Bennett and

West just 6 years apart, both in my first 21 years or so of life (and I had just missed the great Ikeya-Seki) couldn't help but make me feel that another great comet would probably come along before Halley. Intellectually, my study of the past records of comets told me this might not be so, but emotionally I couldn't help feeling that a great comet would appear before I was much older.

The emotional part of me was wrong. The twenty years after Comet West saw many interesting comets. The amazingly successful Australian comet hunter William Bradfield discovered many, some of which came north, and some of which became naked-eye objects and sported tails. There were a number of comets with the names of David Levy and Rodney Austin attached to them. The one of Rodney Austin's three comets which reached its brightest in the spring of 1990 initially had us astronomers thinking it would become a majestic object. But like Kohoutek, it was a first-time visitor that appeared deceptively bright while still far out by the orbit of Jupiter. Fortunately, this Comet Austin's lagging brightness was noticed fairly early, and there was no public relations disaster as there had been with Kohoutek. Still, many of us amateur astronomers were very disappointed.

Of all the comets in the 20 years after Comet West, the most interesting to me—other than Halley, of course—were surely 1994's Comet Shoemaker-Levy 9 and 1983's Comet IRAS-Araki-Alcock.

I never actually saw Shoemaker-Levy 9; its numerous pieces were out near Jupiter and exceedingly faint. But I spent several dozen nights watching the effects of this comet: the spectacular dark spots produced by the explosive entries of its fragments into the Jovian atmosphere. Later in this book, I will have much more to say about the life and death of Shoemaker-Levy 9 and will address the question of how much danger there is of a comet colliding—in our lifetimes or at some far-future date—with Earth.

A Close and Sudden Comet

The closest—and I do mean *closest*—thing that anyone alive has ever seen to a comet's colliding with Earth I saw outstandingly well in 1983.

Comet IRAS-Araki-Alcock was discovered first by the Infra Red Astronomical Satellite and then independently by amateur astronomers Araki in Japan and Alcock (with binoculars out a window!) in England. It was a small and feeble comet. But in early May of 1983 it came closer

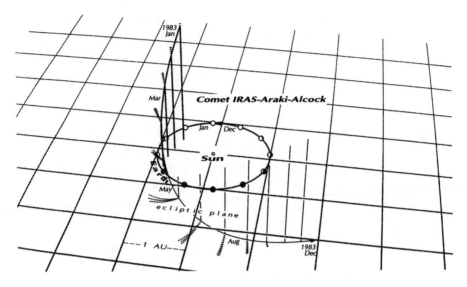

Orbit of Comet IRAS-Araki-Alcock.

to Earth than any comet in over 200 years. The suddenness of this object's arrival, looming, and leaving are captured in an account which I wrote at the time and have never published. This is it, presented in its breathless entirety.

> Received Guy's [fellow astronomy writer Guy Ottewell's] article on comet. Looked at charts but decided not to read article until I have written down my own thoughts.
>
> First word reached me, as it did many people, in the newspaper: new comet, approaching us, due to pass closer than any since 1770, closest encounter in just a few days. Lacking further details, I thought odds were against ideal visibility but knew better than to delay learning more. I was soon on the phone to Jay Gunter, the creator of *Tonight's Asteroids*, that superb and supremely civilized publication which combines finder charts, facts, myths, astronomical history, a forum for amateur astronomers, and sometimes special comet alerts all in a usually 8-page issue that comes to you free in the self-addressed, stamped long envelopes you send to him. The good doctor had received an ephemeris [listing of positions and in this case brightness in terms of "magnitude"] from Fred Espenak, which the latter had calculated from early Brian Marsden orbital elements. The reading of the figures to me over the phone in Dr.

Gunter's rich Southern voice was interspersed with my involuntary exclamations (magnitude 2.1! [as bright as the average Big Dipper star!]).

That was Friday May 6. The next days were frantic with communications to relatives, calls to a few good astronomical friends (wherever they might be) and to local amateur astronomers, and a recitation over the phone of an article for the Atlantic City newspaper (during the running of the Kentucky Derby). But the nights— after the cloudiest (and wettest) March and April on record in much of this part of the country—the nights came, almost cloudlessly, after days of teens and twenties humidities (almost unheard of in the East!), six in a row . . . from first night through climax. With each Earth-turn and daylight flash, a new frame of ever more deeply comet-stained night in the movie of passage.

Friday night. My search with binocs immediately produced a prominent ball of fuzzy light in its still imperceptibly slow roll past the head of the celestial dragon [Draco]. Once located, a roughly magnitude 4.9 object detectable with naked eye. In the 6-inch [telescope] already as large a coma [cloud of the head] as I had ever seen save for Comet D'Arrest in 1976 (which I had also glimpsed naked-eye in superb skies as a remarkably sizable phantom patch of light on the verge of vision low in the south). Yet Comet IRAS-Araki-Alcock ought to improve tremendously on each of at least the next four nights—a period in which it would get almost four times closer and many times brighter—and even those quantities could not really suggest the key thresholds of qualitatively different appearance it would burst through as it rushed us. . . .

Sunday morning. . . . Or so I hoped. Poorer atmospheric transparency permitted no more than yesterday's roughly 15 arc-minutes (about half the apparent size of the Moon) of coma, and its sizability was just slightly more distinguishable with the naked eye. But did those glimpses in the telescope really also show a star-like point midst the coma?

Monday morning. Around 2 A.M. the evening's showers were whisked away at just the right time by one of those cold front sweeps which are a little astonishing and beautiful at even the worst of times. This time, my weary body was instantly reinvigorated; it was a new person running out to greet the coming winds. Last torn cloud curtain edge was withdrawn to reveal a sky clean and sheer to about mag. 7.0 [extremely dim stars] at its summit—and a gleed . . . well, not a gleed, but a handful of strong phosphorescence . . . hung in mid-flight on that sky's north shoulder. We had as good a comet (save for a tail) as the imagination of that word could wish.

A last few exceptionally small cumuli tumbled by on the great winds that still passed silently over me, with still only a fresh rustle here at ground level to herald the certainty of next day's brisk blasts. And the comet seemed an only slightly smaller cumulus, at a some-what higher level, also being blown on some silent wind, a powerful unseen force; or carried by on an unviewable stream which was flowing it toward the aim of Earth's upper axle amongst those made-to-seem-circling constellations as it dove ahead and down before the almost equally rapid rafting of Earth. (Could our modest craft on the river of space shoulder a wave or two to displace the bot-tlecork—or froth-mass?—comet's boisterous course and pace?)

It was like a detached piece of very bright Milky Way (or a smaller but quintillion-or-so-fold closer and mobile Magellanic Cloud [satel-lite galaxy of our own]). It was larger and much brighter than the ad-mittedly much lower Lagoon Nebula was to my naked eye in the south. Through the 8-inch, bad seeing but still a probable (but be cautious!) star-like center and, even with magnification, the coma spreading over half a degree [the Moon's apparent size] at least. (It looked for all the world—or at least for me in that magic clear dark—like another Orion Nebula in size, brightness, and even shape!)

I looked and looked, long enough to know even with the naked eye that it was indeed moving, even as I watched, on its destined course toward the Dippers. Not from tiredness (though tiredness was there) but from peaceful fulfillment and extravagance of good spirits at one and the same time, I lay down on the grass and, after a fine standing bow of thin Moon appeared among the branches in the east, I relaxed there gazing up at the comet fading in morning twilight now just through the gate of two glittering scale-stars of Draco. To my surprise, even its diffuseness did not keep this phan-tom from being distinguishable as a dimmest stain on twilight until within 40 minutes of sunrise.

Monday evening–Tuesday morning. Disappointing transparency as dark fell and the anticipated conjunction with the brighter Guardian of the Pole, Kokab, Beta Ursae Minoris, was already in progress, preventing a good naked-eye view of the comet itself for my visiting friends. It was fascinating viewing the rather bright star cloaked in that slight veil, and the conjunction of the comet's center with the star was surprisingly close.

Later, however, I discovered we were poised on the edge of two weathers and the whole night was a strange swaying between warm and cold (10 degree Fahrenheit changes in 30 minutes beneath a cloudless sky!), still and windy, good transparency and good "seeing" [steadiness of atmosphere] (these latter two never occurred together). In good "seeing" we were convinced of a nucleus (!); in the good transparency later the roughly magnitude 2.8 or 2.9 [approaching Big Dipper-star-brightness] coma was huger than ever, maybe 1 to $1^1/_4$ degrees (2 to $2^1/_2$ times the Moon's apparent width) in diameter to the naked eye.

The night's two greatest realizations: at a magnification of $90\times$ the "nucleus" was perceptibly moving in relation to star after star which it swam (often very close) past; as the comet raced away from Kokab, it occurred to me that I was seeing this ball of glowing mystery being tossed from one Dipper's bowl toward the receptive and bigger bowl of the other—and knew the comet was predicted to just miss the latter bowl's edge . . . unless early orbital elements were off or Earth's tug did pull it aside for the apparent catch!

Tuesday night. A thin cloud blanket swept just far enough south in late evening to win out most of the night (the comet was getting low when the wobbling blanket fully retracted), but it was held off long enough to permit a view of a seemingly larger but not much brighter coma—and one thing else. I stood out in my yard with a modest-sized telescope and beheld, shining out with incredible clarity and brilliance, something which a week before no one in history had ever really seen: the true nucleus of a comet. [*Note:* This supposition of mine and other observers was later disproven. We were glimpsing an incredibly tiny cloud just around the hidden true nucleus, but the belief that we were seeing the true nucleus was a bracing thought.] The true nucleus was the secret glittering dust-shot ice-gem that was the heretofore hidden heart of the solar system's most mysterious kind of object. That beguiling view alone would have made the evening unforgettable, but had the clouds made me miss much more on what might be the best night? Would Wednesday's sights be better—and would weather cause me to miss some or all of them too?

Wednesay night. Clouds all day gave way just in time to another near-perfect night. The great coma nudged past the star Delta Cancri, dwarfing and far outshining the nearby big and bright Beehive star cluster. Nucleus far less visible (because of poor "seeing"?), but sharing with friends at this excellent site we had driven to, and the appearance of yet another marvel, prevented me from doing more than rough estimates of coma size and brightness (I cautiously thought the size a little better than $1^1/_2$ degrees and the brightness about magnitude 2.1. Expert comet observer John Bortle that night assessed at least 2 degrees—four times the Moon's apparent diameter—and magnitude 1.7—bright as the brightest Big Dipper star).

The marvel which wonderfully distracted me was two rays of tail. They were—I think independently—affirmed by my friends. But it was only much later, after the friends had left and before the comet had yet gotten too low, that I concentrated and was able to trace, beyond the unquestionable several degrees of length, far more extent. I believe that each ray was as much as 12 to 16 degrees long (half or more than half as long as the Big Dipper)!

Did I really see these long rays? The gas tail (which I already knew that professional astronomers had photographed) was not supposed to be this long, and it should have been then pointing east roughly along the ecliptic [mid-line of the Zodiac] toward the star Regulus. Was I seeing a dust tail lagging and, due to our nearness— just a bit more than the closest approach distance of 2.9 million miles—greatly displaced in position angle from the anti-sunward direction? Perhaps the rays were merely the edges of a broad fan of dust tail perceptible only where adjacent to the slightly darker surrounding tail-less sky. I thought I could finally see the whole structure of head and nearly the full extent of both rays in one look made with averted [and therefore more dim-light-sensitive] vision. In my favor is the fact that it was an exceptional night with even the zodiacal band (very dim extension of the zodiacal light) definitely detectable across the south sky. Surely some other excellently placed observers got a similar view, or photographed these rays with the proper equipment and exposure? If not [later I learned of at least two expert observers who saw tail features similar enough to mine to lend some credence to my observation], I will be dismayed. But I will still have to believe that what I saw that night was an astounding (more than a million-mile-long?) climax to six unforgettable nights of passage.

Cloud and scarves of comet came by our planet like a thief in the night. And, like the subject of that proverb, the coming was one of glory.

* * *

An event like the passage of IRAS-Araki-Alcock brings to mind many things. First of all, the rarity of such events; second, that astronomy is so rich with categories of objects and events that every year or even season brings us some kind of stupendous rarity that is fascinating to see. When I heard that this was the second closest definitely known encounter of a comet with Earth in history, I immediately remembered that the closest was that of Lexell's Comet (an object more often mentioned in textbooks as a proof of the low mass of comets) because of its passage through the Jupiter satellite system and subsequent orbital changes that—and how is this for celestial billiards?—were what sent it just 1.5 million miles from Earth in 1770. The largest angular extent of its coma was apparently well over 2 degrees [there were, later, reports of larger apparent size for the coma of IRAS-Araki-Alcock], and one of the additional details I find in the old, old book I am looking at is that Lexell is believed to have been shot right out of the solar system. I wonder if that is still thought to have been Comet Lexell's fate—and what will be the fate of Comet IRAS-Araki-Alcock after its close brush with our Earth?

Twice in just over two centuries, comets have come within 3 million miles of Earth, and there have been a number of not much more distant approaches. What if a really large comet were to come as close as I.-A.-A.? The Great Comet of 1861 passed within 14 million miles of Earth, displaying a tail well over 100 degrees long (over halfway across the sky) through which our planet probably passed. Its highly inclined orbit helped place it in a dark sky when its head, perhaps as large as the Moon in apparent size, may have shined as brightly as magnitude −3 (almost as bright as Venus)!

If we go back more than a few hundred years, what still closer amazing passes might we surmise from the limited data? The *Comet Halley Handbook* laconically lists the figures for the most startling pass of Halley's Comet we can determine: The comet's head was apparently less than 4 million miles from Earth on April 11, 1837 when the head, perhaps 4 to 6 (or more) times larger than the Moon in apparent size, burned at an approximate visual magnitude of −3.5 (rivaling Venus). Halley's tail was 93 degrees long at that return.

These are just a few of the marvels of which comets are capable. No one would deny that Comet West of 1976 and Comet Bennett of 1970 were more visually spectacular by far than IRAS-Araki-Alcock. But our close visitor provided some amazing opportunities and appearances which even those giants couldn't match.

Another thing that the I.-A.-A. encounter reminded me about was what a splendid fellowship is that of amateur astronomers. Now, as I am writing this [account of IRAS-Araki-Alcock in 1983], the details come from two members of that fellowship—first that great comet observer John Bortle, and next from Jay Gunter. What's the word? New comet, due to pass within 6 million miles of Earth, excellently visible in a dark sky, magnitude maybe as bright as 3.8 [moderately bright star], closest approach in just a few weeks, may pass directly in front of the Great Galaxy of Andromeda on May 28. What are the odds on this happening? Here we go again!

This next comet, Sugano-Saigusa-Fujikawa, turned out to be so feeble that its light was spread out over too large of an area for it to be readily glimpsed. But the drama of IRAS-Araki-Alcock had been quite sufficient.

The Wait Ends

As Comet Halley approached in 1985, there were indications that it might become very bright, even though this would be one of its least close passes of Earth in history. But the indications proved false. Halley never grew bright enough or sprouted a prominent enough tail to be considered a visually "great" comet at this return. Nevertheless, this most famous of comets not only brought to the world its store of legends and history but also provided professional astronomers with an unprecedented wealth of data (especially from unmanned spacecraft which flew past or through it). Furthermore, Halley's Comet probably kept amateur astronomers with binoculars and small telescopes busy and thoughtful longer than any other comet in this century. Although never truly spectacular to the naked eye, it managed to maintain a moderate brightness long enough to show us an incredible variety of appearances. I'll have much more to say about this return and the story of Halley's Comet in Chapter 8.

My wait to see another visually great comet after West continued until the very final days of my work on this book. In March and April 1996, it was Comet Hyakutake which blazed across the sky, a prelude to what we think may be yet another truly great comet performance—that of Hale-Bopp in 1997.

But there was one series of nights in my watching of Halley which captured an essence of what is most thrilling about comets. It was in late April and early May of 1986, when Halley was a not very bright naked-eye object in the evening sky. Almost everyone had given up on seeing a very long tail from Halley. But a few of us noted that the motions of Earth and the comet would produce a quick change in our angle of viewing the dust tail and that this might result in greatly increased visibility for a week or more.

I was one of the observers who verified this idea. What I saw was like the faint tail "rays" of Comet IRAS-Araki-Alcock—really the edges of the broad dust tail. But there were two differences. First, Halley's tail was much longer. On several nights, I and a few other observers traced the extremely faint tail out to about 30 degrees (one-third the distance from horizon to zenith). On one absolutely clearest night of the year, I may just possibly have traced the edges of the tail out to roughly 50 degrees.

But the second difference between this tail observation and the one I made of IRAS-Araki-Alcock was the difference in the true proportions in space of what I was seeing. Comet I-A-A was just a few million miles from Earth, but Halley in early May 1986 was about ten times farther away, so the true Halley tail dimensions were enormously larger. As I have written elsewhere: "Those late-April and early-May evenings there was the feeling of a giant's form or shadow passing just by and glimpsed by only a few alert and shaken watchers while the rest of the world was sleeping or looking the other way."

I was being passed by a ghost 10,000 times the size of the Earth.

The wraithlike comet which goes by may not always be so vast as this, but there is always, I think, a sense of a monumental passage, the arrival into and departure out of sight by something unique in the heavens, something which hasn't been seen for decades, or thousands of years, or ever, and which may never be seen again.

You think you know what is up there—Sun and Moon, planets and stars, star clusters and nebulae and galaxies. But then a phantom light appears, a comet, and you don't just know that you were wrong, you see that you were, too. After that, the heavens can never be final, can never be taken for granted or be set aside and less than fully regarded. The universe is proven new, always new. For a comet passes . . . and nothing is the same.

COMETS IN GENERAL

Chapter 2
ANATOMY OF A GHOST

Scene for a slightly chilly night in the universe:

We who together represent the Earth, huddle along with the other, nearby planets of the inner solar system—rather close to the campfire we call the Sun. Looking outward we see the steadily if dimly lit faces of the outer planets—Jupiter, Saturn, Uranus, Neptune, and tiny, hardly glimpsed Pluto. We take our leisurely circular strolls (orbits) around the fire, sing our songs, tell our tales, have our meals, conduct our civilization—whatever else we do near our campfire. We think our little camp-out is a pleasantly orderly, eminently understandable affair.

And then suddenly we realize that we and the other planets are not alone.

What has happened?

It's hard to believe but . . . we've just seen a ghost.

What else would you call a hazily luminous, wavering shape, huge but seemingly insubstantial, suddenly floating into view at the outer edge of the circle of bright illumination beyond us? A ghost has materialized beyond Mars. All at once it is flitting in toward the campfire—first it passes above us, then it brightens and elongates as it nears the Sun-flame. Is it real . . . can it hurt us? It rushes right toward us! At the last minute it passes by us . . . is out at the edge of visibility . . . is vanished.

We're spooked for sure now. What are we feeling? Both awe—at least a sense of the uncanny—and fear. Rethinking our memory of the ghost's passage, we realize that we could see through this specter. It seemed like something that was hardly there (yet bright, swift, changing!). Except we could not see through the brightest glowing cloud at its very center. We have a troubling feeling that hidden in that cloud is something quite substantial . . . something which could wreck or painfully disorder us if it ever should hit—and yet (can it be true?) something from whose very substance, multiplied a million times and transformed, we ourselves were originally made?

True or not true, we don't sense much kinship with that strange, otherly thing. And now, as our fire shines on, as our fellow planets bask serenely in its glow, we start to ask ourselves whether the whole experience was just a dream. Perhaps we were falling asleep, as we are now. . . .

We're jolted back awake. There's another ghost floating and wavering out there! It already seems to be bigger, brighter, longer, and more imposing than the first. We look more closely all around us (trying to tell ourselves we are not looking frantically, or seeking escape, for there is no escape and we are in any case truly mesmerized). And now, with a sinking feeling, we discover that a flock of a dozen or more phantoms bright enough to be glimpsed is haunting our inner solar system at this very moment. There's one over there, two above us, one below us . . . now a new one glimmering into view. Each looks different, seems to move differently, wavers and changes. Suddenly we sense (there can be no doubt about it) that dozens more—no, hundreds more—lie just beyond the glow of the campfire we call the Sun and may be preparing to sweep in toward us and the fire. And a little farther out—halfway out into the deepest night between the stars, those other lonely campfires so many miles away—what is out there? Thousands, millions and finally billions of specters, drifting slowly in the night, waiting for the hour when some mysterious impulse will send them flying in to startle, discomfit, puzzle, and . . . delight us.

Our Earth and our solar system are haunted, and what else could we call these flitting phantoms but ghosts?

We could call them *comets*.

If you're like me, you may be more interested in science than seance. But the solar system ghost story I've just told illustrates better than the mere words "mystery" or "bewilderment" or "awe" what comets have been and still, in imaginative analogy, are to us.

For all but the last fleeting second of history (the last 40-odd years in a 2500-year attempt to understand them scientifically), comets have been an utter mystery. Lest you think that "utter mystery" is an exaggeration, or that 90% of everything else in our study of the universe has been an "utter mystery," let me assure you that the mystery of comets has been in a whole different league from anything else in astronomy, at least in our supposedly familiar (almost comfy) solar system. Until 1950, nobody even came close to explaining scientifically what is at the center of comets or what accounts for the spectacular range of features and phenomena they exhibit. Coincidentally, also until that year, no one had any defensible idea of where comets come from—and I don't just mean "where comets come from" in the sense of what their ultimate origins were. I mean that nobody knew exactly where many comets were coming from when they appeared—like the ghosts at our campfire—winging in from the outer solar system. Nor did anybody know how fresh comets could still be appearing after these billions of years.

What is the hidden heart of comets? Where do comets originally, and where do comets now, come from? Those are the howling mysteries which most of the history of comet science was a valiant but almost fruitless attempt to solve. Those are the howling mysteries which the past half-century of comet science grappled with and compelled to start speaking to us—revealing wonders at least as great as any we could have imagined. In this chapter we survey the first few millennia of the scientific quest to understand comets. Then in Chapter 3 we examine the mighty revelations about what comets are and where they come from, which the past few decades of the quest have at last won for us.

Outlaws or Omens?

The people of ancient times believed that Earth was the center of things, so our analogy of the campfire and the ghosts would have to be different for them. But the gist of the analogy would be the same: a feeling ranging from perplexity to outright alarm (always mingled with wonder) about not just the strangeness but also the seeming lawlessness of comets.

Everything else in the heavens more or less seemed to obey laws. Ancient cultures knew that the star patterns—the constellations—always maintained their forms and always appeared in the same place in the sky at their appointed seasons. Admittedly, the "falling stars" were startling, but no important star ever seemed to be missing from the sky, and if you kept watching on a clear night you saw that "falling stars" keep happening. (You couldn't predict exactly when one would occur, but the overall process went on with reasonable regularity every night.) The Sun and Moon would be farther north or south at different times of the year or month, but their movements were easily foreseen. Then there were the planets. These bright and wandering stars seemed almost to have personalities of their own. But the planets (not to mention the Sun and Moon) were always found somewhere in the band of constellations called the zodiac. They never suddenly changed form or suddenly disappeared, and it was possible to predict most of their motions.

Yes, everything in the heavens could be accounted for in terms of predicting where it would go and what it would look like—until a comet showed up.

It was bad enough that comets often appeared as sizable objects, not mere points of light, and that they changed their shape and brightness from week to week or even from night to night. And it was very bad that they could get so bright and show such odd, long, impressive "tails"— such sights were unique and unnervingly strange. But worst of all was the seeming unpredictability of comet behavior: Comets could appear not just in the zodiac but anywhere in the sky, and they could move to anywhere else in the sky before every bit as mysteriously disappearing. Comets seemed to be lawless. There seemed to be no protection against their coming and going, brightening and lengthening and changing color and shape. What else might they do?

Comets were disruptions of the regular scheme of the heavens. And because humans attempt to understand the world by looking for correspondences, people felt that what happened in the heavens must have its reflection on Earth. If a comet appeared, a king would die, an enemy would invade, or a plague or famine would occur, because all these are all disruptions of society and life, just as a comet is a disruption in the heavens. As Shakespeare wrote in *Julius Caesar*,

When beggars die there are no comets seen,
The heavens themselves blaze forth the death of princes.

No wonder, then, that in ancient cultures, comets inspired not only awe but also grave concern or even outright terror.

But notice something interesting: As unpleasant as it was to believe that a new comet might forebode war or epidemic, even these alternatives were apparently preferable to leaving comets unexplained. If the otherwise lawless objects at least performed some supernatural function, they could be partly understood and their effects prepared for.

Aristotle and Seneca

Learned skywatchers of ancient times tried to classify comets on the basis of their resemblance to the shape of earthly things. But the shape of comets is so uncertain and various and chameleonic that one of the Chinese systems ended up with 21 types of "ominous stars" (mostly comets, but maybe some other sky phenomena too) and descriptions of them of this kind: "like a roll of cloth," "with three brooms above it pointing upward," "like a basket 20 feet long with a star at one end." The Roman writer Pliny the Elder did little better with his ten types, which included Pitheus ("figure of a cask, and emitting a smoky light") and Hippeus ("like a horse's mane in rapid motion.")

The only conceivably fruitful scientific debate about comets in ancient times revolved around one question: Did comets occur in the atmosphere, or did they occur beyond it?

We read (secondhand, from a Greek source) that the Chaldean astronomer–astrologers of about the seventh century B.C. had proponents on both sides of this issue. So too had the Greeks and Romans.

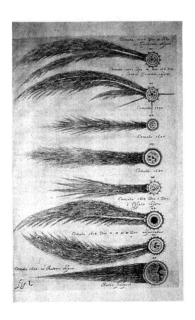

Illustration of different comet types from Hevelius's Cometographia.

The idea that comets occur in the atmosphere was favored by the great Aristotle. It therefore ended up being the prevailing belief for about 2000 years. In his *Meteorologica*, Aristotle rejected the ideas about comets that he attributed to the Pythagoreans, to Hippocrates of Chios, to Democritus, and to Anaxagoras. There was not just one comet, comets were not an elongated appearance formed by two planets when they came close to each other. But Aristotle also rejected the idea that comets were a different kind of "planet"—that is, that they were objects traveling among the planets, far away from Earth and Earth's atmosphere.

In Aristotle's thinking, the world consisted of four elements—earth, water, air, and fire—each in that order less dense and heavy than the previous and thus tending to predominate in a higher sphere. Later, there developed more fully the concept of a fifth essence, or "quintessence," occupying a still higher sphere, that of the heavens. The sphere of fire, just above that of the air, was not flame itself but was potentially flammable. Comets, Aristotle said, occurred when the heat of the Sun (or planets!) caused dry, warm exhalations from Earth, which rose up to meet the lower border of the fiery sphere and ignited from the friction of their motion against it. The comets were then carried partly along by the movement of the fifth, or heavenly, sphere.

Given the limited information he had, Aristotle's theory of comets was really quite ingenious. It even had further details to account for different kinds of comets and for meteors. It seemed to explain all sorts of observed facts. For instance, one would expect that because comets began as dry, windy exhalations from Earth, the weather around the time a comet formed ought to be windy and dry, and Aristotle was convinced that such weather had indeed been noticed whenever a comet formed.

But of course Aristotle and other Greek thinkers placed far too much emphasis on logical extrapolation from theory and far too little on performing experiments and gathering data. They did not have the mind-set of a modern scientist.

Amazingly, however, the other most important writer on comets in ancient times did seem to be advocating something like the use of modern scientific method:

> It is essential that we have a record of all the appearances of comets in former times. For, on account on their infrequency, their orbit cannot as yet be discovered or examined in detail, to see whether they observe periodic laws, and whether some fixed order causes their reappearance at the appointed day.

This writer was Lucius Annaeus Seneca, who survived a sentence of death from one mad Roman emperor (Caligula) only to be exiled and eventually put to death by another mad Roman emperor (Nero, whom Seneca had been assigned the unenviable task of trying to teach and advise). But a few years before the elderly Seneca was forced to bleed to death, he managed to write his *Quaestiones Naturales* ("Natural Questions" or "Natural Investigations").

Seneca supported the idea that comets were celestial objects. He was not able to produce observational evidence to prove this assertion, but his logical refutations of some of the objections to the idea are interesting. Anyone who has ever marveled at comets will also appreciate his response to the criticism that comets can't be astronomical objects because their orbits would have to differ so drastically from those of the planets:

> Nature does not turn out her work according to a single pattern; she prides herself upon her power of variation. . . . She does not often display comets; she has assigned them a different place, different periods from the other stars, and motions unlike theirs. She wished to enhance the greatness of her work by these strange visitants whose form is too beautiful to be thought accidental.

Seneca confessed his limitations when it came to resolving the mysteries of comets. But he wrote a passage in which generations of modern readers cannot help but find a remarkable foretelling of Edmond Halley:

> There will be, some time, some one who will demonstrate in what regions comets run, why they wander so astray from other things, their number and their nature. Let us be content with what we have found; let our heirs also contribute something to the truth.

Another prescient passage from Seneca has been noted by my fellow astronomy writer Guy Ottewell. After pointing out how much more there is to learn about comets, Seneca writes,

> Yet we are surprised if we less than understand these scraps of fire, though the greater part of the universe, God, is hidden! . . . A time will come when our descendants will be amazed that we did not know things so plain to them.

Unfortunately, it would take about fifteen centuries before some of these "plain" things began to be understood.

From Tycho to Halley

For the most part, the later centuries of the Roman Empire and the Middle Ages offered only tired rehashings of the ancient debates about comets. It wasn't until a new group of great astronomers arose in the wake of the Renaissance that major strides were made in our understanding of comets.

Nicholas Copernicus (1473–1543) argued for the existence of a heliocentric (Sun-centered) planetary system, one of the most important insights in the history of human thought. Remarkably, though, comets were not part of his vision of the solar system, because he regarded them as atmospheric phenomena, even as Aristotle had! Ironically, the first important step in combatting this old idea was taken by a man who refused to accept the heliocentric model of the solar system: Tycho.

Actually, Tycho Brahe (1546–1601) was only the most famous of several scientists who proved that the great comet of 1577 must be at a distance greater than that of the Moon. The method was to measure the parallax shown by the comet. Parallax is the change that occurs in an object's apparent position in relation to background objects when the ob-

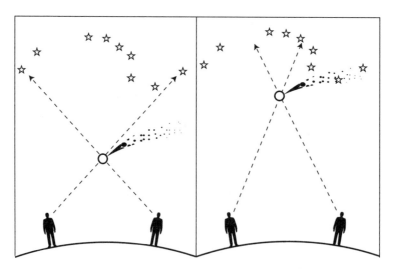

Illustration of parallax.

Three seventeenth-century scholars study diagrams illustrating different ideas about cometary motion. Title page from Hevelius's 1668 work Cometographia.

server views it from different locations. Parallax can be demonstrated by holding your upright finger a foot in front of your face and viewing it with first one eye shut and then the other. Note how the finger seems to change position in relation to a wall or other object considerably farther away than the finger. Now move the finger much closer, just a few inches from your eyes, and repeat the experiment. You have just demonstrated that the apparent movement against the background is much greater when an object is closer to the observer.

What Tycho had to do was make his own careful observations of the 1577 comet's location against the background of the distant stars and then compare these positions with those recorded by observers elsewhere in Europe. Would seeing the comet from a spot a thousand miles away change its position relative to the stars? The result: If the comet showed any parallax, it was far too small an angular change for the comet to lie within Earth's atmosphere. The magnificent comet was, Tycho determined, at the very least more than 4 times farther away than the Moon.

Tycho was famed for being an accurate observer, and his records of planetary positions provided his assistant with the data needed to make a great discovery: the three laws of planetary motion. Tycho's assistant was the mathematical genius Johannes Kepler (1571–1630). After Tycho's death, Kepler's calculations helped rescue the Copernican system by showing that planetary positions could be explained not by circular but by slightly elliptical orbits of planets around the Sun. What were Kepler's ideas on comets? After he observed the 1607 appearance of what was later shown to be Halley's Comet, he wrote a treatise intended for a popular audience in which he claimed that each comet came into being spontaneously from impurities—fatty globules—in the ether (a substance that various philosophers and scientists over the centuries have theorized pervades all of space). Kepler thought each comet had a guiding spirit that passed from existence when the comet did. He also believed that comets were transitory objects, making one appearance only, and that they moved in straight lines—both incorrect assertions. But in one of his speculations he was quite right: There are indeed "more comets in the sky than there are fishes in the sea."

Around 1609, the telescope began to be used for astronomical purposes by Galileo Galilei (1564–1642) and others. The first comet ever viewed in a telescope was apparently the first (not the greatest) comet of 1618. But in the early decades of the telescope's use, neither Galileo nor anyone else made a dramatic contribution to the understanding of comets.

What is arguably the greatest single development in the history of comet science came late in the seventeenth century. Its foundation was laid by the mathematics of universal gravitation worked out by Isaac Newton (1642–1727). But the actual discovery was made by one of the best astronomers of the day, a friend of Newton's who encouraged and even helped finance the publication of Newton's classic work, the *Principia*. This friend was none other than Edmond Halley (the name rhymes with *Valley*).

The basic facts about Halley's prime achievement are simple: Although earlier thinkers had speculated that comets might return, Halley was the first to identify a series of comet appearances and to demonstrate mathematically that they could all be explained as returns of a single object seen at different apparitions. (*Apparition* is the technical term for the set of appearances a comet presents during its entire spell of visibility or detectability. Isn't it appropriate that a "ghost" has an apparition?)

In a letter to Newton in 1695, Halley alluded to his suspicion that the comets of 1531, 1607, and 1682 were apparitions of the same comet. Halley was busy with sea voyages and other activities in the years that followed, but finally, in 1705, he published his work on the topic in "Synopsis of the Astronomy of Comets." There he predicted that the comet seen in 1531, 1607, and 1682 would return again in 1758—a figure he later revised to late 1758 or early 1759.

Halley was right, of course: The comet was first sighted near the end of 1758, and it passed nearest the Sun and Earth early in 1759. The orbit he had determined was a greatly elongated ellipse, kind of like a cigar shape with one end near the Sun (within the orbit of Venus) and the other end very far from the Sun indeed (beyond the orbits of the yet undiscovered Uranus and Neptune).

Halley died in 1742, but not before he modestly staked his claim to fame:

> Wherefore if according to what we have already said it should return again about the year 1758, candid posterity will not refuse to acknowledge that this was first discovered by an Englishman.

Posterity did far more. Edmond Halley became the first person to have a comet officially named after him, and the wonderful returns of his comet make his name a household word once or twice in each century. In addition, posterity has acknowledged Halley as the first person in history to establish that comets do indeed obey the laws of physical science and that important aspects of their behavior can be explained and predicted by rational principles and mathematics.

Short-Period and Long-Period Comets

Now that Halley had unlocked the door, comet science was free to advance. It became more complex and collaborative, so our survey here must expand and concern itself more with subjects and knowledge than with individuals. As we will see in Chapter 3, however, the first successful theories of what comets are and where they come from were definitely the products of two individuals.

One thing astronomers in the eighteenth century quickly realized was that not all comets were going to turn out to have quite the same kind of orbit as Comet Halley. Halley himself recognized this and presented the set of positions some comets moved through as being best represented by the type of curve known as a parabola.

Halley believed (and science today agrees) that these other comets' orbits were really ellipses, not parabolas (a parabola is not a closed curve, so an object traveling such a course would never return). But Halley understood that these ellipses are so immensely elongated that the small section of them on which we observe a comet is indistinguishable from such a section on a parabola.

The diagram below shows the family of forms known as the "conic sections." As you can see, a slice through a cone at increasingly steep angles produces first a circle, then an ellipse, then a parabola, and then the even more wide-open curve called a hyperbola. As astronomers continued to calculate comet orbits from the eighteenth into the twentieth century, they found no comet that entered the inner solar system on a hyperbolic orbit unless it had passed close enough to a planet (usually Jupiter) to distort its path. (If a comet is ever found to approach on an initially hyperbolic orbit, then it will have to have come from another solar system— something that is not regarded as impossible but has never been observed). But astronomers did find what can be considered two major classes of comet orbits, a distinction of tremendous importance in all considerations of the nature and origin of comets.

It turned out that many comets take much less than 200 years to orbit the Sun and that many others take much more. The former are called

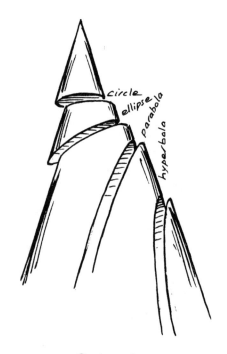

Conic sections.

short-period comets or *periodic comets*, the latter *long-period comets*. The figure 200 years is arbitrary; there are comets with orbital periods of a little less and others with orbital periods of a little more. But the vast majority of observed comets have periods that are either much less or much more than 200 years, and there is a clear distinction in orbital shape and orientation, as well as in brightness, between the two classes.

Short-period comets tend to have orbits which, though seldom circular, are usually quite fat ellipses. Those orbits also tend to be inclined only moderately from the plane of the planets' orbits. Most important, nearly all short-period comets have fairly poor intrinsic brightness (they have to come quite close to Earth to appear even moderately bright in our sky)— and it is not difficult to figure out why. A comet becomes active and may develop a large head or tail when it enters far enough into the inner solar system for the Sun's heat to affect it. Whatever is melting or vaporizing or

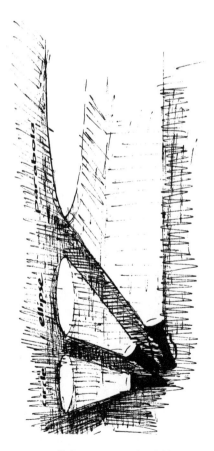

Flashlight beams on a wall demonstrate the different shapes of orbits.

expanding to make a comet brighter and more prominent when it nears the Sun must be finite in amount. Therefore, if a comet has made a great number of passages by the Sun, then it must eventually run low on volatile material. When a comet has been passing the Sun every 10 years for the past 10,000 years, it will have exhausted more of its material than a long-period comet, which may have passed the Sun only once in that time.

All the long-period comets that we've observed have greatly elongated orbits. These comets show no preference as to orbital inclination and may arrive from or depart in any direction. The long-period comets are not all bright. But among their number are many whose intrinsic brilliance greatly exceeds that of any of the short-period comets. These are the comets that can grow brightest, exhibit the most vigorous production of dust and gas, and develop the longest, densest tails.

Halley's Comet is one of the few short-period comets that is intrinsically rather bright. In a way, it is a hybrid of what's typical of the two kinds: Like many long-periods, it is bright enough to put on good shows in Earth's sky; like other short-periods, its orbit is small enough for it to come back repeatedly in human history.

Why are there short-period and long-period comets? The answer seemed fairly simple decades ago. It was noticed in the eighteenth and nineteenth centuries that a majority of the short-period comets have orbital periods of less than 20 years and an *aphelion* (point farthest from the Sun) somewhere near the orbit of Jupiter. Halley himself had revised his prediction to have his comet come back a little later—perhaps in early 1759 rather than 1758—because he believed that his comet was slowed down a little by passing moderately close to Jupiter. By the late nineteenth century, astronomers had seen many examples of comets whose paths were altered—sometimes drastically—by Jupiter. The message seemed clear: All those short-period comets with an aphelion near Jupiter's orbit had been lured into their present orbits by an encounter with the giant planet at some time in the past.

The theory was that every comet was originally long-period but eventually passed close enough to Jupiter—or occasionally to one of the other planets—to get captured into a short-period orbit. Eventually, however, astronomers came to doubt this scenario. Better calculations showed that once it was already short-period, a comet might easily have its orbit distorted and become a member of "Jupiter's family" of comets. But the change from long-period to short-period orbit could not be executed so easily.

There was in addition a raging mystery about long-period comets. How could there still be fresh and bright ones continuing to visit us after billions of years? The supply of fresh comets was somehow being replenished. And it had something to do with the fact that comets were coming in from an aphelion many, many times more distant than Neptune or Pluto. No scientifically satisfying answer to this riddle—the secret of where comets come from—was suggested until 1950, and the issue was not thoroughly worked out until the 1990s.

The Tiny Nucleus and the Vast, Tenuous Coma

Meanwhile, several hundred years of ever better telescopes and lots of bright comets gave astronomers plenty of opportunities to study the structure in both the head and the tail of the phantoms called comets. Scientists were studying the anatomy of a ghost.

The head of a comet consists of two things. The first is the vast cloud of gas and dust called the *coma* (Latin for "hair"; *comet* means "long-haired"). The second is the mysterious, comparatively tiny object (or swarm of objects, scientists once thought) that, when the comet nears the Sun, is hidden deep within the innermost, densest coma. This relatively tiny center is called the *nucleus*.

For at least a few centuries, the term nucleus has also been applied by observers to what is really the "false nucleus"—the densest cloud of gas and dust in which the smaller, true nucleus is hidden. Such a false nucleus might appear as quite a small dot even at high magnification, but there has long been evidence that the true nucleus must be even tinier. One fascinating reason why an upper limit could be set on the size of the nucleus of two particular great comets was that observers failed to see either comet's nucleus in silhouette when they passed directly between Sun and Earth. These two comets were the great comet of 1882 (1882 II) and Comet Halley in 1910. The nuclei of even these mighty comets must therefore, it was calculated, be less than 100 kilometers wide. Any larger nucleus should have been seen as black specks in front of the image of the Sun . . . if the nucleus was a single solid mass.

Actually, the most common conception of the comet's nucleus before 1950 was as a loose collection of ice-covered dust. Even as late as 1953, Raymond Lyttleton updated this basic idea in his "flying sandbank" model of the nucleus.

Whatever the structure of the nucleus, there was another reason why scientists knew it had to be small: the low mass of comets. Comets are influenced by Jupiter and other planets so easily that their masses must be tiny. An object of such low mass could not be very large even if its density were exceedingly low. Modern calculations suggest that the masses range from about 10 million to 10 million million metric tons (a metric ton is a million grams, or 2205 pounds). These figures sound impressive by earthly standards: The largest oil tankers weigh less than half a million tons when full, and the Great Pyramid is about 7 million tons—less than a millionth the mass of a large comet nucleus. But it would take over 70 million large comet nuclei to equal the mass of Earth's Moon, which has only 1/81 the mass of Earth.

And if a comet nucleus is low in mass, imagine how low in mass (and therefore how incredibly tenuous) the coma must be.

A comet nucleus just a few miles in diameter might be able to pass fairly near the Sun and produce a coma thousands of times before exhausting its supply of gas and dust, and the visible coma from such a nucleus might expand to 200,000 miles or more in diameter. Or, to put it even more dramatically, as Guy Ottewell did in *Mankind's Comet*,

> If we take its [the coma's] radius as 100,000 kilometers and that of the nucleus as 5, then the volume of the coma is 8 million million times as large. Yet the matter filling it is merely some dandruff fretted from the skin of the nucleus on this, one of the thousands of passages around the Sun.

A few visible comae have gotten much larger than this. The record holder is the Great Comet of 1811, whose visible coma was over 1 million miles across—larger than the Sun!

What is even more tenuous than a comet's coma? For one thing, the tail that forms from that coma, which may be tens of millions of miles long. It has been said that a long, spectacular tail—spanning a considerable fraction of the sky—could have all its material packed into a suitcase! That is an exaggeration (try an auditorium instead), but a comet's tail is surely the largest (and most glorious) visual use of the smallest amount of material that we can find anywhere in nature, unless it be the stars scattered through the inconceivable immensity of the universe itself or subatomic particles spun around the yawning void within the atom.

What else is more tenuous than a comet's coma? The *hydrogen coma*, or

H coma, which may stretch invisibly for millions of miles beyond the visible coma. But we are getting ahead of ourselves here. The H coma was not discovered until it could be photographed in ultraviolet light from outer space, in 1970. (This feat was accomplished by the OAO, or Orbiting Astronomical Observatory, and the comets whose H comae were thus revealed were Tago-Sato-Kosaka and Bennett—the very first comets I ever saw.)

I have said several times that a comet's coma is composed of gas and dust. The two components can be differentiated in several ways in both the coma and the tail.

For one thing, the gas and the dust of comets shine with different colors. The gas shines blue, the dust yellow or even slightly reddish. And the light from gas and that from dust have different causes. Comets are unique in the solar system in that they shine by both reflected and emitted light. The Sun, like other stars, glows with emitted radiance; the planets, moons, asteroids, and meteoroids all shine by light they reflect from the Sun. But comets have both dust that reflects sunlight and gas that emits its own light—though admittedly the Sun is the ultimate cause of the ionization that sets the gas glowing.

Ionization is giving atoms an electric charge—usually a positive one by stripping them of negatively charged electrons. *Plasma* is a gas—or really a state of matter beyond that of gas, which consists of positively charged atoms (called ions) and the free electrons that have been stripped from them. Scientists knew that solar radiation was somehow ionizing some of the gas of comets, but it wasn't until the 1950s that they understood the role of what we now call the *solar wind*. The solar wind is itself a mighty outflow of plasma from the Sun. Before a comet nucleus is heated enough to produce gas, the solar wind simply reaches the bare nucleus. But when the comet gas becomes ionized (becomes a plasma), it pushes back against the solar wind. The farthest sunward it reaches is called the outer *shock front*, which, as the diagram on the next page shows, is typically about a million kilometers sunward from the nucleus of a large, active comet. About 10 times closer in (though this relationship varies greatly with the strength of the solar wind and the comet's gas production) is the *contact surface*, the boundary at which the outflowing comet plasma is too strong for any of the solar wind to penetrate. In the zone between shock front and contact surface, the solar wind and comet ions are combined and driven laterally around the contact surface, and energy is generated by compression of the comet's plasma.

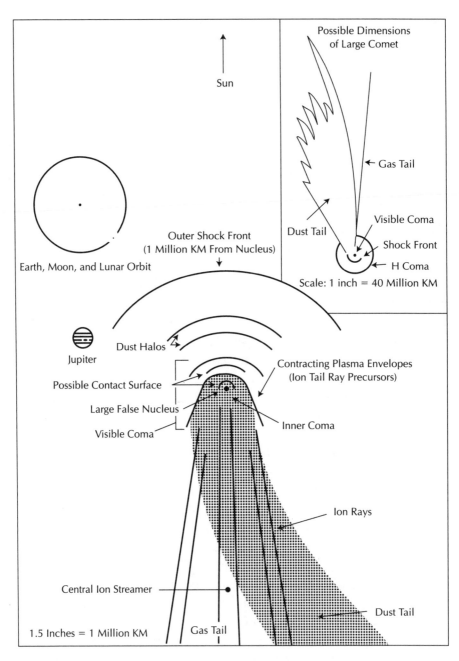

Structure of a large comet.

The visible coma typically reaches its largest physical dimensions in space when the comet is about 1.65 astronomical units (1 a.u. is 1 Earth–Sun distance) from the Sun, about as far out as the orbit of Mars. When a coma gets closer than this to the Sun, its enlargement usually begins to be reversed by compression from the solar wind.

How far out from the Sun does a coma first form? A coma of gases such as carbon monoxide may form when a comet is still beyond the orbit of Jupiter. But the typical coma becomes prominent (in the telescope at least) when a comet gets to within about 3.3 a.u. of the Sun. Because this is the distance from the Sun at which water ice would start to turn to a vapor, the appearance of a coma at this point was long ago recognized as a good piece of evidence that the nucleus must consist of a considerable amount of frozen water. Note that I said the water ice turns to a vapor. In the vacuum of space, the water ice in comets doesn't really melt—it *sublimes*, which is to say that it changes directly from a solid to a gas.

The dust in the coma is neither electrically charged nor atomic in size, so it is not compressed or rebuffed by the solar wind. But there is another force moving the dust particles away from the Sun: solar radiation pressure. Not just light, but other wavelengths of electromagnetic radiation as well, can be considered as consisting of particles called "photons." If a particle is small enough, in the resistlessness of space the momentum it gains from the impact of huge numbers of photons can actually affect its direction of movement. The solid particles in a comet's coma might include some of pebble or even boulder size. But most of what makes up the dust coma of a comet is roughly the size of the particles in smoke. As early as 1836, Friedrich Wilhelm Bessel studied his drawings of dust halos in the 1835–1836 apparition of Comet Halley and was able to do calculations proving that solar radiation pressure could account for the motion of the dust once it had escaped far enough from the nucleus. (Actually, the gravitational pull of the nucleus is so weak that dust and gas do not have to gush away very fast. Solar radiation pressure can quickly take control of the dust's motion.)

Tails of Gas and Dust

The head of a comet is composed of the unseen nucleus and the readily seen coma. But the real trademark of comets is of course the tail.

Perhaps all comets that get close enough to the Sun to develop a coma also develop a tail, but often it is too dim to see. A tail is seen if enough dust or gas has been released into the coma and is pushed away from the Sun by solar radiation pressure and solar wind. The action of those two forces on dust and gas explains the well-known fact—rediscovered by European observers in the fifteenth century and noted by the Chinese perhaps a thousand years earlier—that a tail always points away from the Sun. Like a lot of well-known facts, however, this one needs some qualification.

The biggest qualification is that there are two major kinds of comet tail (or, we could say, separate components of a single tail), and one of them can lag far behind the *extended radius vector* (the radius vector is the straight line from Sun to the comet's head, so the extended radius vector continues that line onward away from the Sun). Both types of tail may be produced by most or perhaps nearly all comets. Usually, we see a comet's tail from an angle such that the two kinds of tail are superimposed and thus difficult to differentiate from each other. But F. A. Bredikhin did so, recognizing that one type—which he called a *Type I tail*—was made of gas, and the other type—which he called a *Type II tail*—was made of dust. (Bredikhin also distinguished a *Type III tail*, but more on that in a moment.)

The gas tail is straight, narrow, bluish, shaped roughly like a cylinder— it lags no more than about 5° behind the extended radius vector. The dust tail is shaped like a curved fan or wing—not very thick in the plane of the comet's orbit, but spreading across an enormous breadth of space in that plane.

The *gas tail*—also called the *ion tail* and *plasma tail*—of a comet can grow outward at incredible rates of speed—millions of miles in as little as hours—with the help of a strong solar wind. So this tail appears straight and narrow and does tend to be directed almost exactly away from the Sun. Gas tails form first and, though less visually prominent than great dust tails, can often get even longer than dust tails.

We discussed above how the plasma of the solar wind and plasma of a comet mixes between the shock front and contact surface and is forced laterally around the contact surface. What happens is that this material forms "contracting plasma envelopes" which emerge behind the coma guided by the lines of magnetic force from the magnetic field carried by the solar wind. These lines of magnetic force are traced by the glow of ionized carbon dioxide (CO^+) to appear as *ion rays*. The rays can sometimes be seen even in fairly small telescopes if not much dust is present.

They range in width from 20,000 or 30,000 kilometers (several times the width of the Earth) down to 1,000 km (only twice the width of Great Britain or Italy)—and perhaps to even finer strands.

In most gas tails there is a central streamer. This is joined at an angle of about 60° by a set of ion rays forming from the contracting plasma envelopes. The rods are like the rods of a partly opened umbrella, but at first quickly and then more slowly, the rods (rays, that is) close on the central streamer, the whole process taking about 15 to 25 hours (and then perhaps beginning all over again).

An even more amazing phenomenon of the gas tail is *disconnection events* (or DEs). Such an event involves a gas tail sometimes many millions of miles long suddenly becoming detached and flying off—sometimes glowing for days, sometimes at combined speed (speed of comet moving onward and solar wind carrying tail away) of more than a million miles per hour. What causes a comet to lose its gas tail—and then re-grow a new one like a lizard? The leading theory is that which Malcolm Niedner and John C. Brandt proposed in 1978: the tail is cut off when the comet crosses a boundary between "sectors" of the Sun's magnetic field. These sectors are immense regions of alternately inward and outward directed magnetic field lines. Passing from one of these sectors to another is passing through a rapid change in polarity, which it is believed could disconnect a gas tail.

The *dust tail* is formed by particles which are pushed out by solar-radiation pressure many times more slowly than the comet's gas. Thus the dust tail of a comet at Earth's or Venus's distance from the Sun would take weeks to grow out to a length of millions or tens of millions of miles. During this time, the comet is moving onward along its orbit, and the dust particles are being left behind to pursue their own orbits. If all the dust particles were of equal size, the dust tail would be a narrow curving line— the farther from the comet's head the particles got, the farther behind the gas tail and extended radius vector they would fall. But the dust of comets comes in different sizes or, ot put it more accurately, different degrees of moveability. Why would one of two particles that were the same size have a greater susceptibility to being moved by solar radiation-pressure? One reason is that one might be darker, which is to say more absorptive of light. And there is indeed evidence of both light and dark dust particles in comets.

In any case, the fact that there may be different degrees of moveability means that a dust tail is not just curved but broad, with its leading edge

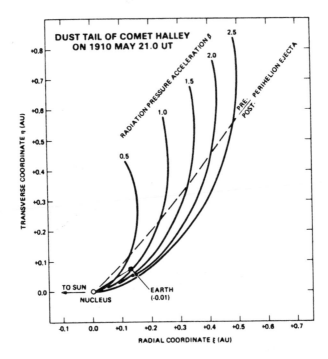

Diagram by Zdenek Sekanina showing "syndynes" (the curved lines) and the perihelion "synchrone" (the broken, nearly straight line) in Halley's dust tail 1.86 days after the nucleus passed between Earth and Sun and 30.8 days after it passed perihelion. The Earth lies 0.01 a.u. below the plane of the paper (the comet's orbital plane). The most movable particles (radiation acceleration pressure 2.5) lie along the longest synchronic curve.

longer than its trailing edge. The accompanying diagram of Halley's dust tail shows this beautifully. The long leading edge consists of dust particles with the greatest moveability or acceleration factor (represented by the Greek letter *beta*), the short trailing edge consists of the dust which is largest and/or most reflective (therefore hardest to move by solar radiation-pressure). The imaginary lines of "equal moveability" (the five curved lines shown in the diagram) are called *syndynes* or *syndynames*. Imagine a comet in which there are several discrete sizes (or size-ranges) of dust particles, all of which are of similar reflectivity. Such a comet, if we saw its tail broadside, would display separate branches in the dust tail which would be natural syndynes—the actual distribution of dust sizes in the tail would be laid out for us like a plan.

But there are other imaginary lines of interest we can draw through a comet's dust tail. *Synchrones* are lines which connect all the particles re-

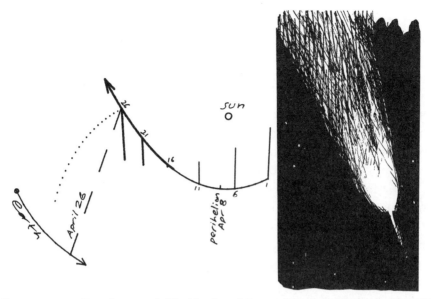

How an anti-tail can be caused: The Earth and Comet Arend-Roland in April 1957. On April 26, the Earth passed through the comet's orbital plane with the comet not far past the Earth-Sun line. The tail was therefore seen (angularly) ahead of the comet, and also some material on the other side of the sightline.

leased from the nucleus at about the same time. Thus in the diagram the dashed line is a synchrone, one of special interest—all the particles along this synchrone were released right at perihelion. In some comets, like West of 1976 and Mrkos of 1957, there may appear streamers in the dust tail which are approximately synchrones—they are the products of particular outbursts of dust from the nucleus. In fact, there can even be secondary synchronic features called *striae*—these occur when outward bound dust in a streamer crumbles into smaller pieces. The synchrones come directly from the nucleus, whereas striae come from a point out along a synchrone. Zdenek Sekanina and J. A. Farrell identified in Comet West 16 striae, derived from 5 dust-emissions, and in Comet Mrkos 26 striae from 16 weaker emissions. In Mrkos there were even two different kinds of striae.

A few pages back, I mentioned that Bredikhin distinguished a *Type III tail*. This appears as a short little brush of light trailing far behind the Type II tail. But in reality the Type III is just especially large dust particles, those that radiation-pressure pushes only very slowly away from the comet. As a matter of fact, it is believed that meteor showers are derived from the large particles in Type III tails (the spectacular connection between comets and meteors is a topic we look at in Chapter 4). Comet West displayed a Type III tail. But in 1957 (the year of Mrkos and its striated tail) Comet Arend-Roland presented Earth viewers with such a tail seen from such an angle that it appeared as a spike pointed right towards the Sun! Such an *anti-tail* is just a product of perspective, visible when we see the thin sheet of dust edge-on and just ahead of our line of sight to the comet's head—a vantage we can get when we pass through the plane of a comet's orbit. The diagram at the top of the preceding page illustrates a positioning of Earth and comet in which an anti-tail can result.

In the several hundred years after Edmond Halley, all of these complex and beautiful details (and more) were learned about the tails and comae of comets. But even as late as the middle of the 20th century there remained a part of a comet's anatomy which was a total mystery: the most important part, the permanent part, the part from which coma and tail arise—the heart or "nucleus" of the comet.

That mystery, along with the other central one—where comets come from—was finally addressed properly by two theories which, coincidentally, both were first announced in papers published in the same year, 1950. Those theories, and the incredible harvest of understandings and new ideas reaped from them in the years since 1950, are the topic of our next chapter.

Chapter 3

DIRTY SNOWBALLS AND THE NURSERIES OF LEVIATHANS

By 1950, comet orbits had been calculated and studied for several centuries, and the complex structure of the coma and tail had been identified and partly understood. But where comets originally came from and what made up the cores that produced the coma and tail in all their intricate, glorious structure—these central secrets were unknown.

The answers came in a flood from the theories of two men.

What Are Comets Made Of?

Surely, if scientists figured out the chemistry of the coma, they could figure out the chemistry of the nucleus it came from.

Ever since the nineteenth century, scientists have studied the spectrum of the light from comae to determine which chemical substances were present. The first photographs of a comet were those taken of Donati's Comet in 1858 by the English portrait artist Usherwood and the American professional astronomer George P. Bond. The first visual spectroscopic observations of a comet were made by Giovanni Battista Donati, not of his own great comet of 1858 but of Tempel's Comet 1864 II. Donati's drawing of the spectrum shows the three *Swan bands* of molecular emissions from C_2 (diatomic carbon), though he did not try to identify them, and they were later studied by, and eventually named after, another early spectroscopist, William Swan. The first photograph of a comet's spectrum, a *spectrogram*, was taken by amateur astronomer William Huggins of Tebbutt's Comet 1881 III. Spectral lines of Na (sodium) were detected in two of the comets of 1882 which passed very near the Sun (this emission is caused by the vaporization of comet dust at these small solar distances, and it adds to the luster of these magnificent comets). Conclusive evidence of the presence of both dust and gas in a comet is the existence of both a *continuum spectrum*, which must be due to sunlight reflected from dust, and an *emission spectrum*, which must be due to the glow from the comet's gas. The source of much of the comet's blue emitted light is ionized carbon monoxide (CO^+). Another constituent of comets that was early identified by spectroscopy is cyanogen (CN).

Scientists have detected several dozen different types of molecules and atoms in the coma by spectroscopy. But can we say with confidence that what is found in the coma must be what is in the nucleus? Not quite. Almost all of the coma consists of "daughter molecules"—the products of sometimes unknown "parent molecules" that were ionized or dissociated (broken down into component parts) by solar radiation soon after their release from the nucleus.

Of course, there has long been little doubt about the identity of some of the parent molecules. The hydrogen and OH (the hydoxyl radical) found in the spectrum of comae are clearly formed by the dissociation of H_2O. However, the abundance of water ice in comet nuclei was not a very helpful clue to the nature of the nucleus. The nucleus could be a single mass of ice or a swarm of ice-coated dust grains and presumably produce a coma with the same spectrum.

But the cometary nucleus is not a swarm of particles.

Whipple's Dirty Snowball

The model of the nucleus proposed by Fred Whipple in papers published in 1950 and 1951 is the *icy conglomerate model*, which is often called the dirty snowball model. This model represents the nucleus as a solid body consisting of ice (mainly water ice) in which dust (very small clumps of mostly silicate matter) is embedded. The model's value for explaining an enormous range of cometary activity eventually became clear.

Why was this model favored over the "flying sandbank" proposed by Lyttleton? The latter had many problems: Not enough ice could be "adsorbed" onto dust grains to produce a phenomenon as long-lasting as a comet; even a rocky body would have to be some meters across (not a collection of smaller particles) to survive passing as close to the Sun as many comets have. Neither could the flying sandbank model explain the strange *nongravitational forces* that seemed to cause some comets to deviate from their predicted paths. But Whipple pointed out that nongravitational forces could be accounted for quite well if they were assumed to be caused by jets of gas and dust shooting out from a dirty snowball.

In the vacuum of space, a large chunk of water ice would change directly from solid to gas as it gained heat from drawing closer to the Sun. Some of the more volatile substances known to exist in comets would be used up after just a few returns if they were not somehow trapped in the ice and so protected from the Sun's heat. As early as 1952, Armand Delsemme and Pol Swings suggested that these substances could be trapped in the crystalline structure of the water ice—they could exist as *clathrate hydrates*. As the ice sublimed into vapor, it would release the imprisoned substances. The occasional release of larger quantities from some parts of the nucleus could lead to the bursts of activity and flares of brightness for which comets are famous. Whipple and other scientists made important suggestions about how a *mantle* of dust covering most of the icy surface of a comet nucleus could help produce the localized jets, nongravitational forces, sunward fans of material, and more. A fresh comet might have much unmantled surface, but at each return some of the dust released by sublimation of the ice would not escape from the nucleus and would eventually settle back on the surface. Eventually, comet scientists came to the conclusion that some of this mantle's dust might be coated with dark organic matter or might itself be dark. The *albedo*, or reflectivity, of a comet nucleus might be very low, causing the nucleus to appear very dark.

The basic concept of the dirty snowball opened up tremendous possibilities for understanding comets. But could a spacecraft be sent to a comet to verify Whipple's model directly?

The Oort Cloud

The greatest single step in understanding where comets come from was taken long before a space probe would get to a comet. In the very year Whipple proposed his "dirty snowball" model, Jan Oort published his theory describing what came to be called the *Oort Cloud* of comets.

We observe such a tiny fraction of the innermost part of a long-period comet's orbit that it is difficult to calculate exactly how far beyond Pluto it goes before reaching its aphelion (far point from the Sun). Not until the time of Oort's work were there enough good observations and orbital determinations to provide a large enough sample (barely) of where the greatest number of comet aphelia lie. The answer turns out to be 50,000 a.u. to 100,000 a.u. from the Sun—a vast spherical shell of space more than 1000 times as far out as Pluto ever gets! Call that shell the Oort Cloud.

If you set up a solar system model in a large field with a soccer ball as the Sun, then to be in correct to-scale size and distance, Pluto would have to be a small pinhead 1000 yards away. And on this scale, a microbe-sized long-period comet at its aphelion would be about 1000 miles distant from your soccer ball Sun! Because 1 light-year equals about 63,000 a.u., and because the star system currently nearest us is 4.3 light-years distant, some long-period comets have aphelia well over a third of the way to the nearest star. How long does it take for comets in this Oort Cloud to complete their orbits? Between 1 million and 10 million years.

If the Oort Cloud is the storehouse for virgin or pristine comet nuclei, then how do we get supplies from that storehouse? Oort and other scientists showed that passing stars (particularly the occasional very close one) could speed up or slow down Oort Cloud comets, sending them flying away from the Sun's grip or falling on that multimillion-year drop toward the realm of the planets. These comets presumably became our long-period comets, until—it was thought—their initial forays near the Sun finally brought them near enough to a planet—usually Jupiter—to get routed into a shorter elliptical orbit (eventually short enough for them to be called short-period comets).

How dense is the Oort Cloud? How many comets are in it? Estimates have ranged from 100 thousand million to 10 million million comets. Whipple and Marsden have even argued that there may be far more: perhaps 10^{15} (a quadrillion) comets. All of these Oort Cloud objects are just chunks of an unusual kind of dusty ice, perhaps usually just a kilometer or two in diameter. But many of them are the seeds of sky-spanning majesty that will someday (or some eon) shine in the heavens of Earth. If all of them were to enter the inner solar system at the same time, the blaze of light, the tangle of countless tails, would be unimaginable and quite unbearable, a million million torches illuminating our world, the Sun itself almost lost in the maze of light. Yet the total mass of all comets in the

Oort Cloud is probably no more than a few times the mass of our small planet and is quite possibly even less than Earth's mass. Which is more amazing, the vast multitudes of comets or their next-to-nothingness? Neither. It is the volume of space across which they are scattered which most astonishes. The density of comets in the enormous region we call the Oort Cloud is incomparably less than even the average density of particles in a comet's tail, which is itself thinner than the best laboratory vacuums achievable on Earth. The region of greatest comet density is almost certainly not the Oort Cloud, where the millions of millions drift, but the comparatively tiny region around the Sun in which a few score sport—and become visible to us—like moths around a flame.

Revelations at Halley

As we'll see in a moment, the idea of a simple Oort Cloud giving rise to long-period comets that eventually became short-period comets turned out to need some modification. Meanwhile, the possibility of the most drastic modification—conceivably even refutation—awaited the dirty snowball model when spacecraft approached Comet Halley.

Would Fred Whipple's snowball model be verified on location? In March 1986, no fewer than five new spacecraft from two nations (the United States was not among them) and one consortium of nations passed

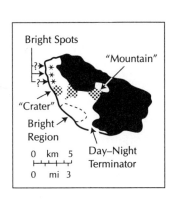
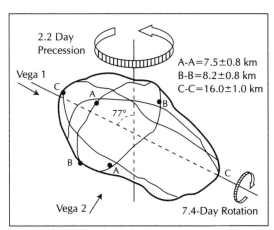

Diagrams of Comet Halley's nucleus based on images taken by the three spacecraft which passed near it in March 1986.

sunward of Halley's nucleus to learn what they could about the legendary object. The two Soviet Vega spacecraft and especially the European Space Agency's Giotto ended up passing right within the coma and, along with their other measurements, obtained photos of the nucleus.

The nucleus was a dirty snowball, but on the surface more "dirt" than snow. It turned out to be somewhat larger and considerably darker than expected—darker than coal! Its potato-shaped form measured about 15 × 8 × 8 kilometers. It was bumpy with hills and craters. Only about 10% of its surface seemed to be active rather than mantled with dark material. Although there were relatively light-colored silicate particles, there were also plenty of dark particles, made of an organic substance called CHON (carbon-hydrogen-oxygen-nitrogen). Being darker and more mantled than had been thought, the nucleus was also hotter—about 80° F, with some notably higher readings. The rotation of the nucleus appeared to be 2.2 days around a short axis with a precession (wobble) of 7.4 days around the long axis, although these results have since been called into question. But whatever the exact timings, it seems likely that the nucleus of Comet Halley travels through space like a "spiral" pass in football—but a supremely sloppy and wobbly one.

And the density of the nucleus? The size of the nucleus is now known, but various ways of estimating its mass produce different figures for its density. One summary article, published a year or so after the spacecraft armada passed, argued that the nucleus could be truly porous—somewhere between 33% and 90% empty space. A later summary estimated the density of the nucleus to be roughly that of bread.

The ice of the nucleus was estimated to be about 80% water ice but (surprisingly) only about 3% or less carbon dioxide and about 17% carbon monoxide, carbon monoxide perhaps being responsible for some of the brightest jets. The nucleus of Comet Halley was spouting like a whale but with seven jets at a time, shooting out an average of about 3 tons of dust per second and from 21 to more than 60 tons of gas per second.

All these data and the many fruitful ideas generated by Whipple's theory still leave us at the beginning—albeit a bright beginning—of understanding the nucleus. As Donald Yeomans explained in 1991, "Before discussing the nature of the cometary nucleus, it should be noted that very few details are known with any certainty. In his review of cometary nuclei in 1988, Michael F. A'Hearn wrote that there are probably more theories describing details of cometary nuclei than there are well-determined properties."

The Kuiper Belt

The theory of the Oort Cloud also needed modification. The composition of comets suggested that they must have formed at a temperature which would have occurred no farther from the young Sun than Uranus and Neptune in the "solar nebula," the cloud of gas and dust from which the solar system formed. How did such enormous numbers of these original comets get 1000 times farther out than Pluto, out to the Oort Cloud? Another problem was that the mechanisms (encounters with the planets) for turning truly long-period comets, coming in from any direction, into short-period comets with mostly low-inclination orbits seemed too inefficient to have produced as many short-period comets as we now see.

What was the solution to these problems? First to imagine an outer Oort Cloud and an inner Oort Cloud, the outer cloud spherical and the inner more flattened. And then to envision, far more flattened into the plane of the planets, nearer than the inner Oort Cloud, a *Kuiper Belt*. This belt had long been suspected. It was originally proposed in 1951 by the famous planetary scientist Gerard P. Kuiper (pronounced KI-per). But was there solid evidence for its existence? And could its existence solve the aforementioned problems?

When the strange object Chiron (not to be confused with Pluto's moon Charon) was discovered between the orbits of Saturn and Uranus in 1977, it looked like an asteroid very bright for its remoteness and therefore quite large. But by the late 1980s, it was starting to show a coma as it headed in for its 1996 perihelion. Chiron is an enormous comet nucleus. Several similar bodies, perhaps smaller and a little farther out, were discovered. Could all these objects have roamed inward from a Kuiper Belt?

In 1992, David Jewitt and Jane Luu, using a CCD (charge-coupled device, an electronic detector) with a large telescope on Mauna Kea in Hawaii, found an object orbiting at more than Pluto's average distance from the Sun. It was just the first of many such objects to be detected in the next few years. Their orbits are not bringing them toward the inner solar system and so may be fairly circular. They are bright enough to be bigger than the comets we're accustomed to— but the even bigger Chiron is a comet, so why not these? And why could there not be more ordinary-size comet nuclei in great numbers out there, just beyond detection? Beyond detection? In 1995, the Hubble Space Telescope took an image which seems to show innumerable such objects in a sample field of sky.

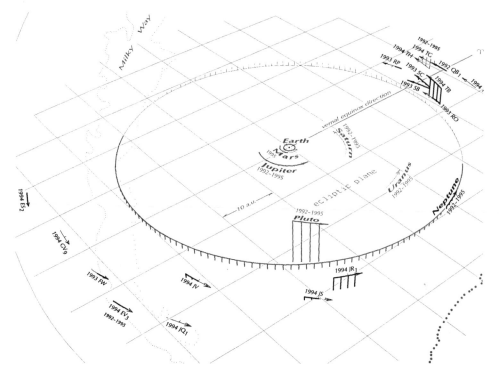

Paths of some of the first Kuiper Belt objects discovered.

This kind of object may need a name. Maybe "Kuiper Belt object" or "Kuiper Belt comet" is not too awkward. Whatever we call them, they may represent the solution to the Oort Cloud–related problems we have noted. And they may open the doors to understanding Pluto, Neptune's moon Triton, and many other mysteries of the outer solar system—not to mention a key episode in our entire solar system's genesis.

Kuiper's theory was that comets formed in the original solar nebula in a disk about 35–50 a.u. from the Sun (this figure was later revised: to 40–50 a.u.—just beyond Neptune) but were then *perturbed* (had their paths altered) by the giant outer planets. In those crowded, collision-filled days of the early solar system, the gas giants sent the comets flying in many directions: right out of the solar system, inward to the inner planets (bombarding the early Earth with water and perhaps organic molecules), and outward to form the Oort Cloud. The amazing thing is that these Kuiper Belt comets may be related to Pluto, to Neptune's moon Triton, and to a few of the apparently captured moons of the outer planets.

Table 1.   **The first Kuiper Belt objects**

		a	*P*	Discovered	by
1992	QB$_1$	43.89	291	1992 Aug. 30	D. Jewitt & J. Luu
1993	FW	43.87	291	1993 Mar. 28	D. Jewitt & A. Luu
1993	RO	39.33	247	Sep. 14	D. Jewitt & A. Luu
1993	RP	39.33	247	Sep. 15	D. Jewitt & A. Luu
1993	SB	39.42	248	Sep. 16	I. Williams, A. Fitzsimmons, D. O'Ceallaigh
1993	SC	39.50	248	Sep. 17	I. Williams, A. Fitzsimmons, D. O'Ceallaigh
1994	ES$_2$	45.27	305	1994 Mar. 13	D. Jewitt & A. Luu
1994	EV$_3$	43.13	283	Mar. 13	D. Jewitt & A. Luu
1994	GV$_9$	42.18	274	Apr. 15	D. Jewitt & J. Chen
1994	JS	36.54	221	May 11	D. Jewitt & A. Luu
1994	JV	35.25	209	May 13	D. Jewitt & A. Luu
1994	JQ$_1$	43.31	285	May 11	M. Irwin & A. Zytkow
1994	JR$_1$	35.26	209	May 12	M. Irwin & A. Zytkow
1994	TB	31.72	179	Oct. 2	D. Jewitt & J. Chen
1994	TG	42.25	275	Oct. 3	D. Jewitt, J. Chen, G. Knopp
1994	TE	40.94	262	Oct. 2	D. Jewitt, J. Chen, G. Knopp
1994	TG$_2$	41.45	267	Oct. 8	O. Hainaut

NOTE: Dates of "discovery" are those when the CCD frames were taken, rather than when the images were found on them. *a*, the mean distance in astronomical units, and *P*, the period in years, derived from it, are uncertain especially for the later bodies in the list. The intention is merely to show roughly how large these orbits are. (Courtesy of Guy Ottewell.)

They may all be *planetesimals*—objects of the kind that originally clumped together to form the planets.

When the cloud of gas and dust that was to form our solar system got dense enough in its center, that center underwent a nuclear fusion reaction and became a new star, our Sun. The other material tended to gather in the equatorial plane of the massive Sun (as particles do to form the rings of planets). The small bunchings of material were going in the same direction at similar speeds and began to clump together with their neighbors—a process called *accretion*—to form larger planetesimals. Eventually, full-sized planets began to dominate and to sweep up into themselves the remaining planetesimals (some of the crashes began to happen at different angles and harder). But small, rocky planetesimals were left everywhere—they became the meteoroids. There was an especially large number of (sometimes big) rocky planetesimals left over at the outer edge of the system of rocky inner planets Mercury, Venus, Earth, and Mars—the asteroids. There was an especially large number of icy planetesimals left over at the outer edge of the system of outer, gaseous planets Jupiter, Saturn, Uranus, and Neptune. What were they? A few of the big, icy planetesimals may have been captured to become moons of the outer planets (most of the moons formed from or with their planets, though). A few of the largest planetesimals, such as Pluto and Triton, had greater proportions of rock with their ice. And the rest of the icy planetesimals just beyond the outer gas giant planets became the Kuiper Belt objects of today.

Thus scientists may have found the Kuiper Belt in these last few years, and it may be of tremendous importance. It would be the place that comets originally came from before many were scattered in incredible numbers to the farthest reaches of the solar system to form the Oort Cloud. But some of the original comets stayed in their original location, in the Kuiper Belt. And although we get long-period comets from the Oort Cloud, we get short-period ones from the Kuiper Belt.

The nucleus of a comet is a dirty snowball—but also tremendously more, much of which we don't know yet. The storehouses of short-period and long-period comets are the Kuiper Belt and Oort Cloud. The Belt and Cloud remind me of this passage from Job; as translated in the Revised Standard Version of the Bible:

> Have you entered the storehouses of the snow, or have you seen the storehouses of the hail, which I have reserved for the time of trouble?

As we'll see in Chapter 5, showers of comets from the Oort Cloud have been proposed as the causes of Earth's greatest episodes of trouble (just ask the dinosaurs). But what I mean to convey in general here by quoting Job is that in learning about the comet's nucleus and the distant sources of these objects we are touching upon the most awesome of matters. And if we think of a later section of Job we are reminded that comets are more than snowballs, they are, when the sun ignites the greatest of them, vast majestic beasts. They are like the Leviathan of Job. And in discovering the Oort Cloud and Kuiper Belt we have found the nurseries of leviathans.

Chapter 4
THE SPECTACULAR DEATHS OF COMETS

A comet must eventually die, at least in the sense of becoming perma-
nently incapable of producing a coma or tail.

The death may occur when the comet nucleus gets smothered in dust
or exhausts its ice, thus becoming a dark, inactive body that can be mistaken for an asteroid (in fact, some of the objects we call asteroids may be
dead comets).

The death may be disintegration into innumerable meteoroids, some
of which can shower in our skies on the same special nights each year.

The death may consist of splittings of the comet nucleus that expose
fresh ice and induce rapid, bright, and entire consumption of the small
fragments.

The death may be the comet's hitting the Sun and getting vaporized!

Or the death of a comet may occur when it hits a planet, such as Earth.
When that happens, many or even most species of life on Earth can be
destroyed and the whole future of evolution radically redefined by a
comet.

The story of comet deaths is a story of comets bright enough to blaze in
broad daylight, of downpours of meteors by the tens or hundreds of thousands per hour, of mysterious lights in the sky racing out of the midnight
to brush past Earth, of the solar system's greatest explosions and the
planet-sized dark clouds they leave, of our entire world burning and

smoking and struggling to be reborn after being doused with acid and shrouded in pitch blackness for an infernal year.

Let's begin the story.

Meteor Showers from Comets

Meteors are commonly known as "shooting stars" or "falling stars." They are the streaks of light that usually last a second or less and occur when pieces of interplanetary rock and iron enter Earth's atmosphere at great velocities and burn up from friction with the air.

In space, these objects are called *meteoroids*; in the atmosphere as the light phenomenon, they are called *meteors*; on the ground (if they reach the ground, which is very rarely), they are called *meteorites*.

Some meteors we see are probably derived from asteroids and have had their far point out in the asteroid belt. These objects might even be considered the smallest of the asteroids, because the definitions of meteoroids and asteroids do not really draw a sharp line between the maximum size of the former and the minimum size of the latter. (In practice, however, we would always call a rocky body a mile or more across an asteroid and a rocky body less than a few dozen feet across a meteoroid.)

Some meteors are derived from asteroids. But the evidence is strong that most have their origins in comets. And it is possible that comets produce all the meteors that are seen in one of the sky's most entertaining and exciting kinds of display—a *meteor shower*.

A meteor shower consists of large numbers of meteors that all appear to diverge from a single point or area among the constellations, which is called a *radiant*. When we see such a display, Earth is traversing a roughly cylindrical ensemble of meteoroids that is spread out all along an orbit and is called a *meteoroid stream*. The source of this stream is thought to be a comet's dust tail, whose particles diffuse into a bundle of orbits near that of the comet. Each year, as Earth comes to the intersection area with a meteoroid stream at about the same date, we experience one of our annual meteor showers. Each year, skywatchers are treated to meteor showers such as the Perseids, the Leonids, and the Orionids. The *-ids* suffix means "children of," which is to say that the meteors of each shower come from a radiant in the "parent" constellation (the Perseids are the children of the constellation Perseus). Some radiants are near a particular star in the sky, and thus we have showers like the Delta Aquarids,

Table 2.   **Comets that have been identified (some dubiously) with meteor streams**

	Meteor stream	Comet	Period
Apr. 21	Lyrids	Thatcher	415
23	Pi Puppids	P/Grigg-Skjellerup	5.1
May 3	Eta Aquarids	P/Halley	76
June 3	Tau Herculids	P/Schwassmann-Wachmann 3	5.5
29	Beta Taurids (daytime)	P/Encke	3.3
30	June Draconids	P/Pons-Winnecke	6.4
July 16	Omicron Draconids	Metcalf (hyperbolic)	
30	Capricornids	P/Denning-Fujikawa?	9
	or	P/Honda-Mrkos-Pakjušaková?	5.3
Aug. 7	Iota Aquarids	P/ Encke?	3.3
13	Perseids	P/Swift-Tuttle	130
	minor with Perseids	P/d'Arrest	6.4
Sep. 1	Aurigids	Kiess (parabolic)	
Oct. 7	Piscids	P/Encke?	3.3
10	Draconids	P/Giacobini-Zinner	6.5
21	Orionids	P/Halley	76
Nov. 5	Taurids	P/Encke	3.3
12	Pegasids	P/Blanpain (lost)	5.1
14	Andromedids	P/Biela (lost)	6.6
18	Leonids	P/Tempel-Tuttle	33.2
Dec. 5	Phoenicids	P/Blanpain	5.1
10	Monocerotids	P/Mellish	145
22	Ursids	P/Tuttle	13.5

(Courtesy of Guy Ottewell.)

which radiate from near the star Delta Aquarii—that is, Delta in the constellation Aquarius.

Some meteor showers last for weeks or months, but these produce low hourly rates of meteors and are clearly the result of broad streams whose members' paths have spread apart so widely that our encounters with individual meteoroids are relatively few. The solitary *sporadic meteors*, the ones that seem to be associated with no meteor shower, may often be surviving members of comet-derived meteoroid streams, streams that have at last diffused too broadly to be identified in any way as discrete entities.

On the other hand, relatively dense concentrations of meteoroids can occur in bunches along a stream that has recently (or even not so recently) been replenished by an active comet. Such a concentration, called a *meteoroid swarm*, can cause intense meteor activity. If Earth

The Leonid meteor storm of November 13, 1833.

passes into a swarm, then an observer in the country might see many more than the dozens per hour visible in the strongest of the annual meteor showers. Such an event, which may feature hundreds or even thousands of meteors per hour, is called a *meteor storm*.

The most spectacular of all meteor storms in modern times have been those of the Leonids (the children of Leo). In 1833, 1866, and 1966, the Leonids were so plentiful that observers in some parts of the world saw "waterfalls" of shooting stars flowing down the sides of the sky and for a few minutes spates at rates of up to 500,000 Leonid meteors per hour! In most years an observer will spot something like five Leonids in an hour under good conditions. But on a single mid-November morning for a few years before, and especially after, the passing of the comet which produces the Leonids, there is a chance of a meteor storm. The next prime opportunities will come before dawn on November 17, 1998, and November 18, 1999, though even on November 17, 1997, we may see a spectacular or at least greatly enhanced display. These events will coincide with the early 1998 perihelion passage of Comet Tempel-Tuttle, which has a roughly 33-year orbit.

The two meteor showers with orbital elements similar to Comet Halley's are the Eta Aquarids of May and the Orionids of October, though storms of either probably occur only within a few hundred years of when a Halley orbital node crosses through Earth's orbit (more on this later in the book). But every year (weather permitting), an observer can see at least a small number of Eta Aquarids and Orionids—and know that these are caused by the incoming flight of debris which once shined in the dust tail of Comet Halley. Thus every year we can watch samples of Halley's Comet flame across the sky.

The first proposed connection of a comet and a meteor shower on the basis of similarity of orbital elements was made in 1866 by Giovanni Schiaparelli, the famous Mars observer (he innocently started the Martian canal controversy when his use of the Italian word *canali*—"channels"—was mistranslated into English as "canals"). Schiaparelli suggested the link between the Perseids and Comet Swift-Tuttle, which had made a bright appearance in 1862. This link was questioned in the 1980s but then was strongly supported in the 1990s when the comet returned and brought with it brief but very strong peaks of Perseid activity.

Remarkably, it was not long after Schiaparelli proposed a connection between Comet Swift-Tuttle and the Perseid meteor shower that nature gave one of the strongest possible corroborations of a link between comets and meteor showers. I don't mean the great 1866 performance of the Leonid meteor shower in association with a passage of Comet Tempel-Tuttle. I'm referring to the Andromedid meteor shower and Comet Biela.

Biela's Comet and the Andromedids

The story actually begins at least as far back as 1826. Just 4 years after the first predicted return of Comet Encke, Captain Wilhelm von Biela was watching for the return of a comet that Joseph Morstadt had told him might be appearing—or rather reappearing if the comets seen in 1772 (Montaigne's) and 1805 (Pons's) were one and the same. Bessel had been the first to calculate the orbits of the two earlier comets well enough to suggest that they might be the same object, but then he had recalculated and retracted his statement. Morstadt believed the original prediction was correct. And so it proved on February 27, 1826, when Biela located a comet well below the naked-eye limit of brightness. Biela's Comet brightened to just better than that limit and was observed for 72 days, enabling its period of about 6.7 years, and its certain identity with the comets of 1772 and 1805, to be established. (Incidentally, Pons located what was to become Biela's Comet on November 10, 1805—just 21 days after he "discovered" what was to become Encke's Comet!)

When Comet Biela was recovered by John Herschel at its next visit in 1832, it became the third comet (after Halley and Encke) to have had a return successfully predicted. But the comet's fame and importance were to come from what happened at the following observed returns, which was quite dramatic.

The 1839 apparition was too poorly placed to see, but in January 1846 observers noticed something remarkable: The central condensation of Biela, and thus presumably its nucleus, had split in two. The true distance between the two remained constant at about 1.6 million miles (over 6 times the distance from the Earth to the Moon). That worked out to an apparent separation of about 14 arc-minutes (a little less than half the Moon's apparent diameter as seen from Earth) when the comet was closest to Earth and was just barely fifth magnitude in brightness. The longer of the two comet tails was only about three-quarters of a degree long (longer than the Moon's apparent width).

Astronomers eagerly awaited Comet Biela's next return. That next return, in 1852, was not a very favorable one for viewing. But sure enough: First the brighter, then the fainter central condensation was spotted. The distance between them had increased greatly.

Unfortunately, the 1859 return was very poorly placed, and the comets (they now deserved to be considered separate comets) were not seen. Astronomers had to console themselves with the knowledge that the 1866 apparition should be in a very favorable position.

What was seen in 1866? Nothing. The general opinion was that the comets had broken up completely and would never be seen again. When Schiaparelli announced the probable connection between the Perseids and Swift-Tuttle, however, there soon followed simultaneous and independent announcements by H. I. d'Arrest and E. Weiss that the Andromedid meteor shower's stream was in an orbit virtually the same as that of Biela's Comet. The Andromedids had been observed at least as far back as 1741, and excellent displays had occurred in 1798, 1838, and 1847. Weiss calculated that a close approach of Earth and the comet(s)—if it (they) still existed—should take place in November 1872.

No Biela's Comet was seen in November 1872. But what was seen on November 27, 1872, was one of the strongest meteor storms of modern history: Rates exceeding 10,000 meteors per hour were observed. There seemed to be little doubt that the display was at least a by-product of some further and fatal breakup of the nuclei of Comet Biela. If any further support were needed, it was provided on November 27, 1885, the next time the comet should have been passing near Earth. That day, shooting stars flew at rates as high as 75,000 per hour, a storm surpassed in all modern history only by the few best Leonid outbursts.

It is important to realize what these facts do *not* indicate. The meteor shower did not replace the comet; remember that the shower had been observed long before the comet was seen to split. Not the shower but rather its enhancement into a storm—in space, the production of a dense meteor swarm—seems to have resulted from the split. The facts also do not prove that the nuclei broke up entirely. There is no doubt that whatever remained of the comet ceased to produce a coma, but one or two dark, perhaps rocky cores probably travel along the orbit and may someday be found—even though the latest attempt to find them, in 1971, was a failure.

After 1885 there occurred some good displays of Andromedids, but nothing greatly better than the ones seen in the decades before the comet split: in 1892 a maximum of 300 per hour, in 1899 one of 100 per hour, even as late as 1940 a peak of 40 per hour. Up until this time, the meteor shower and the comet (or whatever was left of it) had always passed extremely close to Earth's orbit. In fact, in 1832 the announcement that this comet was passing through Earth's orbit—50 million miles ahead of Earth—set off a panic among people who thought the comet was going to hit the planet. And in 1806 (the return during which Pons found it), the comet itself came to within 0.0366 a.u. of Earth, our fifth- or sixth-closest encounter with a comet now known (the comet became about third magnitude, like an average naked-eye star). After 1940, however, the meteors and comet cores (if any) were perturbed away from us by Jupiter. Ever since, what is left has been a very minor and ill-defined shower, with uncertain reports of peaks around November 14–15 and November 23–24.

Split Comets and Dissipated Comets

Why did Comet Biela split, and exactly what happens when a comet splits? In 1976, Zdenek Sekanina was working on a new model for calculating the recession of companion nuclei and had just started applying it to Comet Biela when he received the first reports that the nucleus of Comet West had split. The well-observed further behavior of the four West nuclei gave Sekanina some additional new data with which to test his model. His contribution was to recognize the importance of nongravitational forces (the rocket-like action of jets of dust and gas firing off from the nucleus) and to find a way to take them into account in predicting the paths of the companions and their fates.

Sekanina calculated the possible gravitational interactions of two fragments in the orbit of Comet West and found a wide variety of scenarios, depending on the part of the parent nucleus from which the companion splits off and the initial direction of the companion's motion. According to Sekanina, "the calculated patterns range all the way from the escape of the companion along strongly hyperbolic trajectories (relative to the principal fragment, not the Sun) to pursuance of quasi-stable periodic orbits about the principal mass, terminated in exceptional cases by collision of the fragments." He warned, however, that the fragments are extremely unlikely to achieve a permanent, stable gravitational attachment to each other. Such a pairing or grouping is only remotely possible because of the additional factor that becomes more important than gravitational attraction as the fragments get farther apart: the effects of outgassing from the fragments, especially from freshly exposed surfaces.

The weak gravity of these bodies affects them substantially for only a short time after splitting, and thereafter the rocket action of their outgassing can become the major force in determining their fates. Between two and five fragments have been observed when a nucleus splits (in the special case of Comet Shoemaker-Levy which we will examine later, many more!). But usually only one nucleus fragment survives over time and proves itself the primary. The other fragments tend to be relatively tiny chips off the primary block. But these lesser nuclei are able—even likely—to rival or exceed the primary in brightness for a time (or several times). This is because their irregular shape, their small size, and the large proportion of their surface that is fresh (unmantled) ice make them especially prone to sublimation. They also may tumble chaotically, which keeps bringing fresh areas under the Sun's stare at short (and irregular) intervals. Like wayward, rotating fireworks breaking loose from the main rocket (or perhaps like a balloon that rapidly deflates when one has let go of it), these small fragments will have a wild career as they rapidly expend themselves.

How long do secondary nuclei last? In most cases where they are prominent enough to detect, they remain visible for days or weeks, but only in a few cases for longer. Comet Biela was unusual in that its secondary nucleus survived well enough to come back at the next return as an independent, active—though fainter—object. What has happened to secondary nuclei when they fade from sight even while the primary nucleus remains visible? Sometimes they may have simply exhausted the part of themselves that was exposed ice, leaving a small rocky chunk. More often, the entire body may be essentially a mass of dusty ice that we watch as it evaporates completely, like an ice chip cast into boiling water. The comae of many of these fragments in their last days of visibility begin to expand and diffuse, becoming ever larger but ever dimmer clouds until their surface brightness is too low to detect—a last puff of vapor spreading ever more thinly into nothingness.

Imagine how difficult it must sometimes be for observers to keep track of the identity of various fragments. Not only must they consider the angle at which we see the fragments' paths projected onto the sky. Mutual gravitational attraction and erratic rocket effects must also be factored into the interpretation of which fragments are which on various dates. And if that weren't enough, the fragments' freshly exposed surfaces may make them flare and dim erratically.

Sekanina believes that the actual splitting need not be violent. In most cases, fragments may separate at speeds of only a few meters per second. The secondary nuclei may often be slabs of ice which have slid off the parent, possibly when the rotation became too swift or the dust mantle on that particular slab too heavy. The splitting of a nucleus may be less a breaking up than a peeling off.

Such mechanisms would help explain why splitting seems to occur as frequently far from the Sun as near—in one case, at as much as 9 a.u. beyond (and before) perihelion. Only a few of the 21 cases of split nuclei that Sekanina originally studied could be explained by tidal disruption by the Sun or Jupiter (the sungrazers 1882 II and Ikeya-Seki are examples of the former, periodic Comet Brooks 2 of the latter). In the sungrazer 1882 II, the splitting must have been more violent, for in addition to the more enduring fragments that stayed in the coma, there appeared for just a few days a number of remarkable other fragments that had gotten much farther away and looked like additional smaller comets, each with its own tail! Other split nuclei of a tamer sort have produced streamers within the coma or trailing back from it (which might also justly be called tails).

Do nucleus splittings cause increased activity, increased gas or dust production, and greater brightness? Perhaps they always do. All that can be said confidently is that comet fragments sometimes get stirred enough for the activity to be observed. The evidence is certainly clear for Comet West in 1976, when the splitting of nucleus D from nucleus A was accompanied by an increase in both the total brightness of the comet and its dust emission (especially several dust streamers in the tail).

In their final days, secondary nuclei resemble what might be called dissipating comets. There are a number of comets that have approached the Sun and, either before or after perihelion, have faded, grown diffuse, and simply vanished once their comae became too spread out to detect.

Are there any split comet nuclei that have separated far enough to come into view years apart, seemingly as separate comets, even if still on very similar orbits? In 1971, Ernst Opik argued that the chance of comets' *not* being associated in these kinds of genetically related groups was 10 to the minus 39th power and that groupings of two to seven members make up 60% of all known comets. But in 1977, Fred Whipple noted that some of Opik's assumptions were incorrect and that only one genetically related group could be proved. Fifteen supposed groups had been designated A to Q ("I" and "J" designations were not used), but only the members of group M are now widely accepted as related.

The members of group M have a special name, whose very mention sends chills of awe down the spines of comet observers. These comets are the Kreutz sungrazers.

Splendor of the Sungrazers

The Great Comet of 1680 that fascinated Halley and Newton was the first comet calculated to have passed close enough to the Sun to justify the appellation "sungrazer." But in 1843, the Great March Comet passed even closer: within the Sun's inner corona, a mere 131,000 kilometers—0.19 of the solar radius—from the Sun's blinding surface, the photosphere. Unlike the 1680 comet, this one was observed in broad daylight, in fact right up to within 1 degree of the Sun's disk. Soon after, its tail

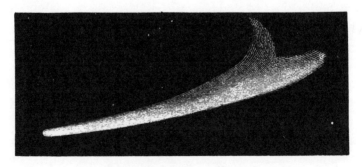

The Great September Comet of 1882.

shot out to the staggering length of 300 million kilometers—across the orbits of Mercury, Venus, Earth, and Mars and beyond!

The orbit of the 1843 comet was different from the 1680 comet's but remarkably similar to that of the lesser yet still spectacular comet of 1668. Doubt remains about whether the 1668 comet belongs to the Kreutz group with the Great Comet of 1843, but if so, it may be the only Kreutz member that is not a sungrazer, having passed "only" about 5 times closer to the Sun than Mercury does. That perihelion distance does not place it in the top 20 of smallest perihelion distances; the Great Comet of 1843 came about 12 times closer to the Sun.

In the decades after the 1843 comet, there were several grand comets but no sungrazers. Then, in seven astounding years from 1880 to 1887, there came three sungrazers in orbits remarkably similar to that of the 1843 comet. The orbit of the third remains imprecise because the comet was observed so briefly, but the second comet was dramatically visible for months and had its orbit accurately calculated—by Carl Kreutz. This was the Great September Comet of 1882 (1882II), whose maximum brightness may have been more than 100 times greater than the full Moon's!

How could these comets survive passing so close to the Sun? Only their hairpin turns at speeds of up to several million kilometers per hour prevented them from being pulled into the Sun, boiling away, being torn to rubble by tidal stresses, blowing apart under steam pressure, or vaporizing completely—even rock and metal vaporize easily at temperatures of 5000 K.

Consider for a moment the speed of a sungrazer comet. If the Moon orbited Earth at such a speed, we would see it complete its orbit and go through its entire set of phases in less than an hour. If Earth traveled around the Sun at this velocity, each season would last about 3 days, and the year would be complete in less than 2 weeks. Light itself, the fastest thing in the universe, travels about 500 times faster than a sungrazer, so these comets do not undergo significant changes in time, length, and mass as a result of relativistic effects. But no other enduring, discrete, macroscopic object in our solar system travels anywhere near so fast.

Although the sungrazers' great speed gets them away from the Sun very quickly, they do suffer from the tremendous expenditure of their materials—surely only the largest sungrazers could survive more than one or two perihelion passages. And the tidal and other stresses are likely to cause splitting of the nucleus. Two of the three greatest sungrazers of modern times, the comets of 1882 and 1965 (Ikeya-Seki), suffered dramatic splits that made them all the more interesting.

Table 3. **Sungrazers and first sunstrikers**

Sun's surface		T		km 696,000	radius 1.00	a.u. .004653	P
(Aristotle)	372 B.C.		I?				
	1106	Feb. 4	II?				
	1668	Feb. 28	??	9,964,000	14.32	.066604	
		Mar. 1	*??	1,500,000	2	.01	
(Kirch, Newton)	1680	Dec. 18	—	931,000	1.34	.006222	
(Richaud)	1689	Nov. 30	—	9,639,000	13.8	.064430	
		Dec. 2	*??	1,500,000	2	.01	
(Jacob)	1695	Oct. 23	—	6,328,000	9.1	.042297	
		Oct. 23	*??	1,500,000	2	.01	
	1702I	Feb. 15	*??	1,500,000	2	.01	
Great March Comet	1843I	Feb. 27	I	827,000	1.19	.005527	513
Great Southern Comet	1880I	Jan. 28	I	822,000	1.18	.005494	
Object Towfick	1882	May 17	??				
Great September Comet	1882II	Sep. 17	II	1,160,000	1.67	.007751	759
Great Southern Comet	1887I	Jan. 11	*?	723,000	1.04	.004834	
Du Tolt	1945VII	Dec. 27	II?	1,124,000	1.62	.007516	
Pereyra	1963V	Aug. 23	I	758,000	1.09	.005065	903
Ikeya-Seki	1965VIII	Oct. 21	II	1,165,000	1.67	.007786	880
White-Ortiz-Bolelli	1970VI	May 14	II	1,328,000	1.91	.008879	
Howard-Koomen-Michels (= Solwind 1)	1979XI	Aug. 30	II?	245,000	.35	.00164	
Solwind 2	1981I	Jan. 27	*?	730,000	1.05	.00488	
Solwind 3	1981XIII	July 20	*?	639,000	.92	.00427	

Listed are the comets (known or possible) going within 0.1 a.u. of the sun; and one other, the comet of 1668 (at least 9 unlisted comets went nearer than it), T is perihelion date, P is estimated period in years. Membership of the M or Kreutz group of comets is shown by: "I" or "II" (subgroups), "?" (subgroup uncertain), "??" (membership in the whole group uncertain), "—" (not a Kreutz member). "*" indicates that perihelion date and distance are as given only on the assumption that the comet is a member of the Kreutz group. For some comets there are two entries, the second being the possible Kreutz-sungrazer solution of its orbit. Perihelion distance is given in three ways: kilometers from the sun's center; fraction of the sun's radius; astronomical units (a.u.). The last three comets were sunstrikers, their perihelia being (probably) within the sun. (Courtesy of Guy Ottewell.)

Ikeya-Seki was clearly a less impressive object than the 1882 comet. Yet it was a marvel to those people, especially at southerly latitudes, who got a good view of it before it faded. And there is no telling when another great sungrazer might come. But speculation about the splitting of sungrazers brings us to consider the origins of the Kreutz group. The favored interpretation of the known facts is the most marvelous of all ideas about the Kreutz objects: the idea that all of these comets, which include some of the brightest and longest-tailed in history, are probably just pieces of a much larger parent sungrazer which was visible not many thousands of years before the beginning of recorded history.

The Parent and Evolution of the Sungrazers

What was this parent of the sungrazers like? Its nucleus was probably many times larger than most comets—perhaps as much as 120 kilometers across. A body of that size coming so close to the Sun almost certainly would be an object of absolutely staggering brightness. Though some comet scientists question this, Zdenek Sekanina has told me that the Sungrazer Parent's absolute magnitude could have been on the order of -5. The *absolute magnitude* of a comet is the brightness in the magnitude system that it would have if it were 1 a.u.—one Earth–Sun distance—away from both Earth and Sun. The parent of the sungrazers could have passed Earth's orbit when 90 million miles from Earth and still, even at that early stage, have easily outshined Venus. When you consider that this absolute magnitude of the parent is 100 times greater than that of its mighty 1882 II fragment, you begin to glimpse the possibilities.

It seems possible that the Sungrazer Parent could have been visible in broad daylight for weeks and outshined a half Moon (how much more intense it would have been!) even when far enough from the Sun to be visible well outside of twilight. Were lesser comets wrenched from it that were themselves large enough to form impressive naked-eye comets with it as it departed the Sun? Was the tail of the main comet bright enough for tens of degrees of it to be visible in broad daylight?

Putting aside our speculations about what the Sungrazer Parent may have looked like, let's consider when it made its last appearance as a whole object. Brian Marsden did a study of the Kreutz members and, among many interesting findings, calculated that the parent object probably first split up between 10 and 20 revolutions ago. With an orbital period of very roughly 1000 years, the original Kreutz comet thus was probably seen between about 18,000 and 8000 B.C.—a very long time ago, but certainly a time of considerable culture in some places in the world. Part of the period was Ice Age, so this stupendous comet may have blazed over a wintry world.

The way Marsden envisions the evolution from the first major split of this comet to the current assortment of lesser sungrazers—now spread over at least several hundred years (a sizable fraction) of the original orbit—is also fascinating. He believes the two halves of the original comet may have persisted until the years 371 B.C. and A.D. 1106, at which times each probably divided into at least three pieces, which have been returning as the sungrazers of the past several centuries. The 371 B.C. comet was the one mentioned by Aristotle, which apparently even naked-eye observers thought was split in two and whose tail stretched at least one-third of the way across the sky (60 degrees) according to the crude observations of the time. The A.D. 1106 object was discovered only 2 degrees from the Sun on February 4 from Europe. After this daylight visibility, it appeared in the southwest after sunset on February 7, and on February 10 was observed in China, Japan, and Korea in the west sky with a tail perhaps 90 degrees long and up to 5 degrees wide. It remained visible until at least mid-March.

Marsden thinks that the 371 B.C. comet could have been the progenitor of what he calls Sub-group I and the A.D. 1106 comet the progenitor of Sub-group II. The two sub-groups are distinguished by a difference of about 20 degrees in the "line of the nodes" for the members. In Sub-group I the major member is the Great Comet of 1843; in Sub-group II the major members are the Great Comet of 1882 and Ikeya-Seki of 1965. Marsden argues that the 1882 and 1965 comets were the largest pieces of the 1106 comet's breakup.

Only a very violent split could put two fragments of a comet on orbits different enough to bring them back a human lifetime apart by the time of their return seven or eight centuries later. Other appearances of sungrazers just a few years or even months apart would be derived from more gentle separations. A split of the former kind—in fact, a split of an even extremely violent nature—might be necessary to send fragments into orbits markedly unlike those of the other Kreutz members. Yet such splits may be possible, as is evidenced by the strange independent little comets apparently spawned and ejected far from itself by the great 1882 comet. Marsden makes what he calls "a wild speculation" about the Great

Comets of 1680 and 1843. These two, he says, may have been the major fragments of the split of the 371 B.C. comet. The 1680 comet now has an orbit very different from those of Kreutz members, but Kreutz himself pointed out that its orbit passes within 0.0005 a.u. of the orbit of the 1882 comet—intriguing, though possibly just an unusual coincidence.

Are there any theories other than the breakup of a giant parent comet that could explain the Kreutz sungrazers? Seargent mentions Sekanina's proposal about "a collision of proto-comets at a heliocentric distance of about 1 a.u." But presumably this is just a speculation which Sekanina finds far less likely than the breakup of a parent comet.

By the way, the main argument of Marsden's plan for the genesis and evolution of the Kreutz group would require returns of the A.D. 1106 and 371 B.C. comets in the early centuries of the Christian era. There do seem to be some poorly reported but compelling candidates—certain objects which seem to have been comets observed near the Sun in daylight.

Could-Be Sungrazers

Another interesting enterprise is the search for records of Kreutz members that may have been visible only briefly because of a poor line of sight from Earth. What would the record of such a comet be like? Sometimes it might be a report of a bright object seen only once near the Sun in the daytime or just after sunset. Among Kreutz candidates of this sort are two that were sighted the same year as the Great Comet of 1882 (1882 II).

The second of these was a "star" seen at Broughty Ferry in Scotland around 11 A.M. on December 12, 1882. The tantalizing piece of information is that the "star" looked crescent-shaped with optical aid. Could that have been the parabolic shape of a bright comet's inner coma? Unfortunately, however, its position may rule out its being a Kreutz member even if it was a comet.

That was only 3 months after 1882 II reached perihelion. But 4 months before the great comet's perihelion, during the total solar eclipse of May 17, 1882, a very bright comet was observed right beside the sun with a half-degree-long, strongly curved tail. The key observations were made from Tewfick in Egypt, so this comet is often known as Object Tewfick. It was never seen before or after the eclipse, but that is perhaps not surprising considering the time of year: Kreutz members coming to peri-

Comet during total eclipse of the Sun in May 1882.

helion between about May and August approach from the opposite side of the Sun and stay hidden in its glare on their journeys in and out.

The aphelion point of the Kreutz members is in the direction of the brilliant star Sirius. Thus in summer, we see them (or, rather, fail to see them) approach from almost behind the Sun and curve just in front of the solar disk (or almost so) before receding again in the same section of sky as the Sun. This also means that we see them from an angle which greatly foreshortens their tails except for a day or two around perihelion. The dust tails of Object Tewfick and Ikeya-Seki at perihelion were much longer than they appeared, because they were strongly curved as a result of lagging behind the rapidly accelerating comet as it approached its blistering encounter with the Sun. Talk about strongly curved: The head of the Great Comet of 1843 must have swung through an arc of 292 degrees (out of 360) in a single day around perihelion (imagine how curved the lagging dust tail must have become).

Incidentally, much of the dust is vaporized right around perihelion in a sungrazer perihelion passage. As a result, even if our viewing angle on the tail is favorable, it is only after perihelion that a stupendously long and magnificent tail forms.

The Sunstrikers

Up until 1981, the Kreutz sungrazers' almost literal skimmings of the surface of the Sun were the closest encounters known of any object with the

Sun. There was only one way to do better. It was achieved by a series of comets thought to be members of the Kreutz group, albeit small ones. These objects are the first of what we could call "sunstrikers."

All were very small objects, with nuclei probably under a kilometer in width, and were never observed visually. Their images were found on photographs of the Sun's corona taken by an instrument aboard the U.S. Defense Department satellite P78-1. The device used for the photographs, a coronagraph, creates an artificial eclipse of the Sun's blinding photosphere with an occulting disk so that details of the pearly, structured "atmosphere" of the Sun can be detected.

The first and best-studied of the sunstrikers was also the one with the smallest perihelion distance. Comet Howard-Koomen-Michels was the first comet ever discovered by a spacecraft, and the interval between its passage and the time it was found by examining photographs was over 2 years. Not until September 1981 did Russ Howard have an opportunity to look at the photos of August 30, 1979, and discover the amazing sequence of images recorded with the instruments designed and operated by Martin Koomen and Don Michels.

The photographs show the area from a distance of about 10 solar radii inward to 2.5 solar radii from the Sun's center. Everything closer in is blocked by the occulting disk. The comet and its tail are clearly seen approaching the Sun, and the head going behind the occulting disk. But as the hours pass in which a sungrazer should reemerge, what is revealed instead is a dramatic spreading of the tail into first a fan and then a kind of semicircle, brightening parts of the corona before expanding further and fading from view.

What happened? The calculations of Brian Marsden and Zdenek Sekanina leave little doubt that this comet's course was that of a Kreutz member and was carrying it to a collision with the Sun's surface. Sekanina's figures show that the perihelion distance from the Sun's center was going to be about 0.35 solar radius—in other words, within the Sun. The comet must have traveled about 3 million kilometers in the last 40 minutes before collision! Whether the comet's nucleus entirely disintegrated shortly before the moment of collision cannot be determined, but according to Sekanina, the rate of sublimation was so great that in the last 3 hours, the mass loss would have been equivalent to a layer of "dirty" water ice 50 meters thick or a layer of CO_2 ice 200 meters thick (compare this to the meter-thick layer of ice which Halley's Comet would lose if it lost it equally all over its surface during all the many months and years of an entire apparition).

Table 4.   **Timing and location of some points on the path of Comet Solwind 1**

Predicted time[a] 1979 Aug. 30 (UT)	Predicted offsets from center of sun (solar radii[b])		The comet's head was to
	in right ascension	in declination	
21^h09^m6	−2.44	−0.55	disappear behind the occulting disk
21 30.1	−1.74	−0.36	reach perihelion distance of 1970 VI
21 36.7	−1.49	−0.29	reach perihelion distance of 1965 VIII
21 41.8	−1.28	−0.23	move normal to the line of sight
21 48.3	−0.99	−0.15	begin to transit across the sun's disk
21 49.3	−0.94	−0.14	reach perhelion distance of 1843 I
21 51.7	−0.83	−0.11	reach perihelion distance of 1963 V
21 53.8	−0.72	−0.09	enter the photosphere
22 10.9			reach the perihelion

[a]As observed from the earth, i.e., including the light time.
[b]Instantaneous solar radius was equivalent to 15.87 arcmin; coordinates refer to eq. 1950.0.
(Courtesy of Zdenek Sekanina.)

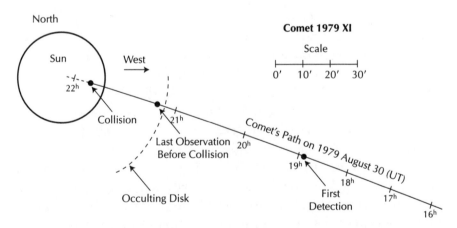

Path of SOLWIND 1 into the Sun (adapted from a diagram by Zdenek Sekanina).

Sekanina also studied the tail, which remained visible for over a day after the comet had hit the Sun. He was able to prove that the surviving tail was dominated by submicron-size dust grains which were launched toward the Sun by the powerful sublimation but then drastically deflected away by the intense radiation pressure pushing outward from the Sun—and, incidentally, toward Earth. The forward scattering of light from these particles some time after the comet hit could account for the observed brightening of the tail. The extent of the tail to the northeast shows that some of the dust could have survived to as close as 1.4 solar radii from the Sun's center before being vaporized and that such particles are likely to be made of a very refractory substance, perhaps of carbon-based composition.

Are sunstrikers rare? There has been a suggestion that we look for evidence of them in other solar systems in the spectra of stars. In our own solar system, the P78-1 satellite imaged five more little Kreutz members before it was destroyed in a "Star Wars" anti-satellite weapon test on September 13, 1985. (About this destruction, comet enthusiast Joseph Marcus wrote, "Contrary to the assertion of the U.S. Defense Secretary, Caspar Weinberger, that P78-1 was 'burnt out,' the satellite's coronagraph in fact was returning useful images to the very end.") Because more than three Naval Research Laboratory scientists per discovery were involved in the finds after Comet Howard-Koomen-Michels, the practice of naming the comets Solwind 2, Solwind 3, and so on was instituted. Were these all sunstrikers? Apparently not. (Although Solwind 3's perihelion distance was less than the Sun's radius, Solwind 2's was calculated by Marsden as 1.05 solar radii.) But none ever emerged from its close encounter with the Sun.

The same can be said of the next batch of satellite-discovered mini-Kreutz objects. The Solar Maximum Mission satellite—better known as SMM or Solar Max—was the first satellite ever repaired in orbit, an achievement accomplished by the Space Shuttle astronauts in 1984. Images taken by Solar Max from 1987 through 1989 turned up nine more little Kreutz comets that didn't emerge from the Sun's vicinity. Several of these rivaled Venus in brightness briefly before their destruction and were almost certainly fragments of what was once a larger object; if these had hung together, they could have emerged to give us a fine sungrazer display. Solar Max itself met its end, almost 10 years after launch, when it reentered Earth's atmosphere and crashed in the Indian Ocean on December 2, 1989.

The Solar Maximum Mission comets were named SMM-1 through

SMM-9. The first of the P78-1 comets was officially and retroactively named Solwind 1—the first object ever recorded on its deathflight to the surface of the Sun.

Can Comets Become Asteroids?

There is an even more dramatic way for comets to die than suncrashing—more dramatic, at least, from our biased point of view. I refer, of course, to comets hitting the Earth. However, before we consider the question of possible Earth–comet collisions, we should note that by all estimates, the chances of Earth's being struck by an asteroid are far greater than the chances of its being hit by a comet.

Now it is true that the asteroids in the main "asteroid belt" seem to be in very stable orbits, which never bring them near Earth. There is only a certain number of small asteroids of a special kind that can cross over, under, or through Earth's orbit and which are subject to major changes in their orbits in the course of a few million years—or in far less time when they pass very close to one of the inner planets. These latter asteroids are the ones which can end up targeted for Earth, and which resemble comets in enough respects to raise a provocative question: Could at least some of the objects we call asteroids be the extinct cores of comets?

The asteroids with orbits that bring them farther in than the main belt can be divided into several classes. *Apollo asteroids* have a semimajor axis (average distance) of more than 1 a.u. and a perihelion distance of less than 1.017 a.u. (the approximate aphelion distance of Earth). *Amor asteroids* have a perihelion distance of more than 1.017 a.u. but less than 1.3 a.u. (the approximate perihelion distance of Mars). *Aten asteroids* are those that have an average distance (a "semimajor axis") of less than 1 a.u.

All of these classes include members which can be called Earth-approaching. The closest known approach of an asteroid until recent years was that of Hermes, which in 1937 passed only 780,000 kilometers away (about twice the lunar distance). But in the past decade, better observation has revealed a number of bodies smaller than Hermes which have come much closer. Four of these objects have come closer to Earth than the average distance of the Moon. The current record holder is asteroid 1994 XM1, which on December 9, 1994, passed 0.0007 a.u. (112,000 kilometers, or 70,000 miles) from Earth—only about 9 Earth diameters from our world! To judge from its dimness, this object is probably

Comet Ikeya-Seki photographed by Dennis Milon on October 31, 1965, from the Catalina Mountains near Tucson, Arizona.

Comet Hyakutake as photographed at about 2:30 a.m. on March 27, 1996, by Ray Maher near Jake's Landing in New Jersey.

The white dust tail and blue gas tail of Comet West at 4 a.m. on March 7, 1976, photographed by Betty and Dennis Milon at Sullivan, New Hampshire.

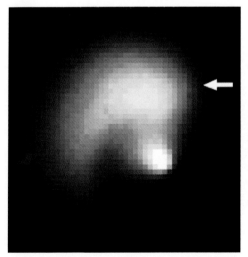

This image of Comet Hale-Bopp obtained by the Hubble Space Telescope on October 5, 1995, shows a giant counterclockwise jet and enshrouded possible fragment (indicated by arrow) coming out of the cloud-hidden nucleus. (Courtesy of Space Telescope Science Institute.)

Hale-Bopp, photographed on May 22, 1996 (soon after it was first sighted with the naked eye). Notice the trail of a satellite also recorded on this image taken by Gordon Garradd.

"The Adoration of the Magi," a fresco by Giotto in the Scrovegni Chapel in Padua, shows over the nativity scene what is probably Halley's Comet as viewed at its 1301 return.

Phil Hudson of Fort Wayne, Indiana, took 2-minute exposures of Comet Hyakutake near the North Star each hour from 9 p.m. to 4 a.m. on the night of March 26–27, 1996, and combined them to form this image.

The parabolic shape of Comet Hyakutake's coma, photographed by Richard A. Keen on March 27, 1996.

Comet Hyakutake and the Big Dipper, photographed on the night of March 24–25, 1996, by Paul Ostwald.

Comet Hyakutake photographed on the night of March 24–25, 1996, by Paul Ostwald.

Comet West photographed March 1976 by Dennis di Cicco.

The forked tail of Comet Hyakutake photographed on April 16, 1996, by Johnny Horne near Kelly, North Carolina.

Comet Hyakutake, Venus, and the Pleiades, photographed on April 17, 1995, by Ray Maher at East Point, New Jersey.

Halley's Comet photographed March 11, 1986 by William Liller.

only about 6 to 13 meters across, about the size of a very small house, so maybe we should call it a meteoroid. Its discoverer, James Scotti (who was using the Spacewatch Telescope at Kitt Peak in Arizona), told *Sky & Telescope* that although he did find this object, he "missed the other 40 or 50 objects of similar size that probably also passed within the Moon's distance that day."

The Earth-approaching asteroids are suspected of including some dead or inactive comet nuclei among their ranks. What suggests that this could be true? Foremost is the similarity to some comet orbits. Most of the Earth approachers have rather elongated orbits that cross the orbits of one or more planets and could not be stable in these paths over astronomically long periods of time. These asteroids are also assumed, on the basis of their estimated albedoes (reflectivities) and their observed brightnesses, to be very small objects—mostly just a few kilometers or less wide—similar to the size of comet nuclei or perhaps to comet cores if such exist.

From what is known of active comets' composition and from what scientists can haltingly venture to surmise about their origins, is it possible for comets to have nonvolatile cores? Some scientists have suggested that such cores could exist only if comets formed far enough inward in the solar nebula (perhaps within the orbit of Saturn) or if they contained enough short-lived radionuclides (such as 26A1) to generate the heat needed to melt their interiors. But current theory favors comet formation at least as far out as Uranus and Neptune. Whether comets could have contained enough radionuclides to form a core remains unknown.

Another way to try to determine whether comets can become asteroids is to investigate those objects of either class that suspiciously resemble objects of the other class and thus may represent a transition phase between the two.

On the asteroid side of the fence there is 944 Hidalgo. Until the discovery of Chiron, in 1977, no object defined as an asteroid was known to journey so far from the Sun. Hidalgo travels out as far as Saturn on a highly inclined orbit that exhibits a greater average distance from the Sun than Jupiter. If Hidalgo is dark like the outermost belt asteroids and the Trojan asteroids (which travel in two groups at the stable Lagrangian points ahead of and behind Jupiter on its orbit), then its diameter may be about 60 kilometers. That is certainly much larger than most comet nuclei, but it puts Hidalgo in a class with the periodic comet Schwassmann-Wachmann 1, which follows an orbit with average distance between

Jupiter and Saturn, may be about 75 kilometers in diameter, and may be covered with dark (probably carbon-based) material. This comet's orbit is actually more circular and less inclined—hence less "comet-like"—than that of the asteroid Hidalgo. Furthermore, Schwassmann-Wachmann 1 goes through quiescent spells when, in several respects (including its spectrum), it resembles an asteroid. Only when it flares up, becoming hundreds of times brighter, does it produce a coma which shows it to be clearly cometary.

Might it be that both Hidalgo and SW-1 were originally giant Kuiper Belt comets?

Because the momentous question we are working toward concerns the chance that a comet or asteroid will strike Earth, we should be most interested in this context in comets and asteroids which roam around the inner solar system. The least active comet which comes at least in to the orbit of Mars seems to be the periodic comet Arend-Rigaux. After showing coma and tail at its discovery apparition in 1951, this comet appeared to be shining only by reflected sunlight at the returns of 1957, 1964, and 1971. Several scientists predicted that although the 1978 return would be the most favorable for viewing since its discovery, the comet would still appear "starlike" (which is, by the way, a literal translation of the word *asteroid*, for the asteroids were so named because all but the largest are too small to show as anything but points of light in a telescope). What did Arend-Rigaux show in 1978? It fooled the experts: At the 1978 apparition it displayed both coma and tail, albeit weak ones. Nevertheless, this comet seems on the verge of entering an inactive phase in which it will look—for a long time? forever?—like an asteroid in the telescope.

An even more interesting story is that of Comet Wilson-Harrington. This object was discovered on photographs in 1949. It showed coma and a tail. The amazing thing was that when L. E. Cunningham calculated the orbital period it turned out to be 2.31 years—much shorter than the record-holder, Comet Encke. There was however an uncertainty of more than 2 years in this figure, for the comet had been photographed for too short a time to permit a confident determination of its orbit. Scientists thought that this comet might never be seen again (or at any rate connected with this apparition) because even in 1949 when it passed only 0.16 a.u. from Earth its brightness was only about magnitude 16—very dim indeed.

Now we flash ahead to 1979. Eleanor Helin discovered asteroid 1979VA, which later received the official asteroid number 4015. But it wasn't until 1992 that Ted Bowell calculated this asteroid's motion back through the years in an attempt to find pre-discovery photographs and learn more about the object. He found a 1949 image of it—with a tail! Brian Marsden naturally remembered Comet Wilson-Harrington and subsequent studies proved that it and asteroid 4015 are one and the same object. It is now comet 108 on the list of periodic comets—but which is it, asteroid or comet? Apparently at its 1993 perihelion passage, no sign of cometary activity was detected. Clearly, this is a transitional object. (By the way, is its orbital period shorter than Comet Encke's? No, Wilson-Harrington travels in an orbit which averages about 4.3 years, compared to Encke's 3.3).

What other objects in the solar system could be inactive (whether exhausted or pristine) comets masquerading as something else? The outer satellites of Jupiter and Saturn are prime candidates. Their sizes are mostly small (typically that of a very large comet nucleus), and their orbits around their respective planets are variously eccentric, highly inclined, and even retrograde (clockwise as seen from the north, unlike all the planets and most of the moons). No asteroids have retrograde orbits around the Sun. Many comets do. Whatever they are, it seems likely that these moons could not have formed with the planets they now circle, but instead were captured. And although the capture process is apparently not well understood, it may have occurred in the very early days of the solar system, when the vastly extended atmospheres of the proto-Jupiter and the proto-Saturn (helped) snare these bodies. An interesting question is whether all of Jupiter's known moons will remain captives of the planet. Some of their orbits can change drastically, and one of the moons can pass to more than 30 million kilometers from Jupiter. And here again there is a fascinating blurring of distinctions, for a number of comets have passed right in among Jupiter's satellites and been temporarily captured. The periodic Comet Brooks 2 came closer to Jupiter than any comet known up until that time and yet managed to escape. The periodic comet Gehrels 2 has been captured by Jupiter on at least three occasions: from 1783 to 1786, from 1833 to 1835, and from early 1967 to late 1974. On this last occasion, it was in an elliptical orbit around Jupiter for almost 8 years!

Perhaps the most interesting object that still looks like an asteroid but seems very likely to be an extinct comet is Phaethon, discovered in 1983. This object has a small orbit that takes it well within that of Mercury: If it was once an active comet, no wonder its ices have been exhausted.

Phaethon's cometary origin is suggested by its orbit, which is about the same as that of the meteoroid stream that produces one of the greatest of the annual meteor showers, the Geminids. There is a chance that true asteroids—rocky objects that have never been part of a comet—could produce meteor showers. But such showers and their meteors ought to be substantially different (for instance, more enduring) than those which occur in a typical meteor shower. There is still debate about whether Geminids are similar enough to the other showers whose connection to comets has been established.

Are all the Apollo and Amor and Aten asteroids really extinct comet cores? One important question is whether there is enough chance for asteroids to be perturbed out of the main belt to become Apollos, Amors, and Atens. Most researchers had concluded that such a change is possible but that the process is not efficient enough to have produced anything like the number of known Apollos, Amors, and Atens. But in the 1980s, work in the new field of chaos theory offered an alternative interpretation.

Jack Wisdom concentrated first on one of the Kirkwood gaps—regions in the main asteroid belt which are clear of asteroids because of gravitational interference by Jupiter. The Kirkwood gaps exist because the orbital period of any object in a roughly circular orbit in these regions would be a submultiple of Jupiter's orbital period. In other words, Jupiter takes about 12 years to circle the Sun, and if an asteroid took about 4 years (in a circular orbit), then every third time the asteroid came to a certain point in its orbit, Jupiter would be out there beyond it, exerting immense gravitational force. Such an arrangement slows or speeds up a small object, enlarging or shrinking its orbit.

But the matter turns out not to be so simple. Wisdom calculated that objects entering the Kirkwood gap at 2.5 a.u. from the Sun could stay in this region in nearly circular orbits for as much as a million years before their repeated encounters with Jupiter every third orbit would suddenly result in their being kicked out into a radically different orbit. The "stable" circular orbits would have been a deception, even if sometimes a long-lasting one. The slightest differences in initial velocity and direction between two bodies entering this region could eventually result in their flying into radically dissimilar orbits. Could any of these objects be sent inward toward Earth? Wisdom calculated that one of every five asteroids in this particular gap is kicked out into an elliptical Earth-crossing orbit within 500,000 years! George W. Wetherill consequently concluded that the most abundant type of meteorite, the chondrite, must be derived from Wisdom's mechanism.

Wisdom's work provided a way for "genuine" asteroids (ones that never were comets) to get into Earth-crossing orbits. It does not, of course, rule out the possibility that some of the Earth-approaching objects we call asteroids are comet cores. And what of the possibility that some meteorites—meteoroids which have reached the ground—are derived from comets? No meteorite has ever been proved to be derived from a comet. Many of the members of meteor showers, derived from comets, may be very small and fragile, incapable of surviving entry through Earth's atmosphere. But fireballs—meteors brighter than Venus—have been observed in comet-derived showers, and some of this type could conceivably reach the ground. Another question is whether any ice from a comet has ever survived being vaporized before reaching the ground. What might happen if a sizable fraction of an icy comet nucleus broke off and entered our atmosphere?

That is a question we will explore in our next chapter. For we have looked at the other ways a comet can die but there is one way which must necessarily hold the greatest interest for us: what happens when a comet dies by colliding with the planet Earth?

Chapter 5
SWORDS OF DAMOCLES

Comets may have been crucial agents in the creation of life on Earth.

In the earliest days of our solar system, over 4 billion years ago, collisions between astronomical bodies were very common. The young Earth must have been bombarded by immense numbers of comets flung into the inner solar system by the gravity of the giant outer planets. Scientists believe that these impacting comets were a vital source of water on the early Earth. This conviction was probably only temporarily shaken by the Galileo probe's December 1995 finding of less than the predicted amount of H_2O in Jupiter's atmosphere (which comets are also supposed to have watered).

But comets may have furnished Earth with more than water. They could even have brought to Earth organic molecules, which are the very building blocks of life. That's not quite like saying that comets seeded the Earth with life . . . but it begins to approach being such a statement.

So comets, it is proposed, have been our great benefactors. We could even go further and claim that we ourselves—or at least important ingredients of ourselves—are the descendants of comets.

But there's a darker side to the story. What is the single most destructive force which threatens the long-term survival of life-forms on Earth? Unless it is human beings themselves, it is probably the collisions of large astronomical bodies with our planet. This would include asteroids and comets, but there may be reason to think that comet impacts, though rarer, are worse.

"Waiting for the end of the world." Drawing by R. Jerome Hill from Harper's Weekly, June 4, 1910.

Comets may be swords of Damocles dangling over us.

We are understandably fascinated and concerned by the destructive possibilities of comets. We have also in recent years made tremendous breakthroughs in understanding the mass extinctions of the past. So this chapter is devoted to the darker subject of comets as destroyers. But even in the telling of these fearful tales we should remember that life has been made stronger by hardships. And there are great ironies in the history of extinctions and survivals. If the dinosaurs had not been destroyed, would human beings exist today?

The Impact Theory Is Born

The great breakthroughs in explaining how the dinosaurs were destroyed, and in understanding the roles of comets and asteroids as agents of destruction and evolution on Earth, have occurred in the last 20 years. The main trunk of the theory gave rise to some branches that have not borne fruit. But these ideas involved comets and were so colorful that I want to discuss some of them here, too.

The start of this whole revolution in our understanding came in 1977, when geologist Walter Alvarez and his father, Nobel Prize-winning nuclear physicist Luis Alvarez, found unusually large amounts of the terrestrially rare element iridium in a clay stratum half an inch thick. The age of deposit for the stratum was about 65 million years, the boundary time between the end of the Cretaceous Period and the beginning of the Tertiary Period. That time featured one of the greatest dyings in the history of life on Earth, when perhaps 75% of all species perished—including the dinosaurs. This great dying at the Cretaceous–Tertiary boundary, which of course is the very reason for there *being* a boundary, is sometimes called the K-T event.

Theories about what caused the K-T event and killed the dinosaurs had stirred the fancy of scientists and the general public alike over the years. But now the Alvarezes seemed to have a solid lead to understanding exactly what had done the killing. They concluded that the over-abundant iridium found in this layer in many places around the world must have a source beyond Earth. Rejecting the kind of exploding star called a supernova as a cause, the Alvarezes decided that both the iridium and the circumstances which caused mass extinction must have come from the impact of an asteroid . . . or possibly a comet.

From the beginning, the Alvarezes recognized that the impact itself could not be directly responsible for most of the killing. Instead, the dust that the impact raised would fill our atmosphere and block out sunlight, lowering temperatures in most places by tens of degrees and putting a stop to photosynthesis. Plants would die, animals which fed on plants would die, meat-eating animals which fed on plant eaters would die. This scenario and the Alvarezes' calculations provided one of the first models for scientists studying the issue of "nuclear winter"—the possible destruction of most life on the planet by cold, dark, and starvation under smoke-blocked skies following a nuclear war.

The Alvarezes' theory met with immediate controversy. Opponents argued that the source of iridium need not be extraterrestrial (it could, for instance, be volcanic eruptions), that the iridium deposits could not be precisely correlated with the time of the great mass extinction, and that

Sketch of Meteor Crater in Arizona, by Doug Myers.

the mass extinction did not occur so abruptly as the Alvarezes and allies supposed and as the disaster they hypothesized would require.

Other critics pointed out the lack of an astrobleme (giant impact scar) that could be dated as 65 million years old. A dramatic rejoinder was the suggestion that a major impact in a geologically sensitive area might set off (or become involved in) surface processes that would heal or disguise the wound. The best candidate area was the North Atlantic, where Iceland was heaved from newly upthrust volcanoes at just about the right time. The picture: Iceland is pushed up or erupted out to fill the scar that the comet or asteroid caused, while the dinosaurs die in the cold and foodless dark.

Furthermore, if the object were a comet, its nucleus might be too fragile to survive the trip through Earth's atmosphere intact. Could the dust released from the ice fill the atmosphere as effectively as material thrown up from the Earth by a rocky body's impact and cause the great darkness and dying? (As we'll see later, this explanation proves untenable for several reasons.)

It would take until the 1990s to find the crater that is probably the "smoking gun" in the case of the murder of the dinosaurs. But in the early 1980s, more attention was given to an offshoot of the Alvarezes' idea that, in its various forms, was even more sensational. This offshoot idea was that the mass extinction, the K-T event, was caused not by one asteroid or comet but by a barrage of comet collisions with Earth . . . and that such an event recurs at regular intervals. The proponents of this theory argued that roughly every 26 to 28 million years Earth is bombarded with as many as two dozen comet impacts in a million-year period.

Showers of Comets?

Even while Walter Alvarez was digging in Gubbio, Italy, in 1977, Alfred Fischer and Michael Arthur were pointing out evidence that mass extinctions occur every 32 million years. But it was not until 1983, 3 years after the Alvarezes finally published their impact theory, that the paleontologists David Raup and John Sepkowski discovered a 26-million-year period between mass extinctions and made the connection with the Alvarez theory. After all, no terrestrial cycle was as long as 26 million years, so there had to be an astronomical cause for the extinctions. Before we go any further, I hasten to add that as far as I know, no claim of any periodicity in extinctions has ever been proved. But it's easy to look critically on these efforts with hindsight. Let's pursue the story and see where it took the researchers involved.

The early 1980s saw another claim about periodicity—periodicity in the amounts of cratering on Earth. Richard Muller and Walter Alvarez then used statistical methods to match the ages of known craters to the extinction record. Although no one giant crater had been found to match the age of the K-T event, Alvarez and Muller pointed to what seemed like greater amounts of cratering at that time and roughly at the times of 38-million-year-old and 11-million-year-old mass extinctions (both obviously less severe than the Cretaceous–Tertiary dying). They concluded that in each of these periods, Earth was struck by perhaps as many as two dozen large bodies over a million years or less.

This scenario answered the complaints that mass extinctions were not abrupt enough to have been caused by one impact. It would also seem to establish whether the bodies were asteroids or comets. In this theory, comets must be the culprits for several reasons. The Oort Cloud comets are stupendously more numerous, mysterious, and precarious in their orbits than any known or even theorized belt of asteroids. Along with other major uncertainties about comets, there is above all the fact that one prolonged gravitational nudge at regular intervals could indeed send Oort comets inward by the myriad to ensure a number of collisions with Earth.

The problem was to find a workable mechanism whereby a very long gravitational nudge could happen at rare but regular intervals.

The phenomenon has been called a "comet shower," and if such a thing ever really occurs it is not just deadly but perhaps more fantastically awesome and beautiful than any sky display ever seen by human eyes. If devastating impacts are to be guaranteed and the mass extinctions thus explained, then possibly a thousand million comets would have to rain on the inner solar system in under a million years.

What would a comet shower be like? Every night the twilight glow in the west would descend to reveal an awesome fan of the tails of several dozen bright comets, some even overlapping and blended in the sky. A few would stretch across most of the span of the heavens—heavens fringed and filigreed with still more dozens of tailed comets and crowded with luminous puffs of close-passing but tailless comets. When the tail ends of the longest dusk comets began descending toward the west, those of the dawn comets would be near them at the meridian or at least high in the east. The total effect might resemble the fingers of two hands extended and partly interlocked to enfold the Earth, or it might look like a half-woven cocoon of silky light strands from which the Earth must burst to be freed as a vast and weighty butterfly. By day, every month would feature a few days in which a comet or two with short intense tails would guide the Sun across the sky. By night—astonishing thought—there might be more comets visible to the naked eye than stars.

A comet shower would supposedly last for many thousands of years, perhaps for a million years, so an ever-changing, comet-filled sky would become as much an accepted part of nature as the weather. Conscious beings living under such a display might become jaded by sights that we in our few-cometed sky would regard as the most awesome splendors imaginable. Yet perhaps not: The variety of comets is so great that perhaps a comet shower would not be subject to loss of wonder. Perhaps no week would pass without ushering in the once-in-a-lifetime beauty of some new comet's colored, curved tail or nucleus-starred head or without heralding the arrival of comets in breathtaking combinations.

We return from this vision as if from a dream, but it may not be a dream. Perhaps this wasn't the sky that the final dinosaurs lived under. But there probably ought to be major variations in the number of comets entering the inner solar system over vast spans of time. Even if nothing like more-comets-than-stars occurs, there may have been times in Earth's history when the heavens were far more comet-rich than they are now—and perhaps other times when they were more comet-poor. There certainly must have been enormous numbers of comets in the skies of Earth at times in the earliest days of the solar system, though clouds of various sorts may have blocked all view of them from the surface.

At irregular intervals in our revolution around the center of our galaxy, there must be times when our solar system passes near exceedingly rich associations or clusters of stars. And at these times, there are more nearby stars to muster a herd of comets, or set loose an avalanche of comets, our way. Whether alien stars or star clusters could ever cause a full-force comet downpour I do not know. But astronomers apparently think it out of the question that stars or clusters could produce comet showers with any precise periodicity.

What alternatives for producing regular comet showers are available to the scientists who argue for their existence? Three fascinating sources: large interstellar clouds, a strange companion star of our Sun, and a planet beyond Pluto disturbing a disk-shaped array of comets much closer than the Oort Cloud. Let's look at each in turn.

Clouds and Nemesis

Clouds of interstellar gas and dust have been variously invoked as the source of all comets and as a source for both replenishing and stripping out comets from the Oort Cloud. But in the 1980s, Richard B. Stothers and Michael R. Rampino proposed that the solar system's rare passages through such clouds disrupt the Oort Cloud enough to cause comet showers, impacts, and mass extinctions. Rather than arguing that comets are set loose each time our solar system enters a spiral arm of our galaxy, where the clouds are mostly located, Stothers and Rampino suggested that this happens each time our solar system oscillates up or down through the galaxy's equatorial plane, where the clouds are also thicker. Apparently, what chiefly recommends this version of the basic idea is that the time intervals for these plane passages are better. They are thought to occur about once every 33 million years plus or minus 3 million years, a close match with the 31 million years between extinctions Stothers and Rampino found in the fossil record.

Eugene Shoemaker and other critics, however, counter that because our solar system is now near the equatorial plane of the galaxy, the 33-

million-year cycle would be well out of phase with the periodic extinctions. One problem in such discussions has been that the solar system's location with respect to the plane has itself been debatable—although some new research may be on the verge of pinning the location down.

The second object proposed as the sender of comet showers is a hypothetical companion star to the Sun. Such a star may have first been seriously proposed by Gerard Kuiper, decades before the Alvarezes and the comet shower controversy. Its existence is not at all unreasonable. A majority of the stars that astronomers have studied are members of double- or multiple-star systems. Detecting the "duplicity" (doubleness) of many stars is often difficult, so there must be even more of these systems than we yet know, and the arrangement may be far more common than solitary stars. The question is whether our own Sun is a loner.

If a companion star to our Sun shone at all, then to remain undiscovered it would have to be a very low-luminosity object in an orbit immensely farther out than that of Pluto. Its gravitational effects on the solar system would be difficult to detect from our vantage point near the Sun. (From many light-years away, it would be easier: we could look back and measure a slight wiggle in the path of the Sun across the heavens, proof that some sort of companion body was pulling on it.)

The best way to distinguish the companion from innumerable other faint red stars would be by its large parallax (the apparent change in its position in the sky in comparison to much more distant stars) when we see it from the left and right sides of Earth's orbit. In the 1980s, Richard Muller conducted a survey to try finding the companion star by its giant parallax.

Muller's interest was not surprising, because he was a member of one of the two teams of researchers that independently proposed the existence of a companion to explain comet showers. His team, which included Marc Davis and Piet Hut, was the one that named its hypothetical companion sun Nemesis, after the Greek goddess of retributive justice. The other team—Daniel P. Whitmire and Albert A. Jackson IV—spoke of the object as "the death star."

If the companion exists and sends comet showers that cause the great dyings in the history of Earth, then of course these names are both apt.

Let us first consider the Muller-Davis-Hut version, Nemesis. They propose the star is a red dwarf having only about 1/3 the mass of the Sun and just 1/1000 the luminosity. The greatest difficulty with their theory is that Nemesis must be placed in an orbit unlike that of any companion star ever determined. It cannot lie on the near side of the Oort Cloud most of the time, it must usually be on the far side (that is, beyond it). Only near perihelion, at about 30,000 a.u. from the Sun, would it pierce all the way through the main concentration of the Oort Cloud. The aphelion, which Nemesis should be near now, would have to be 2 to 3 light-years away. Eugene Shoemaker is one critic who points out that such a star would have an unstable orbit and would be very likely to have escaped the Sun altogether by this time in the history of the solar system.

Whitmire and Jackson came up with a different version of the companion star. In their theory the "death star" is likely to be a black or brown dwarf, a kind of sun that emits little if any visible light. Such an object could not be seen even on an orbit that brought it to within 2000 to 9000 a.u.—at least 50 times farther out than Pluto, but almost as many times farther in than the Oort Cloud. To get the required 26-million-year period, this orbit too must be elongated so that at least the aphelion is out in the Oort Cloud. A study by Jack G. Hills calculated that a star passes within 3000 a.u. of the Sun once every 500 million years. Whitmire and Jackson figured that their "death star" could shower as many comets as such a passing star, even if it was further out, because it would be moving far more slowly with respect to the comets.

The orbit suggested by Whitmire and Jackson presumably had its own stability problems. But in 1985 a third orbit was proposed for a companion star, this time by a leading authority on comets and on the basis of a further line of evidence. Armand Delsemme took a sample of 126 dynamically "new" comets that are thought to have left the Oort Cloud within the past 20 million years and whose orbits might therefore show signs of the perturbation that set them loose. That perturbation, said Delsemme, could be the gravitational nudge of an object traveling only about 0.2 or 0.3 km/sec—far slower than an interstellar gas cloud or a passing star. Delsemme felt that a companion star was a good explanation. Assuming a 26-million-year period, he calculated that in order to produce the motions of the comets in his study, the companion's orbital plane would have to be almost perpendicular to Earth's. Delsemme even went so far as to predict that this "death star" should be near aphelion at a point about 5 degrees from the pole of the ecliptic (in Draco the Dragon) right now!

A decade has gone by and no death star or Nemesis has been found in the region Delsemme suggested, or in Muller's search for a star of large parallax, or in any study of the data from IRAS (the Infra-Red Astronomical Satellite).

The IRAS data were also considered a likely source of evidence for another object that could conceivably cause comet showers and mass extinctions: a tenth planet.

Planet X

The Planet X theory of mass extinctions, published in 1985, was originated by John Matese and one of the authors of the black dwarf "death star" theory, Daniel P. Whitmire. According to them, the undiscovered tenth planet of our solar system would travel in an orbit about 100 a.u. out (twice as far as Pluto's aphelion) and would require approximately 800 to 1000 years to complete one revolution. The trigger for comet showers would come from a precession (a kind of slow rotation) of the orbit, now inclined about 45 degrees to that of Earth (more than twice as tilted as Pluto's). The entire rotation of the planet's orbital plane about the Sun would take 56 million years, so twice in this period—every 28 million years—the orbit would precess Planet X into the midst of a disk of comets lying beyond the orbit of Neptune.

Whitmire felt that this scheme had several advantages over his death star. First, the orbit of Planet X would be far closer to the Sun and more stable, though still unprecedentedly strange. Second, unlike the death star, a Planet X of this kind (with a highly inclined orbit at about twice the distance of Pluto) and an ultra-Neptunian belt of comets have been proposed for other reasons. Indeed, we have already seen that strong evidence for a Kuiper Belt of comets has been gathered in the past few years. On the other hand, one piece of evidence originally produced in support of a planet in such an orbit was a seeming irregularity in the motion of Comet Halley, and that has now been otherwise accounted for.

The original reason for positing a Planet X, though questioned, still endures: There may be an unknown body perturbing the orbits of Uranus, Neptune, and Pluto. The problem is that the orbits of these planets is not known with sufficient precision to establish for certain whether such a perturbation is occurring. The tenth planet has not been found in the data from IRAS. Nor have the paths of the Pioneer and Voyager spacecraft leaving our solar system yet given any indication that such a planet exists.

Strangely, one way to prove or disprove a tenth planet's existence might be by studying centuries-old observations of other planets' conjunctions (meetings) with each other. In 1979, Steve Albers published a few of the results of a pioneering computer program he had written and run several years earlier. The article added many more entries to the list of the two previously known "mutual occultations" of the planets (one planet appearing to pass directly in front of the other—the closest of all planetary conjunctions). In an article in *Sky & Telescope*, Albers specifically suggested looking for observations of such occultations that involved Uranus and Neptune, in the hope of finding prediscovery sightings of those planets. The idea was that if someone had sketched the star field surrounding, say, Jupiter around the time of a Jupiter–Uranus or a Jupiter–Neptune mutual occultation, the sketch might contain an object demonstrably not in the position of any star and therefore obviously Uranus or Neptune—in a slightly different place than predicted even by Albers's ultra-accurate program. To find out exactly where Uranus and Neptune were hundreds of years before they were discovered would give us additional data points with which to compute their orbits—and maybe prove or disprove whether they are being perturbed by any major unknown world out there.

Albers's suggestion was apparently taken up by Charles Kowal and Stillman Drake, who made the wonderful discovery that Galileo himself had made sketches showing Neptune as what he thought was a star near Jupiter in January 1613. The Galileo sighting of Neptune provided a position for the planet much earlier than ever before—over $2\frac{1}{4}$ centuries before its discovery. But any additional refinement of Neptune's orbit made possible by this piece of data has not been enough to solve the Planet X mystery. Perhaps another such finding would suffice. Or perhaps our continued tracking of the outbound Pioneers and Voyagers will provide the answer.

Comets Versus Asteroids

Neither Nemesis nor Planet X has been found. And one of the very reasons for our originally invoking them—a periodicity in mass extinctions and cratering—seems not to have held up. Another reason to consider them was the lack of a K-T event crater. But now we must leave these colorful but unsupported theories behind and quicken our pace with the exciting details of calculations that do bear fruit, mysterious but undeniably real events of tremendous explosive power that have happened in our own century, and the discovery of what is almost certainly the K-T crater, the "smoking gun" for the impact theory of the dinosaurs' death.

First, the calculations that bear fruit.

In the early 1980s, in the midst of rampant speculation about luminous or nonluminous death stars, Planet X, fossil and cratering records, interstellar dust clouds, and the like, two of the world's preeminent comet scientists produced a study that seems not to have received sufficient attention in this context.

The study was made by Zdenek Sekanina and Donald Yeomans in 1983, after the close and visually rather dramatic passage of Comet IRAS-Araki-Alcock and the fairly close but visually disappointing passage of Comet Sugano-Saigusa-Fujikawa. The study was based on a computer search these scientists made for Earth-approaching comets among the entries in Brian Marsden's comprehensive catalog of cometary orbits. Sekanina and Yeomans were able to estimate the degree of incompleteness of their survey and were able to draw some interesting conclusions with a high degree of confidence.

They found, for one thing, a collision rate far lower than that which had emerged from an earlier study by Everhart and were able to show the defect or limitation of Everhart's work. Sekanina and Yeomans's figures indicate that, on average, collisions between Earth and an active comet should

Destruction of the Earth by a comet expected to arrive on June 13, 1857 (from L'Illustration, *March 21, 1857).*

occur only once every 33 to 64 million years, assuming that the present rate of comet visits to the inner solar system is typical. They cite a study (by Shoemaker, Williams, Helin, and Wolfe) which found that Earth-crossing asteroids brighter than absolute magnitude 18 (those which have masses of 10^{15} grams or more) should strike Earth once about every 300,000 years—roughly 110 to 220 times more often than active comets do.

What conclusions can be drawn from Sekanina and Yeomans's figures and from those of Shoemaker and his colleagues? First of all, we might speculate that a vast majority of asteroid impacts do not cause mass extinctions, but that it is possible every collision with an active comet does. Perhaps no asteroids large enough to destroy most species of life on Earth have hit in hundreds of millions of years? Maybe the dust accompanying an active comet nucleus which explodes in our atmosphere is much more effective than the debris thrown up by an asteroid strike at blocking sunlight and setting off the attendant chain of disasters and deaths? Perhaps there are no comet showers, but the rare collisions of individual comets with Earth cause mass extinctions? If the last of these statements is true, then no precise periodicity of mass extinctions exists—and no need for a death star or Planet X.

Sekanina and Yeomans, in discussing the K-T event, point out the difficulty of assessing the situation. "Whether the massive projectile was a comet or Earth-crossing asteroid," they write, "the consequences of such a collision are extremely complex and depend strongly on the unknown circumstances of the fall (impact on land versus in the ocean; mechanical strength, initial velocity, and angle of incidence of the projectile)." Conversely, it is extremely difficult to reconstruct which kind of object entering at which angle could have caused a specific disaster. This is true not only for cataclysms that occurred millions of years ago but also for some that have happened in modern human history.

The Tunguska Event

The outstanding example of a modern collision with an object whose type we cannot yet certainly identify took place a little after 7 A.M. local time on June 30, 1908, in central Siberia. Observers reported an object resembling "a piece broken off the sun" hurtling across the cloudless sky. When the body was about 8.5 km over a remote wilderness area near one of the Tunguska rivers, it exploded, producing a "tongue of fire" which may have extended up to about 20 km in altitude and was visible from as far as 400 km away. Trees were felled radially away from the blast for about 50 km. The heat wave was felt in the village of Vanavara, 60 km away. People and horses were knocked to the ground as far as 160 to 240 km distant. Earth tremors were felt (even though no major piece of the object reached the ground) as much as 900 km away, and the sound was heard at a distance of over 1000 km. Seismographs and barographs continued to register the shock waves long after they passed beyond human perception. In fact, the atmospheric shock circled the world twice before it became undetectable by instruments. That night, the skies across northern Eurasia, all the way west to Great Britain and the Atlantic, were filled with intensely bright clouds which may have been enhanced noctilucent clouds composed of ice-covered dust—the dust of meteoroid, asteroid, or comet.

What caused "the Tunguska event"?

The controversy has continued now for almost 90 years. It has been calculated that if the object had arrived just over 4 hours earlier, it would have destroyed the city of St. Petersburg, and the Bolshevik Revolution might never have happened. But the social turmoil leading up to World War I and the Soviet revolution which followed helped prevent scientists from coming to seek answers about the mysterious reports of the great explosion in the Siberian wilderness.

Willy Ley (in *Watchers of the Skies*) tells how Russian astronomer Leonid A. Kulik was stimulated to learn about the Tunguska event when in the winter of 1920–1921 he read about it on the back of a calendar page that had been stuck in a library book years earlier, probably as a bookmark. The calendar publisher had reprinted stories of interest from newspapers on the backs of the calendar pages—in this case, eyewitness accounts from the week after the Tunguska event!

Kulik made his first visit to the region in 1921 but was not able to get to the almost inaccessible "ground zero" site until 1927. The famous film footage of his expedition shows him garbed in a suit and hat with netting to protect him from hordes of insects. The damage Kulik found to the still demolished forests, and the meteorites he did not find, showed that the Tunguska object must have vaporized almost completely in an explosion which has since been estimated as equivalent in force to a 12.5-megaton bomb. (Indeed, there has been concern that another Tunguska explosion could be mistaken for a nuclear blast and set off a nuclear war.)

Kulik's findings seemed to dictate against the object's having been a giant meteor, and the decades that followed have been witness to a preposterous series of speculations about its nature. Among the suggestions are that the object was a piece of anti-matter, a nuclear-powered alien spaceship in distress, and, of course, a "mini black hole"! Throughout the 70 years since Kulik's findings, however, the debate among serious scientists has focused only on whether it was a meteor or, of course, a comet.

By the 1980s, the debate matched Lubor Kresak's argument for a very small comet nucleus or piece of a comet nucleus against Zdenek Sekanina's case for a large meteor. In 1978, Kresak had proposed that the Tunguska object was related to the Beta Taurid meteor shower, which occurs at the end of June and has itself been derived from Comet Encke. According to Kresak, the Tunguska object was a very small fraction of the nucleus of Comet Encke and was not observed approaching Earth because it came at us from the direction of the Sun (an old fighter pilot trick for sneaking up on one's enemy). The cometary hypothesis explains the lack of a crater, for if it were predominantly ice, the Tunguska object would presumably have been too small to survive passage through Earth's atmosphere.

But in 1983 Sekanina argued that the Encke and Tunguska orbits were too badly misaligned, that the Tunguska object did not come in at cometary speed, that a fragment of comet nucleus would not have survived to as low an altitude as 8.5 km. Instead, he suggested, the Tunguska body was a meteor whose size and angle of trajectory were in the right range to produce neither a fragmenting object (whose pieces would have burned up less violently) nor a ground-reaching object. (Even fairly large meteors are slowed enough by the thicker air when they reach an altitude of about 12 km that they stop glowing and travel the final distance to the Earth at a few hundred kilometers per hour—no faster than a skydiver from a great height before the parachute opens—and reach the ground as cool, dark meteorites). The Tunguska object was certainly not of the exceedingly rare kind that are so massive they are little impeded by Earth's atmosphere and so reach the ground at hypersonic speed. The Tunguska object, claimed Sekanina, was a meteor whose size and entry angle and speed were just right to cause it to vaporize almost instantly in the lower atmosphere. Similar cases of giant but instantly vaporizing meteors have been observed—though here "giant" still means far smaller than the Tunguska object must have been. If the Tunguska object was a meteor, then its diameter may have been about 100 meters and its mass 1 million tons (R. Ganapathy suggests a revision to 160 meters—big enough to sit in a sports stadium like an egg in a cup!—and 7 million tons).

If other large, though far smaller, meteors have disappeared so quickly, with a flash, there still remains the question of how this occurs. As Andrew Chaikin put it in a review of the Tunguska mystery in 1984, "The means by which a million-ton object destroys itself in an instant are not understood." Of course, not quite all of the Tunguska behemoth was turned into vapor or energy: The material that caused the display of bright night clouds is likely to have been fine dust particles (more than could have been derived from the dustiest comet tail, says Sekanina), and some of the vapor seems to have condensed into submillimeter-size particles found at the site of the destroyed forests. Also, Ganapathy's estimate of a larger Tunguska object is based on his finding of submicron-size grains in ice laid down around 1909 in Antarctica. If they are from Tunguska, then certainly a lot of the object's debris managed to circle the world.

Another finding in relation to Tunguska may be the most alarming of all. In 1981, Richard Turco and his colleagues applied to Tunguska a model they had developed to study the effects of atmospheric nuclear explosions on Earth's ozone layer. It is now well known that in recent decades, depletion of ozone (mostly from the use of chlorofluorocarbons in such products as aerosol sprays) has permitted greater levels of ultraviolet radiation from the Sun to reach our lower atmosphere—mostly, but not only, in the polar regions—and thereby increased the danger of skin cancers and cataracts. Nuclear explosions can produce large amounts of nitric oxide (NO) in the upper atmosphere, also depleting ozone. What about the Tunguska object? In 1978, C. Park had calculated that Tunguska formed 30 million tons of NO in the atmosphere between 10 km and 100 km in altitude: 5 times as much as the total amount of NO, NO_2, and HNO_3 normally found in Earth's entire atmosphere! The Turco team found that similar to the amounts created by a nuclear war and calculated that the ozone layer in a 10-degree-wide latitude band centered on Tunguska (61° N) must have been virtually destroyed (over 80% depletion) for several months. Worse yet, the average ozone depletion over the entire Northern Hemisphere during the first year after the Tunguska event was about 45%—enough to triple the dosage of ultraviolet radiation getting through to ground level. The depletion would have remained greater than 10% for several years.

Might we hope that the model of Turco and colleagues, or Parks's calculation, is in error? Unfortunately, actual observations support the theoretical work. As part of a long-term study of variations in the so-called solar constant (the Sun is a slightly variable star), measurements of atmospheric transmission at certain wavelengths were made on Mt. Wilson in California, and those measurements can be used to determine the abundance of ozone over the site (it is at about 34° N). There are large day-to-day variations resulting from the patchiness of the ozone layer, but the average annual concentrations in the period from 1909 through 1911 show a 30% depletion of ozone.

In light of these findings, it is sobering indeed to hear Eugene Shoemaker's estimates that a meteor with energy equivalent to Tunguska's arrives about once every 300 years (within a factor of 2) and that meteors with 4 times as much energy occur once every 1000 years. According to these odds, there is a 12% to 40% chance of a meteorite with energy equivalent to that of Tunguska in 75 years. But wait: surely this does not mean that such an object will necessarily produce all the effects of a Tunguska, because the combination of mass, entry angle, and speed may not be in the correct range to cause such a devastatingly effective and quick release of energy. If this were not so, then even the vastness of Earth's surface and its unexplored areas before modern times would probably not be sufficient to prevent other Tunguska-like (or worse) events from having been recorded in some form in history. Also, of course, if Tunguska was a very tiny comet nucleus or a piece of Comet Encke's nucleus, then the chances of another Tunguska event in the next few centuries may be exceedingly small.

Chicxulub

Could the astronomical object that caused the K-T event, which wiped out the dinosaurs and most species on Earth, have been a Tunguska-like meteor or comet that therefore did not leave a crater for us to find?

No. If one object caused such a worldwide cataclysm, then it must have been large enough to pass through Earth's atmosphere. As early as 1980, the Alvarezes and their colleagues calculated that if a rocky asteroid delivered the amount of iridium they found in the K-T layer, then it must have been about 6 miles in diameter. Other methods for estimating the size of the asteroid or comet have confirmed that calculation: 6 miles wide if an asteroid, 12 miles wide if a comet nucleus. Most of the matter in Earth's atmosphere is concentrated in the lowest 5 or 6 miles of it. But if the object that caused the K-T event was that wide or wider, then Earth's tiny atmosphere would hardly have slowed it down.

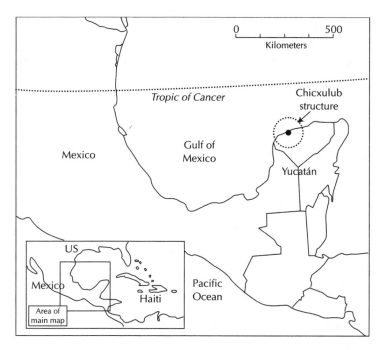

Map showing location of the Chicxulub crater.

Where then is the crater from the K-T impact? Sixty-five million years is a pretty long time, even in the lifespan of our planet. The rearranging of Earth's surface by plate tectonics could well have covered or pulled apart the K-T crater, making it impossible for us ever to identify, especially if the impact was out in the deep ocean. But in 1984, geologist Bruce Boher and colleagues argued that the shocked minerals in the K-T layer must have been formed from continental rocks, suggesting an impact either on dry land or on a continental shelf in relatively shallow water. Other than these minerals and the tremendously overabundant iridium, the remarkable ingredients found in the K-T layer included immense quantities of soot and, in some parts of the world, "turbidite deposits": deposits of tumbled rocks indicative of tsunamis. Tsunamis are giant waves set in motion by earthquakes—or by other, rarer shocks to the Earth. Geologist Alan Hildebrand discovered that the 65-million-year-old turbidite deposits were thickest in the Caribbean. He also studied the thickness of what would have been ejecta from the impact, such as shocked quartz. Hildebrand found the greatest thickness of them in Haiti, where they were 25 or more times thicker than at other sites around the world. Again the evidence pointed to the Caribbean. Was the crater there?

In 1990, Hildebrand came up with a candidate formation. During the course of 65 million years, it had been covered over with sediments. But gravity and magnetic field studies indicated the existence of a circular structure partly on the north shore of Mexico's Yucatan Peninsula and partly in the Gulf of Mexico. Further study has left little doubt that this is the crater resulting from the impact that brought an end to the Cretaceous Period and the reign of the dinosaurs.

The circular formation is as much as 200 kilometers in diameter, and its name is Chicxulub.

Chicxulub is variously pronounced CHEESH-oo-loob and CHICKS-oo-loob.

The Mayan word *Chicxulub* means "devil's tail."

Let's imagine, if we can, what happened on the fateful day sometime between about 64,930,000 and 65,030,000 years ago.

The world is still ruled, as it has been for over 150 million years, by the largest land animals ever to live on Earth, the dinosaurs. But their doom is approaching from space at tens of thousands of miles per hour.

The object approaching Earth may be a giant comet or asteroid. If the former, it has been a spectacular sight in the skies of Earth for several days. Whatever it is, it is so big that our atmosphere will not shatter it or even slow it down before it reaches the surface.

It hits! The impact brings to Earth a kinetic energy roughly equal to that of 100 million 1-megaton nuclear bombs. The force of impact is

"If you fear the comet, live on a mountain-top." Drawing by Arnold Moreaux, from Sketch, April 27, 1910.

comparable to an earthquake of about 11 on the Richter scale—about a million times stronger than the 1989 San Francisco earthquake. The spot hit will someday be the north shore of the Yucatan Peninsula (which some of Earth's largest hurricanes now ruffle—what a gentle puff compared to the K-T event!) but is at this time under perhaps several hundred feet of water. But even the deepest ocean is shallow compared to this enormous object. In fact, a 2-mile-thick layer of limestone here is immediately vaporized, producing enough carbon dioxide to raise Earth's temperature by maybe 20° F for years.

The wave of vaporization spreads out and up. A crater roughly as big as New Jersey is formed. Whether the ocean waves will be up to several miles high or not depends on a variety of factors, but significant waves will bounce back and forth in the region of the Gulf of Mexico for days.

What danger is there to creatures thousands of miles away? Plenty. Remember the soot in the K-T layer? It indicates that most of the forests in that luxuriously vegetated world burned up. This probably happens when debris cast into space by the impact falls back through the atmosphere as a rain of meteors so numerous and bright that the heat all over the world is literally broiling. Perhaps only creatures protected by heavy rain or snowstorms, or cloistered underground or undersea, survive this searing heat and the resulting worldwide forest fires.

But the nitrogen oxides produced by the energy of the impact create an acid rain many times worse than today's that falls all over the world for years. The ozone layer is perhaps entirely compromised. And the greatest killer of all now comes: impact winter.

Volcano Winter, Impact Winter, and Nuclear Winter

What a chain of terrors a single comet impact could set off on our planet! But to the initial survivors would fall the dreadful hardship of trying to live in a world so cold and dark that we can hardly imagine it.

What would this "impact winter" be like? One way to try to conceive of it is to visualize a tremendously more severe version of a phenomenon the world has experienced a number of times in human history: volcano winter.

When the volcano Pinatubo in the Philippines erupted in 1991, the explosive force of the event was not so great as that of some other volcanoes in modern times. But Pinatubo was remarkable for lofting high into the atmosphere many millions of tons of sulfur dioxide. Like the clouds of the greater Krakatao in 1883 and the lesser El Chichon volcano of Mexico in 1982, Pinatubo's sulfur dioxide cloud spread in the stratosphere all the way around the world. For several years after each of these eruptions, the normally colorful total eclipses of the moon were darkened to deep blood-red or gray or black, and the twilights of the world glowed bright with intense color, the final crimson band lingering in the west until the Sun was as much as 18° below the horizon and the faint stars were all aglow in a black sky overhead. After El Chichon's eruption, scientists learned (or relearned—a few had figured this out in the 1960s) that the bright twilights and dark eclipses occurred because the gas was combining with water vapor to form a high-altitude haze of sulfuric acid. This haze was far more effective than volcanic "ash" (silicate particles from volcanic eruptions) at catching and reddening sunlight as high as 10 to 20 or more miles in the atmosphere. It was also more effective at blocking light that is usually refracted by Earth's atmosphere into Earth's shadow and onto the Moon during a total lunar eclipse.

Flaming twilights and dark eclipses—ominous sights. But the more consequential effects of these great sulfuric acid hazes from volcanic eruptions are a lessening of the intensity of daylight and a reduction in temperature around the world.

A friend of mine who is a landscape artist noted with disapproval for many months after Pinatubo the more diffuse shadows and milky skies on even those days when humidity was low and the air mass in the lower atmosphere was clean and clear. Weather clouds, even the comparatively high-altitude cirrus, could be seen sharply in front of this white haze, which was strongest in vast areas of sky around the Sun. Did scientific measurements confirm her (and my own) observations of such effects? Yes. Half a world away from Pinatubo, the intensity of sunlight was reduced by 12%, surely a significant amount.

How significantly did the reduction in sunlight affect Earth's temperature? In the year or two after Pinatubo's eruption, there was a measurable drop in the world's average temperature. The drop was less than 2°F, which might not seem like much. But in a worldwide average temperature, it can be quite significant, leading to much greater individual temperature variations in some places and to major disruptions of weather patterns. It is especially remarkable that Pinatubo had this much cooling effect in a world that had been steadily warming (the global warming is due, most scientists now agree, to the "greenhouse effect" from the vast increase in the amounts of gases such as carbon dioxide and methane produced by our burning of fossil fuels and other practices). If the Pinatubo cooling had not had to first offset the strong global warming, think how much greater it would have been.

Volcanic eruptions of the past have lowered Earth's temperature and darkened its skies far more than Pinatubo did. The year 1816 was known as "the Year Without a Summer" in Europe and North America: June snow and frost troubled New England; crop failures led to famine and indirectly to the spread of epidemics elsewhere; and the chilly weather in Europe gave rise to images of cold and darkness in the lines of poets such as Byron and the pages of a book by young Mary Shelley, *Frankenstein*. The Year Without a Summer came in the wake of the mighty eruption of the Indonesian volcano Tambora.

The first person to associate a cold winter and milky, dull skies with volcanic eruptions elsewhere in the world was probably Ben Franklin, who suggested a connection between bitter weather in North America and Europe and two Icelandic eruptions in 1783 and 1784.

How extreme can the worldwide effects be of an eruption which produces vast quantities of sulfur dioxide? According to Carl Sagan and Richard Turco, "The largest eruption in the last 10,000 years whose effects have been quantitatively determined seems to have been Mt. Rabaul on the island of New Britain in the present Republic of Papua New Guinea" in the year 536. They mention records in China of frosts and snows in July and August that year and resulting famines that killed 70% to 80% of the people in some regions of southern China. And how about this? One Middle East source was said to have recounted, "The sun was dark, and its darkness lasted for 18 months; each day, it shone for about four hours, and still, light was only a feeble shadow." From my own personal experience of the volcanic-hazy days after El Chichon and Pinatubo, I can understand how a more extreme version of such days would indeed give this impression.

None of these worldwide sulfuric acid hazes has dimmed light levels as much as the more localized ash or smoke pall from a volcanic eruption or extensive forest fire. The "Dark Day" of May 19, 1780, in New England was perhaps almost as dark as night at noon. The cause is believed to have been smoke from western forest fires. On September 26, 1950, people in New England and parts of the mid-Atlantic states were treated to views of a blue and green Sun and Moon and of a purple (red plus blue) Moon during an eclipse that night). But in the process the daylight was so dimmed that in the middle of a perhaps otherwise clear day, a professional baseball game had to be stopped and canceled on account of darkness. The cause was again the pall from forest fires, this time in Alberta, Canada (the pall drifted across the Atlantic and also affected British skies). As much as a few hundred miles downstream of a tremendous ashy blast like that from Mt. St. Helens in 1980 (and even more so from mightier eruptions such as Krakatao and Tambora), day can be so darkened that it is literally hard to see objects a few feet away.

To imagine impact winter, imagine the darkening right under one of the heaviest ash or forest fire smoke clouds, covering not just a few hundred square miles but the entire Earth, and lasting not just hours or days but many months. Such conditions would produce drops of tens of degrees in average temperature—far more than enough to offset the warming from increased carbon dioxide released from the 2-mile thickness of limestone vaporized across the 200-km crater of Chicxulub.

The K-T impact set so much of the world's vegetation on fire that the soot from that alone, not to mention the dust blown directly up from the blast, could easily have shrouded the world in Stygian darkness for months. Calculations suggest that for about 1 to 2 months after the K-T impact, it was too dark to see even a few feet in front of you—anywhere in the world. About 2 to 3 months after the event, noontime was about as bright as a night of the Full Moon. And for perhaps an entire year, it was too dark for plants to perform photosynthesis.

Thank goodness that no impact of this size is likely to occur again for tens of millions of years, you might well say. But there's bad news. To reproduce some of the effects of the K-T impact is far beyond the capability of any much smaller impactor or of anything the human race can yet do. But to produce something close to the severity of impact winter may not be nearly so hard. Apparently, nuclear explosions are a particularly efficient means of setting vast areas on fire and casting an ashy pall over the entire Earth. And a much smaller object than the K-T comet or asteroid might be able to produce an impact winter.

For that reason, we cannot so easily rule out the worldwide danger from a small piece of comet nucleus or large meteoroid in the next few hundred years. In addition, in just the past few years, there have been several events which at first glance might suggest that Earth's collisions with comets are far more common than the Sekanina-Yeomans survey implies. The first event was the 1992 recovery of Comet Swift-Tuttle, which calculations for a while had suggested had a small chance of colliding with Earth at its next return in 2126. The second event was the monumentally powerful 1994 collisions of pieces of Comet Shoemaker-Levy 9 with Jupiter. The third event was the close passage in early 1996 of the bright Comet Hyakutake. Do any of these events suggest that the Sekanina-Yeomans calculations are amiss?

I don't think so. There was only a remote possibility of Swift-Tuttle striking us in 2126, and now we know that in that year it will actually pass much farther away than Hyakutake did in 1996. A comet collision with the massive and gravitationally persuasive Jupiter is tremendously more likely than one with Earth, so the Shoemaker-Levy 9 impacts, extremely rare events of human history, were no indication that Earth gets hit more often. Finally, Comet Hyakutake may have looked impressive, but the more than 9 million miles between Earth and its nucleus was an immense margin of safety.

All three of these fascinating comets and their displays receive extensive attention later in this book. And we will also consider the benefits, dangers, and difficulties of trying to deflect a comet if we were to calculate that it was headed straight at Earth.

But I want to end this chapter with an emphatic reminder that a genuinely serious danger which does exist in our contemporary world is that of nuclear winter.

The case for elimination of virtually all the world's supply of nuclear weapons has been made with eloquence in Sagan and Turco's book *A Path Where No Man Thought* (New York: Random House, 1990). That book was written just before the break-up of the Soviet Union and the general consensus that the Cold War was over, before the United States and Russia agreed to a new, swifter schedule of major nuclear diasarmament. But the threat of nuclear winter lives. We would only need to explode a small fraction of even the greatly reduced arsenal planned for the year 2001 to plunge the world into deadly darkness and untold human suffering. Meanwhile, the capability of many more nations and even international terrorist groups to manufacture nuclear weapons is close at hand. It is imperative that we demand our governments to work harder to head off these dangers. We must all work much harder at being peacemakers.

And if learning the facts about a comet or asteroid's crash 65 million years ago has helped in the slightest to inform and motivate us to prevent nuclear winter? Then a stupendously great good will have come from our study of that horrific event.

Chapter 6
OBSERVING AND
DISCOVERING COMETS

To see a vast and glowing cloud of mystery hurtling through the night, coming to light for perhaps only 1 year in a million—this year in which you are seeing it—what is that like?

Just observing a comet is a moment of discovery, not least because no one in the world may be seeing it exactly as you are at that moment. I am not making some obscure point about the philosophy of perception. Comets are clouds within clouds, their borders forever uncertain, trailing off toward infinity. Those clouds are fed by a spinning, spouting mountain of ice honeycombed with the ingredients for fresh home-made solar system from an old—$4\frac{1}{2}$ billion-year-old—family recipe. The recipe is still a secret, the spinning may incorporate complex wobbles, the spouting depends on what rapidly changing time of day and season and year and "weather" the latest spot on the comet ready to steam is experiencing.

Thus every observation of a comet is a discovery, the details of which depend on the quality of your sky, the combination of eyepiece and telescope you use, your experience as an observer, and that incredibly complex thing you are looking at.

Now try to imagine being the first person ever to see a particular comet. Usually, such a discovery is the culmination of hundreds of hours of patient and persistent, alert and sky-loving searching. And that searching results in having something as old as the solar system, and temporarily much larger than Earth, being named after you, possibly for as close to forever as humankind can manage to make it.

This chapter offers a small selection of observers' tips and some of the many moving (but also instructive) stories about comet discoverers. Beginners wishing to apply these methods should get themselves a general guide to learning the sky and selecting and using binoculars and telescopes. Anybody wishing to read more stories of comet discoverers is encouraged to check the bibliography for the books by Leslie Peltier and by David Levy and the huge section of discoverers' stories in William Liller's *Cambridge Guide to Astronomical Discovery*. Liller's book as a whole is the most comprehensive presentation of ideas and techniques for the dedicated amateur who wants to find comets or novae.

Seeking Dark Skies

Most of us must visually observe comets in order to discover them, so let's start out with observing. Many of the following points are for those who wish to look for faint comets with optical instruments, but even if you are just a skywatcher who hopes to enjoy an occasional great or good naked-eye comet such as Hale-Bopp, some of these pointers will help you to see and understand bright comets better.

First of all, you need to know what kind of observing site is good for comets.

The answer is pretty simple: the place with the darkest sky.

What I mean by this first of all is a place where you have the least possible light pollution. *Light pollution* may be defined as excessive or misdirected artificial outdoor lighting. If you've tried looking at the starry sky anytime or almost anywhere in recent years, you probably know what a problem this has become. It's not just the nearby glare—wasted light going directly in your eyes—from a neighbor's yardlight or streetlight. It's the nearby or even distant *skyglow*, the total light going up from a city to be reflected in the atmosphere. A typical city of 50,000 may have a major impact on how many stars you can see in its direction from a distance of 10 miles or more. A typical city of 500,000 might have an effect 30 or 40 miles away.

In the United States, several billion dollars a year is wasted on light that goes directly off into space, and about 3/4 of this waste could be eliminated through good lighting fixtures and practices. Amateur astronomers are finally beginning to get legislation passed to help control light pollution and are starting to work on the problem with environmentalists, lighting engineers, lighting manufacturers, and electric utilities. If you want to learn more about light pollution and about the effort to literally save our view of the stars (while saving huge amounts of money and energy), send a self-addressed, stamped long envelope to IDA (the International Dark-Sky Association), 3545 N. Stewart, Tucson, AZ 85716.

You may eventually help to get light pollution in your area reduced. But for now, you need to find the darkest present sites for comet observing. Although it may be inconvenient to observe away from your home, you should compare how well the stars, the Milky Way and other celestial sights appear from home and various locations which are different drive-times away. When you want to look for a dim comet or see the very faintest extension of a naked-eye comet's tail, you may be willing to drive farther to the darkest site you know. Under ordinary circumstances, your home site may be dark enough to get adequate views of comets.

Light pollution is a problem every night, but you should often be able to avoid brightly moonlit nights and hazy weather; both of these, like light pollution, hinder good comet observing. By the way, don't forget that humanmade skyglows are often considerably reduced late at night or in the hours before dawn.

Preparation and Your First Comet Observations

Of the many tools for comet observers, none are more useful than your eyes themselves. What can you do to make them more effective?

Several things. First of all, assuming that you have managed to get to a reasonably dark location, you must allow for *dark adaptation*. The chemistry of your eyes changes in dark conditions, permitting the rod cells to become much more sensitive to dim light sources after about 10–20 minutes in darkness and a little more sensitive the longer you stay out. But this is true only if you don't get a bright light suddenly shining in your eyes to set back the process. As experienced astronomers know, one good way to prevent getting your dark adaptation spoiled is to use a red flashlight (or to cover an ordinary one with red cellophane)—red light does not much disturb the dark adaptation.

Another technique to help you see fainter objects (for instance, faint comets in a telescope or binoculars and more of a bright comet's tail with the naked eye) is *averted vision*. There are more of the light-sensitive rods a bit off from the center of your retina. Therefore, if you try directing your gaze just a bit away from a faint star or comet, you will note that it appears brighter, and you will be able to detect it more easily.

Of course, other tools to enhance vision are telescopes and binoculars. The hard fact is that if you want to see comets more often than once every few years, you will need optical aid. Contrary to what beginners might think, however, a large telescope is not always best for comet observing. Much of comet watching is done with small telescopes and even binoculars, if the object is not too faint. The reason for this is that with a larger telescope, you generally cannot use very low magnifications. High magnification is good for trying to see the structure of the brightest innermost parts of the coma (if your comet is not altogether too dim), but it tends to spread the available light of a comet's faint outer parts over too large an apparent area. Much of a comet is like fog: Get too close to it (or seemingly so, by high magnification), and what seemed visually dense from a distance simply thins out and disappears.

Let us assume you have the positions for a comet you want to observe, you have those positions marked on a star map (ideally, a star map showing even stars that are dimmer than the comet will be), and you are able to find the comet in the correct position. What is the first thing you should try to do once you have the fuzzy patch of light that is a comet in your field of view?

The first thing is to enjoy it. You've got one of the solar system's most mysterious objects before you, so relish its mere existence there—and then its appearance. Study its shape (if it has any!), consider sketching the shape and any details of its structure you can detect. (Sketching doesn't just provide you with a record of your endeavor, by the way. It also trains your eye, eventually enabling you to see far more.)

More Advanced Observations

There are several features to look for in comets and several properties to rate or estimate.

Naturally, you want to determine how bright the comet you are seeing is. The rule here is to use the lowest-magnification instrument or instrument–eyepiece combination with which the comet is still plainly visible. Once a comet gets to seventh or sixth magnitude, it is generally not quite visible to the naked eye, but binoculars will show more of the outer coma than a telescope and will therefore provide a greater total brightness—and therefore allow a more accurate assessment of how bright the comet is. After it brightens to fifth or fourth magnitude, you should start trying to make brightness estimates with the naked eye.

The tricky thing about brightness estimates of comets, however, is that a comet's coma is an "extended object," a fuzzy patch whose light is spread out over a considerably larger area than the point of a star. How, then, can we accurately compare the brightness of a comet to that of a star?

Of the several methods used, the most common and easiest is the *Bobrovnikoff method*. In this method, you defocus the binoculars or telescope until the images of the comet and nearby stars are similar in size, and then you make comparisons. Is the out-of-focus image of the comet about midway in brightness between star A and star B? Then if star A shines at magnitude 6.0 and star B at 7.0, the comet's brightness should be about 6.5.

How do you defocus the comet and star images when the comet is bright and you are using your naked eyes? If you wear eyeglasses, removing them may help defocus the comet and stars by roughly the right amount. Other means (such as comparing them in a reflective sphere—for instance, a garden globe) can be employed if necessary.

You can also estimate the size of the coma or of any tail that is visible. This can be done roughly if you know the size of your telescopic eyepiece's field of view, and it can be done more accurately by comparison with the known angular distance between two stars. (Much better than either of these is using the "drift method," a measuring reticle, or a filar micrometer—but these are advanced techniques beyond the scope of this book.) With a long tail, what you see can be plotted and measured on a star atlas. Or if the tail is really long—say 10 degrees or more—the start and end point of the tail are noted and a formula is employed to calculate the length.

Another property of the coma which can be estimated is the DC, the *degree of condensation*. This is how concentrated toward its center a comet's coma is. The range is on a scale of 0 to 9, where 0 is a completely diffuse object with no concentration toward the center and 9 is a star-like coma.

Instead of observing an entire coma looking like a (slightly fuzzy) star, you may see a star-like point of light in the midst of a fairly strongly concentrated coma. This *apparent nucleus* (a dense cloud around the always much tinier and unseen true nucleus) may look larger and fuzzier on some nights because of its own expansion but also because of poorer "seeing"— that is, images being less sharp images as a result of turbulence in Earth's atmosphere. A central cloud not so small and concentrated as an apparent nucleus could be called the *central condensation*.

Even when you don't see a tail, you may notice that the coma is elongated along a certain axis. It is helpful to be able to note the PA, or *position angle*, of any cometary feature in relation to the center of the comet. You've got to know which direction in your field of view is north. Position angle is measured from 0 degrees as north to 90 degrees as east, 180 degrees as south, and so on back around to 360 degrees, or rather 0 degrees again, as we return to north.

A feature sometimes observed even in fairly faint comets, especially if they are periodic comets, is a *sunward fan*. If a tail is seen, try to determine whether it is a straight and narrow gas tail or a broader and fuzzier dust tail.

Features in Bright Comets

You sometimes have to study a faint comet carefully to see any structure in its coma or any trace of tail that may be present. But a bright naked-eye comet can show numerous complex features unlike anything viewed in dimmer comets.

Look for the blue of ionized gas in the coma if you are using an optical instrument with enough light-gathering power. People's ability to see colors in faint light varies, but in some comets I've found the blue becomes distinctly noticable with an 8-inch or 10-inch telescope and a fifth- or sixth-magnitude comet. In comets as bright as first magnitude, the blue of gas and the gold of dust start to become apparent to the naked eye. (And then there is red in comets. See Chapter 13 about Hyakutake, and notice references to red comets in the chapters about comets of the past.)

If enough gas and dust are produced by a comet and it gets close enough to the Sun, the coma takes on a striking parabolic outline. This can be seen in a number of comets, even some of barely naked-eye brightness in a relatively small telescope.

Additional features are easily observed with the parabolic form only if a comet gets quite bright. Between the two arms of the parabola, behind the central condensation, either a *shadow of the nucleus* (a narrow strip of

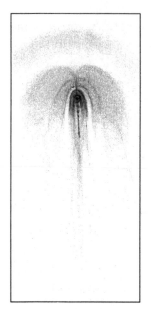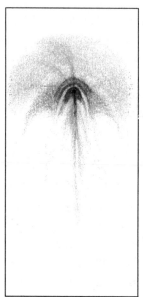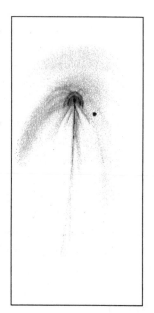

At Volcano, Hawaii, Stephen James O'Meara made these sketches of the intricate structures in the head of Comet Hyakutake seen through a 4-inch Tele Vue Genesis refractor at 177X. The three views are at March 23, 14:00 UT; March 24, 15:00 UT; March 26, 14:00 UT. The dot near the comet in the third sketch is a star.

reduced brightness) or a *spine* (a very thin line of greater light intensity) may appear and extend more or less back along the mid-line of the tail for a short distance. Presumably, "the shadow of the nucleus" occurs because less sunlight gets to the dust behind a relatively dense central condensation. But I begin to tread into territory I'm not sure of here. Little has been written about modern observations of some of these amazing features, and on the rare occasions when one gets a great comet bright enough to show them plainly, they can be bewildering (but awe-inspiring).

A bright comet which is dusty may produce *envelopes* or *haloes* of dust, expanding out at first sunward from the parabolic outline. Examination of the innermost coma at higher magnification can reveal tuft-like *jets*, geysers of dust or gas shooting up from the compact active regions on the hidden nucleus. Linear or ring-shaped active areas on the nucleus could produce the haloes; if seen edgewise, they would mimic true jets. But all such dust-producing activity ends up in haloes if it is vigorous and plentiful enough.

You may see some of these cometary features in no more than a few comets in your entire lifetime. But because this book is devoted to great comets, it is well worth describing these features—and better yet illustrating them.

Some of the most exciting structures, such as ion rays and synchronic bands, are found in the tails of comets—but these were described in Chapter 2, and you will learn more about them as we tour the individual great comets of history in chapters ahead.

Names and Designations

If you discover a comet, then you get to have it named after you. But what are the exact rules governing this process?

Through much of history, comets received the title of the year or ruler's reign in which they occurred. More rarely, a comet might be so closely associated with a ruler's death that it became that ruler's comet. As the scientific study of comets took over in the eighteenth and nineteenth centuries, some comets were still named after the year they occurred, but more and more of them were named for their calculator (the four calculators' comets to date: Halley, Lexell, Encke, and Crommelin) or their discoverer. Independent discoveries of the same comet sometimes resulted in several names linked by hyphens, as in Swift-Tuttle. Recoveries of a lost or long-gone comet sometimes got the recoverer's name added (Herschel-Rigollet). Eventually, the official rule became that the first three independent discoverers of a comet could get their name attached to it (as in Comet Kobayashi-Berger-Milon and Comet Tago-Sato-Kosaka). If the official announcement of the new comet has already been made, you are out of luck; and if you are the fourth independent discoverer, you are out of luck.

The three-name system has produced some of the more glorious and preposterous mouthfuls of euphony and cacophony ever to pass human lips: Comet Pajdusakova-Rotbart-Weber . . . Comet Bappu-Bok-Newkirk . . . Comet Bakharev-Macfarlane-Krienke . . . Comet Whipple-Fedtke-Tevzadze . . . Comet Van Gent-Peltier-Daimaca. Now I understand that as part of an official change in comet designations instituted in 1995, the attachment of a third name is going to be "discouraged"—whatever that means.

In addition to names—and perhaps more and more, in scientific papers and communications, in *replacement* of names—comets have official designations. If you are going to be reading a lot about comets in the days and years ahead, you really need to learn the new system. And because you will see plenty of use of the old system in books written before 1995, you would do well to know how the old designation system worked, too.

There's a lot to say about the designation systems, old and new. My friend and fellow writer Guy Ottewell has explained it in the most congenial way imaginable in his *Astronomical Calendar 1996*:

Comet designations have changed. They are, like much else in astronomy, administered nowadays by the International Astronomical Union's Central Bureau for Astronomical Telegrams, or Minor Planet Center, a function of the Harvard–Smithsonian Center for Astrophysics, located in the Smithsonian Astrophysical Observatory, in Cambridge, Massachusetts—a body itself in need of handier redesignation! How about the Astronomical Clearinghouse? One thinks of it as "Brian Marsden and crew," but has to take a deep breath before each attempt to refer to it.

The previous system (Marsden describes its evolution in the *International Comet Quarterly*, Jan. 1995) was that each comet on discovery or recovery was given a provisional designation such as "1994b," meaning the 2nd to appear in 1994; if, as began to happen in 1987, more than 26 appeared, there had to be "$1987a_1$," etc. (the record was a 34th, "h_1," in 1989 and 1991). When the perihelion dates were safely enough determined, a year or two later, they were given permanent designations in perihelion order, in the style "1994 I" "1994 II" ... (The record here was XXXVII, for 1987). Thus these could be in different years: Halley's Comet at its most recent visit became 1982i and then 1986 III, because recovered in 1982 and reaching perihelion in 1986. A "P/" could be placed before the name of a comet if it was periodic, that is, with a period under 200 years.

The system decided during 1994 and imposed from the beginning of 1995 is more like that used for asteroids, and is more convenient for the administrators. It is applied to new comets and retrospectively to all those of history, as in the tenth (1995) edition of Marsden's *Catalogue of Cometary Orbits*.

A comet gets a designation like "C/1995 A1." In this, "A1" means that it was the first to be reported in the first half of January, "D3" the third in the second half of February, etc. "I" is not used, and it presumably becomes possible with practice to remember that

"N" is for the first half of July, and "Y" for the latter half of December (so "Z" isn't used either).

"C/" is changed to "P/" if the comet proves to be periodic. Thus "C/" is used to mark the non-periodic comets. Periodic comets whose orbits are regarded as established, as after they make their first predicted return, are numbered: "1P," "2P" . . . in the order in which their periodicity was recognized, starting with Comets Halley and Encke; the latest is "119P" (Parker-Hartley, upon its first return in 1995). For these comets the "/1995 A1" part of the designation is dropped: there is no new designation at new appearances. For a periodic comet that has long been unrecovered, such as 3P Biela which disintegrated during its 5th appearance, "P" is changed to "D" (which could connote "doubtful," "disappeared," "deceased"). The prefix can be changed to "X" if no orbit can be computed; or to "A" if what was at first taken for a comet turns out to be an asteroid, leaving the designation to fit in among those for other asteroids. In this system, then, though there is some changing with changed status, there is no changing from "temporary" to "permanent" designations; and perihelion dates become irrelevant.

The asteroid and comet systems are still not quite parallel, partly because so many more asteroids turn up each month. Comet P/1982 A2 Helin-Roman-Crockett became 111P Helin-Roman-Crockett. But asteroid 1983 RD2 became 5050 1983 RD2 became 5050 Doctorwatson. (I ignore parentheses, a gratuitous complication to make both systems look even more nerdish than they already do.)

So now you know about comet designations.

Tips for Would-Be Discoverers

There is no object in the universe which an amateur astronomer has a better and richer chance to discover, or can use simpler means to discover, than a comet. Perhaps there is no new entity of any sort in nature which the amateur can try to discover more simply or with greater likelihood of success. Nor would an amateur in any other field stand a chance of making a find that might be more widely celebrated or might have a more rich and potent impact on both society and science.

Roughly half of the discoveries of long-period comets are still made by amateur astronomers. Comet seeking began with a few French practitioners in the wake of Comet Halley's first predicted return in 1759. It grew in popularity (after the death of the most prolific of all comet discoverers, Jean Louis Pons) in the mid-nineteenth century, partly as a result of the second (and maybe never-to-end) wave of asteroid discoveries. It rose to new levels of success with a few Americans in the late nineteenth century. It remained important in spite of the frequent—and usually incidental—photographic finding of comets by professional astronomers looking for other objects in the twentieth century. In the 1980s, it was rumored that satellites like IRAS would soon detect all the new comets before amateurs had a chance, thus ending the sport, hobby, avocation, and passion once and for all. But this did not happen, partly because obtaining funding for new satellites has not been easy. Now, in the October 1995 issue of *The International Comet Quarterly*, Dan Green writes,

> A noticeable lack of discoveries occurred from January until July, when C/1995 O1 (Hale-Bopp) was found—undoubtedly due to the termination of regular, monthly photographic hunting with the 18-inch Schmidt telescope at Palomar by Gene Shoemaker *et al.* and Eleanor Helin *et al.* at the end of last year. Comets are surely being missed, and it is unfortunate that the Palomar program terminated before other search programs could be in place for continuation of coverage. Meanwhile, the amateur hunter will surely have a good opportunity during the next few years to find comets, until all-sky CCD-scanning programs can be put into place.

Let us hope that this latest threat to amateur comet hunters does not end the game forever. But in any case, the prospect of less high-tech competition in the next few years should give amateur hunters an incentive to get out there and start searching while they can.

During the past few centuries, comet hunters have had an increasing variety of telescopes to use. But whether reflector, refractor, or rich-field telescope or binoculars are chosen, the basic requirements remain the same. The most important need is for a large field of view, which usually goes hand in hand with fairly low magnification. Searching the prime areas of the sky takes too long with small fields of view, and the higher magnification may spread out the frequently diffuse, low-surface-brightness image of a comet to imperceptibility. The distinguished veteran comet observer John

Bortle wrote in the early 1980s that for finding comets, a magnification of 20× is best on telescopes with aperture (diameter of main mirror or lens) of 10 centimeters (about 4 inches) or smaller. For larger telescopes, twice as much power as the aperture in centimeters is best. Bortle judged that an aperture of 25 centimeters (about 10 inches) is about the maximum for a good comet-hunting telescope, but he acknowledged that some observers had succeeded with larger telescopes. In the early 1980s, Canadian amateur astronomer Rolf Meier discovered his third comet with a big, 40-cm reflector. In 1984, David Levy was just getting started at making comet discoveries with his 16-inch telescope named "Miranda," but he has rung up seven more finds as a visual searcher since then.

Where are new comets most likely to be found? On the professional astronomer's photographic plates of the opposition (opposite the Sun) part of the heavens—where many comets with perihelion outside Earth's orbit are nearest to us and brightest. But most of these objects never get stirred much by the Sun, and they are often short-period comets exhausted by the repeated (though feeble) stirrings they have had. "Brightest" for them may be only sixteenth magnitude or even dimmer, so they are not discernible to virtually any visual observers, even with giant telescopes.

Amateur comet hunters usually look for long-period comets, which sometimes become spectacularly bright and are usually discovered in a quite different region of the sky. The preferred areas for visual comet hunters are in the west during and after late evening twilight and in the east before and during early morning twilight. Bortle suggests that morning observers begin vertical or horizontal sweeps about 45 degrees above the horizon and work downward for about an hour and a half so as to reach the horizon near the start of morning twilight (about 90 minutes before sunrise, depending on your latitude and the time of year). The pre-dawn hours are significantly better than the post-dusk ones for comet discoveries, especially because retrograde comets will be approaching the Earth at this time and place in the sky. If you want to increase your chances of discovering a comet in this way, you have to get up very early.

Calculations by Lubor Kresak have shown that for comets brighter than tenth magnitude—almost the limit at which new comets are found visually—the greatest number can be located between about 30 degrees and 35 degrees from the Sun. Comets are at their brightest when close to the Sun, and not infrequently a comet enters the inner solar system on the far side of the Sun from us and then suddenly rushes out, full-blown and relatively bright, into the twilight.

Equivalent to those periods when a comet moves out from behind the Sun and is fair game for anyone are those periods just after the full Moon when suddenly the evening sky is moon-free again. Similar but lesser examples are those times when, by coincidence, many of the geographical locations that host a lot of comet hunters are clouded over for a number of nights. Not clouds but the body of Earth itself may block the view of a new comet from many of your fellow comet hunters if you happen to live in the Southern Hemisphere. In the twentieth century, the person who has discovered by far the most comets visually is William Bradfield of Australia. He has great skill, but he also has far less competition than his counterparts in the Northern Hemisphere.

How long must one hunt before discovering a new comet? The average time appears to be somewhere between about 200 and 600 hours, but Donald Machholz observed for 1700 hours before making his first find (in that period, he says he was "probably the most active comet hunter in the world" at 500 hours per year). Then he observed for 1742 hours before finding his second comet. The hours you search must be attentive ones, of course, with properly overlapping sweeps at speeds of less than 1 degree per second at a good site in good conditions. The greatest problem is that so many star clusters, nebulae, galaxies, and faint double stars resemble the hazy patch of a comet at low magnifications. That was what the first great comet hunter, Charles Messier, found out and did something about: His famous catalogue of "Meissier objects" was compiled as a list of objects a comet hunter could safely ignore. It's ironic that Messier has become far better known among amateur astronomers for this handy list of bright and striking deep-sky objects than for his many comet discoveries.

David Levy and some of today's other leading comet hunters feel that the sights seen in the course of the search are perhaps as rewarding as the comet at road's end. Is there any better way to learn richly and deeply the fields of the heavens? If you are so impatient to find a comet that the joy of the road has no charm, you're probably better off giving up. But just in case you find that new comet tonight, how do you report it? The detailed answer appears in Appendix 7.

The Photographic Comet Finders

Comets are such individuals it seems appropriate that they be named after the human individuals who discover them. And those people form as diverse a group as one could imagine.

We could break the group into two large parts: the professionals who find comets on photographic plates (and now CCD images) and the amateurs who find comets through their telescopes. (Even these broad categories may be blurred. Amateur David Levy has gone to work with professionals who obtain and study photographs on which they have found numerous comets. And on January 27, 1996—just a few days before the great Comet Hyakutake was found—amateur Edward Szczepanski came across a new comet on his photograph of the galaxy M101.)

The discovery of faint comets on photographic plates may suggest less adventure, romantic spirit, or individualism than the amateur findings. But this professional work requires knowledge and skills of its own—without which professional astronomers would not be in the position to examine these images in the first place. Fred Whipple once said that you have to scrutinize—carefully, knowledgeably—an area of photographic plates equal to a city block to find a new comet. Elizabeth Roemer and her colleagues were responsible for 77 comet recoveries between 1953 and 1976. Writing in 1982, Lubor Kresak said that the 48-inch Schmidt telescope at Mt. Palomar had made 32 photographic discoveries of comets.

And then there is the team of Eugene and Carolyn Shoemaker, joined by David Levy (and other assistants such as Henry Holt and Phillippe Bendjoya). Other members of the team play their role in taking the photographs, but apparently it is Carolyn who searches the images. She is the first to witness each comet on them. At the time of this writing, the Shoemaker name graces (by my count) 33 comets—a record. According to David Levy, even armed with her skill and swiftness, it takes Carolyn about 100 hours of searching to come up with each comet. And that, Levy adds, doesn't include the time spent exposing and developing the film and preparing it for scanning. In addition, the Shoemakers must drive 500 miles each month to the observatory.

The Eighteenth- and Nineteenth-Century Hunters

For understandable reasons, the stories of visual discoveries of comets are often more varied and colorful than those of the professional's photographic finds.

The sport of visual comet hunting got started after the first predicted return of Halley's Comet. Charles Messier thought he was the first person to recover the comet in January 1759, but he then learned that a German farmer, Johann Georg Palitzsch, had beaten him by almost a month. This disappointment must surely have been a factor in inspiring Messier to start hunting for comets in the 1760s. At first, he had the field to himself, but his countrymen Jacques Montaigne and Pierre Mechain soon became his rivals. Messier's fierce competitiveness is suggested by a famous (and it is to be hoped apocryphal) story. Shortly after Messier's wife died, he heard that Montaigne had found a new comet. A friend, seeing Messier wracked with grief and thinking he was upset over his wife's death, said "I am sorry." Messier allegedly replied, "Alas! Montaigne has robbed me of my comet!"—and then, trying to recover, "Poor woman."

Meissier's desire for comets was so great that King Louis XV called him "the comet ferret." In 1775, Messier's friend Lalande tried to invent some new constellations (none of his configurations is still in use), and one of them he named *Custos Messium*, "the Keeper of the Harvest" or, in French, *Le Messier*, in honor of his friend the reaper of the harvest of comets. Messier was involved in the discovery of at least 15 (maybe 21) comets, and he was the sole discoverer of 12—though 11 of them were either long-period comets (which won't be back anytime soon) or had orbits that could not be calculated. (The other one was named for its calculator, Lexell.)

In 1801, Messier's last comet was independently discovered just before he spotted it. It was the first comet discovered by the person who became perhaps the greatest comet hunter of them all: Jean Louis Pons. Pons started out as the doorkeeper of the Marseilles Observatory. He has his name on 26 comets, the first of which he discovered at the age of 40. The number of comets that Pons independently found (including some he discovered after other observers) may be as high as 37.

Between 1786 and 1797, Messier had competition from an English woman who in that relatively short time ran up a total of eight new comets. She was none other than Caroline Herschel, sister and assistant of the musician-turned-telescope-maker-turned-astronomer William Herschel. William Herschel became famous by discovering in 1781 what he at first thought was a peculiar new comet. The fuzzy little disk turned out to be the planet Uranus, the first planet discovered since prehistoric times. Later, William Herschel's son Sir John Herschel became an important astronomer and was a key observer of Halley's Comet at its 1835–1836 return.

Between Caroline Herschel and Carolyn Shoemaker there have been, by the way, many female comet discoverers. Maria Mitchell of Nantucket launched her career with her award-winning discovery of Comet

Mitchell in 1847 (one of the three later independent discoverers was Madame Rumker, wife of the director of the Hamburg Observatory). Other women who have discovered comets (some more than one) include Ludmilla Pajdusakova, Margaret Vozarova, Eleanor Helin, Pelageja F. Shajn, and Liisi Oterma.

I mentioned Mitchell's award-winning comet. Medals and monies have occasionally been offered to those who could discover a comet. From 1831 to 1848 Denmark's Frederick VI offered a gold medal to each such discoverer. Then the Vienna Academy of Sciences took up the task and supplied prizes to each comet finder. But the most famous of comet prizes was the $200 that H. H. Warner offered in the late nineteenth century to any U.S. discoverer of a comet.

The person who benefited the most from this award was the eventually famous Edward Emerson Barnard. Barnard used his first $200 Warner prize (a great deal of money in the 1880s) to buy a lot and eventually—with the help of more comets—to build and pay off a little cottage for himself, his wife, and his mother. Barnard wrote,

> Those were happy days, though the struggle for a livelihood was a hard one, working from early to late, and sitting up the rest of the twenty-four hours hunting for comets. We looked forward with dread to the meeting of the bills which must come due. However, when this happened, a faint comet was discovered, and the money went to meet the payments. The faithful comet, like the goose that laid the golden egg, conveniently timed its appearance to coincide with the advent of those notes. And thus it finally came about the house was built entirely of comets. This fact goes to prove the great error of those scientific men who figure out that a comet is but a flimsy affair after all, infinitely more rare than the breath of the morning air, for here was a strong compact house, albeit a small one, built entirely of them. True, it took several good-sized comets to do it, but it was done, nevertheless.

Barnard went on to discover a total of 16 named comets. In this endeavor he was outdone by another enterprising American observer of his time, William R. Brooks of upstate New York, who amassed 22 named comets, second only to Pons among visual observers. But Barnard went on to become one of the great figures in the history of American astronomy and one of its most eagle-eyed observers. The variety of heavenly objects other than comets that are named for him testifies to his accomplishments: Barnard's Star, Barnard's Loop, Barnard's Galaxy, and Barnard's many dark nebulae (B1, B2, and so on).

We still do not know what learned person perpetrated a famous hoax on Barnard. Barnard was not a victim in the sense that he believed the hoax. No, he was the astronomer who an 1891 newspaper article claimed had invented an automatic comet finder—a machine that found comets and set off an alarm to awaken the sleeping searcher. It took Barnard several years to get the newspaper to retract the story.

A marvelous and more pleasant tale involving Barnard is that of his dream of comets. Barnard had been eagerly observing the Great September Comet of 1882, the mighty sungrazer, and no doubt went to sleep with his mind filled with comets. He dreamed of seeing comets all over, bunches of them. He woke up before dawn on the morning of October 14 and hastened out to observe the great comet, then to start scanning the sky with his telescope for new ones. But he had gotten only 6 degrees or so from the great comet when he beheld a sight never seen before or since that month: a group of five small comets, presumably spawned from the big one! Other observers confirmed Barnard's sighting, and other such apparently secondary comets were seen, including one Brooks spotted a week later which, unlike the others, was observable for more than just a single night.

The whole nineteenth century, but especially the second half, was a marvelous time for comets and comet hunters. Two other great American comet discoverers, about 20 years before Barnard and Brooks got their start, were Lewis Swift and Horace Tuttle, whose fates were quite different. But their story is given in the profile of Comet Swift-Tuttle in the next chapter.

A World of Comet Hunters

The twentieth century has witnessed the spread of comet hunting all over the world among an even more varied assortment of observers. In 1968, a 16-year-old amateur astronomer, Mark Whitaker, began hunting for comets as a summer project with a 4-inch reflector and after only three nights found a new comet. (It was independently discovered two nights later by professional astronomer Norman Thomas and became known as Comet Whitaker-Thomas.) In 1983, the 71-year-old veteran observer George Alcock discovered Comet IRAS-Araki-Alcock with binoculars while looking out his window in his pajamas. (Alcock is the discoverer of five comets and five novae, and in 1989—the year he turned 77—he spent 293 hours and 23 minutes searching the skies!)

No country has produced more devoted or successful comet hunters in recent decades than Japan. One of the Japanese astronomers' great inspirations was Minoru Honda, who told his moving life story as a sky lover and comet hunter in William Liller's book *The Cambridge Guide to Astronomical Discovery*. Until William Bradfield passed his total, Honda had discovered more comets (I count ten, but a reliable source I consult says twelve) than any other living person. After 1968, he concentrated his efforts on novae and proceeded to find five of *them*. Honda died in 1990, just a few weeks after William Liller received his essay about his life. According to Liller, "Honda is survived by his widow, whose name, Satoru, in Chinese characters is the combination of 'comet' and 'heart.'"

Honda said that a book he read in 1930 stated that only two comet discoveries had ever been made from Japan, and both of these were really recoveries of periodic comets. He set out to change that—and his example encouraged generations of Japanese comet hunters to follow.

Honda's is not the only inspiring story of a Japanese comet hunter. Consider the great Kaouru Ikeya, who took a factory job at the age of 14 to help support his family but eventually saved enough money to build his own inexpensive telescope. He searched the pre-dawn skies before work for 335 hours over 109 nights in more than a year before finding his first comet. His co-workers were so proud of him that they put together a gift of $300 (this was over 30 years ago, so quite a sum) to help him continue his comet hunting. That first comet reached third magnitude and displayed a tail as long as 19 degrees on photographs. Ikeya proceeded to discover another comet the next year (1964) and then, in 1965, one of the great comets of the century—the sungrazer Ikeya-Seki. In 1966, Ikeya found a rather dim comet and then, at the end of 1967, another comet that also ended up being called Ikeya-Seki. In 1965, Ikeya had found the great comet just 15 minutes before Tsutomu Seki; in 1967, he beat Seki by no more than 5 minutes!

The many Japanese amateurs' skill and intensity is demonstrated by the events of October 5, 1975. That night, three of them independently (in different locations) discovered Comet Mori-Sato-Fujikawa within 70 minutes of each other. Later that night, five of them independently discovered Comet Suzuki-Saigusa-Mori within half an hour of each other. Hiroaki Mori is the only person ever to have discovered two comets visually in one night.

There may be far fewer dedicated comet hunters in the Southern Hemisphere than in Japan alone. But one man in Australia has proved so skillful in discovering comets that there have been few left for others. This is the man once called the "wizard of Dernancourt," William Bradfield. When Roger Sinnott of *Sky & Telescope* visited Bradfield in April 1977, he described the 6-inch refractor that the century's greatest visual comet-hunter uses.

> Close inspection shows that this seeming hodgepodge of unfinished boards, rusting bolts, and baling wire is in fact superbly suited to its appointed task. No off-the-shelf commercial telescope I know of quite equals this instrument's combination of manageability, ruggedness, light grasp, and wide field for comet sweeping.

Bradfield has discovered two comets in four different years. He has found two in 13 days, and three in less than 3 months. He discovered his first ten comets in less than 8 years, had only one cometless year in a period of 7 years, and finally, after a dry spell of $2\frac{1}{2}$ years, completed his dozen in less than a dozen years. In the 10 years from 1985 to 1995, Bradfield's discovery rate slowed (perhaps more from light pollution problems than age), but he managed to find five more. In the summer of 1995, the word went out: Bill Bradfield had done it again, this time finding a comet just visible to the naked eye (it was dimmer, but interesting, when it reached Northern Hemisphere skies). Even in the Northern Hemisphere, about one in every eight of the comets I've observed in my life has been a Comet Bradfield. Every one of his comets has borne his name alone. That comet he found in the summer of 1995 was the seventeenth "Comet Bradfield."

At the time of this book's writing in early 1996, the ranking of visual comet discoveries by living observers reads like this:

William Bradfield, 17
Antonin Mrkos, 11
Don Machholz, 9
David Levy, 8

The finding rate of Machholz has increased greatly since the more than 3400 hours it took him to find his first two comets! David Levy has his name on at least 20 comets (by my count), having assisted the Shoemakers in many of their photographic finds—including the last of their periodic comets, the amazing Shoemaker-Levy 9 whose pieces collided with Jupiter in 1994.

David Levy is a Canadian-born comet hunter who moved to the American Southwest specifically to improve his chances of finding comets (he estimates that 300 nights a year are suitable for comet seeking in this climate). Levy is also the gentle and eminently likable author of a number of engagingly written astronomy books (he obtained his Master's degree in English, doing a thesis on the poetry of Gerard Manley Hopkins—and its allusion to a particular comet!). He also writes a column for *Sky & Telescope*, profiling the lives of (mostly amateur) astronomers and the "observing experience."

Levy's own words tell his stories best. And that might also be said for the man who was his greatest inspiration, Leslie Peltier of Delphos, Ohio, an unassuming farmer who was "the world's greatest non-professional astronomer." Earlier in this century, Peltier discovered a dozen comets and made 132,000 variable-star observations. He earned $18 by picking 900 quarts of strawberries in order to buy his first telescope, "the strawberry spyglass." Later, his skillful observations stirred Henry Norris Russell to send him a 6-inch refractor which Zaccheus Daniel had used to discover three comets and that Peltier was to use for his dozen. Peltier tells his life story in the wonderful book *Starlight Nights*. (I still remember my delight at coming across it in a marvelously disheveled old bookstore over 20 years ago, but the book has been reprinted much more recently.)

The stories of men and women peering into the night patiently for the tell-tale (and tell-tail) stains of moving light warrant a book of their own (Levy's fine book *The Quest for Comets* comes close to this ideal). But perhaps the joy of discovering a comet is best described in a chapter of Peltier's book that recounts his first comet find. Peltier wrote that the Daniel comet seeker "had come back to life again," and he was able to carve in its mahogany below the years of Daniel's finds, the name PELTIER and the year 1925. Many more inscribed years of comet discoveries were to follow, but perhaps no carved legend could hold as much wonder and delight as that first. The carved name and number still fix the memory of that gentle young man on his way to send his first discovery telegram, clattering on his old childhood bike through the dark Ohio country night of anchored dog barks and the stars he loved.

HALLEY AND THE SHORT-PERIOD COMETS

Chapter 7

HALLEY AND ITS PERIODIC KIN

"Of all the comets in the sky, there's none like Comet Halley.
We see it with the naked eye, and periodic-ally."

The above lines of doggerel explain with admirable precision why Halley's is the most famous of comets. What they say is that Halley is the only comet that becomes conspicuous to the naked eye repeatedly—indeed, consistently, even unfailingly—for at least as far back through history as we have fair records on comets. In nearly every lifetime there is likely to be at least a few long-period comets much brighter than that lifetime's appearance of Halley. But these are generally comets which have not been seen for thousands of years and will not be seen again for as long. They appeared unexpectedly, so we could not prepare for them and anticipate them for years or decades—as we can our old friend Comet Halley.

I interrupt this calm and rational dissertation to exclaim from the heart. So beautiful *and* so faithful . . . what more could one ask?

Let us try, however, to arrive at a reasoned explanation of Comet Halley's greatness. John Bortle once pointed out the most important factor in creating Comet Halley's unique record of outstanding visibility: Only Halley is consistently (at almost every visit) bright at rather large angular elongations from the Sun. Along similar lines, Guy Ottewell gave the following more extensive explanation of why Halley has been able to weave its sight into all the centuries of the human race for at least thousands of years:

The fact is that Halley has caught the attention of mankind so often because only it has long durations of visibility, *and* great brightness outside twilight and often at large elongations from the sun, *and* only brief interruptions of visibility by the sun's glare, *and* occasional spectacular approaches to the earth. For all this to be possible its natural adequate brightness is requisite but not sufficient (some of its comrades may have more of it); the real key is a combination, unique to it, of orbital features.

This is from *Mankind's Comet,* the book that Guy and I wrote about Halley's Comet. Following that passage is the most thorough and trenchant discussion (by Guy, though I hope my collection of facts and my suggestions helped) that you'll ever read of the merits of Halley's orbit. It comes close to being a "recipe" for the characteristics you want in your ideal comet.

You first want an orbit that cuts something like half or two-thirds of the way across Earth's orbit and is retrograde (the comet moves in the opposite orbital direction from that of Earth). These two qualities ensure a long period in which the comet is fairly close to Earth and Sun with the possibility of good encounters with the comet both on its way in and on its way out.

You next want a comet orbit inclined moderately from Earth's (more inclined gets the comet too far from Earth, less inclined fails to keep the comet away from our line of sight with the Sun).

Third, if more of the land and people of your world are in its Northern Hemisphere (as on Earth), you want the perihelion position of the comet to be near the northernmost part of the orbit (or, more generally, you want the part of the comet's orbit that is within Earth's to be mostly north of the plane of Earth's orbit).

Finally, you want the size of the orbit to be such that the comet on it does not come back so often that it is taken for granted or so rarely that most people never see it.

Of course, Halley's Comet is only one of well over a hundred comets that are known to be periodic. A few of these have an intrinsic brightness that can rival Halley's. A few of them come back in a period close to Halley's magic "once in a lifetime" (an average of about every 75 years). But not one of them has both of these properties, together with the other merits of Halley's orbit. Only two periodic comets other than Halley are recorded as having been seen in pre-telescopic times. They were seen at

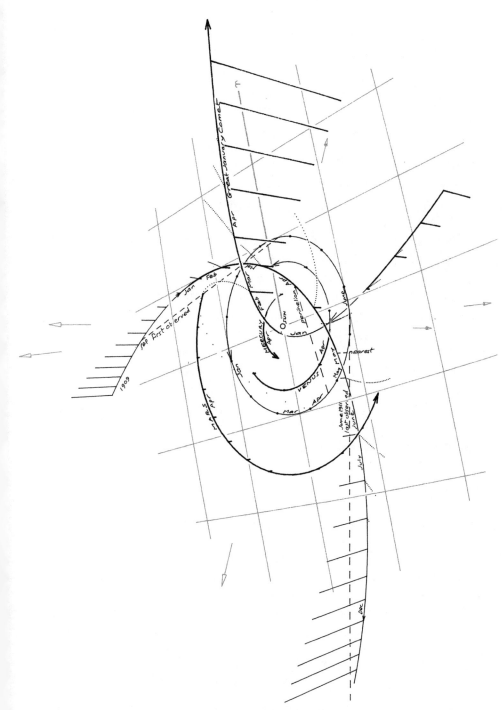

Diagram of the orbits of Halley's Comet and the Great January Comet in 1910.

135

a few returns. Halley's appearance was recorded at each of its last 30 returns, and it is possible that the comet has been visible to the naked eye at every one of its hundreds of returns for at least tens of thousands of years.

Chapter 8 of this book is entirely devoted to Comet Halley. Then the next section of the book opens with a chapter entitled "What Makes a Comet Great?" There we will see that there are indeed long-period comets brighter and more active than the august Halley. But we'll find that most returns of Halley result in it being a "great"comet (about one out of every seven or eight "great" comets in the past 500 years has been an apparition of Halley) and that no other periodic comet has had a "great" apparition in all the historic record. Thus Halley is the only repeatedly—almost unfailingly—great comet, even if it is seldom the "comet of the century" in brightness or overall grandeur.

In the meantime, having demonstrated why Halley is the king of the periodics, let's examine, in this chapter, many of the other fascinating periodic comets. The following selection of interesting periodic comets is by no means exhaustive. It is merely meant to be a generous sampling.

BIELA   See Chapter 4, "The Spectacular Deaths of Comets."

BROOKS 2   This remains the most impressive example of a comet which passed near a planet and was broken into pieces but survived and escaped (Shoemaker-Levy 9 is the spectacular example of a shattered comet that crashed into the capturing planet).

At its discovery apparition in 1889, this comet's nucleus was observed to consist of five pieces (the most ever seen up until that time). Calculations later showed that the fragmentation was probably a result of tidal disruption during the comet's encounter with Jupiter in 1886, the closest nonfatal approach of a comet to a planet in history. Brooks 2 came to within 0.001 a.u. of Jupiter, passing among the major moons of the planet for several days and even within the orbit of Io, Jupiter's innermost large moon. The result of this close encounter was not only a fractured nucleus for the comet but also an altering of its orbital period from 29 years to 7 years and a decrease of perihelion distance from 5.48 a.u. to 1.95 a.u. Ever since, the period of Brooks 2 has remained about 7 years, but the comet seems to have lost brightness. It has never approached anywhere near the brightness of eighth magnitude that it achieved at the discovery apparition, nor has there been any sign of the four lesser pieces of nucleus.

CHIRON   The largest known comet nucleus, Chiron (pronounced KI-ron) is believed to be about 200 kilometers wide—but when first discovered it was thought to be an asteroid, perhaps the first of a new belt beyond the orbit of Saturn. Chiron was discovered photographically at Palomar by Charles Kowal in 1977. It was assigned asteroid number 2060 and Kowal suggested naming it "Chiron" after the wisest of the centaurs (half-man, half-horse) of Greek mythology. The idea was that this might be the first example of some kind of transitional or hybrid object, and that other such objects could be named for other centaurs (a few similar objects have in fact been found and named, and have their aphelion even farther out than Chiron's). Chiron ranges from a little closer to the Sun than Saturn's orbit almost out to Uranus's orbit. It reached the perihelion of its currently just over 50-year-long orbit on February 14, 1996, but its visual magnitude in 1996 has lingered at only 15. This is true despite the fact that since 1989 there has been detected around Chiron a coma—first of gas then also of dust—proving that this object deserved to be classified as a comet.

The orbit of Chiron is unstable in the long term: close encounters with Saturn and eventually perhaps Jupiter should occur at intervals of thousands or tens of thousands of years, at which times the orbit is altered radically. Chiron and objects like its fellow "centaur" Pholus appear to be working their way ever so slowly down a ladder towards Jupiter from their original locations in the Kuiper Belt of comets. The ultimate fate of Chiron may be to get ejected from the solar system altogether. If it passed near the Sun on its way out, what a grand cometary display it might produce!

D'ARREST   Five years after he and Galle got the first looks at Neptune knowing what it was, Heinrich Louis d'Arrest sighted this comet, which since then has been observed at numerous returns on an orbit that varies between 6.4 and 6.7 years in length. In 1976 the comet's perigee (closest approach to Earth) and perihelion (closest approach to the Sun) occurred on the same day, and at this extremely favorable apparition, Comet d'Arrest was seen with the naked eye for the first time ever, by a number of observers (including myself).

The comet reached a maximum magnitude of 4.9 but was a good deal harder to see than this suggests because its coma was spread out to a half-degree or more with a 1-degree tail extension (the comet was also rather low in the sky for observers around 40 degrees north latitude, though it was visible nearly all night long). The minimum distance was 0.15 a.u. Jupiter has since then perturbed Comet d'Arrest, and its increased perihelion distance (now out to 1.30 a.u.) may prevent any return as favorable as the 1976 one for many decades to come.

ENCKE   This is the comet of shortest known period and the most closely studied of all. It is one of only four comets named for the person who calculated its orbit rather than for its discoverer. It was not recognized as the same comet at the favorable apparitions where it was first spotted by Mechain (1786), Caroline Herschel (1795), and Pons (1805). The second time Pons discovered the comet, in 1819, Johann Franz Encke was able to determine its 3.3-year elliptical orbit, to recognize the earlier apparitions for what they were, and to predict correctly the next perihelion passage in 1822.

Comet Encke's slightly early returns and slightly shrinking orbit have been attributed to "nongravitational forces"—the rocket-like effect of jets of dust and gas from its nucleus. It has been observed at far more visits than any other comet: The 56th observed return to perihelion occurred in 1994, and the 57th should be witnessed in 1997. Since the "Eureka!" apparition of 1819, this object has been missed only at the 1944 perihelion passage, when men were waging war around their planet and had little time to look for comets.

Although this comet's orbital period is the shortest that has been firmly established, the orbit is more elongated than those of most short-period comets, carrying it from perihelion near Mercury's orbit out to an aphelion of 4.10 a.u. (the outer edge of the asteroid belt, well within the orbit of Jupiter). The perihelion distance is interesting, because a close pass of Mercury by Encke in 1835 proved that the previously determined values of the planet's mass were too large (in 1848, Encke again approached Mercury, this time to within about $5^1/_2$ million kilometers). Encke's aphelion distance, about 0.5 a.u. less than any other comet's, inspired Barnard in 1912 to suggest that astronomers attempt to photograph it through aphelion passage. That is precisely what was done in 1913 (making this the first comet ever observed throughout its entire orbit) and thereafter at almost every aphelion since 1972, when the magnitude of its presumably bare, frozen nucleus was just 20.5.

Encke's aphelion is so comfortably far inside the orbit of Jupiter that the comet's path remains stable. In fact, no one has yet been able to figure out how Encke might originally have gotten into its present orbit!

Encke's orbit is not only the most steadfast but also has perihelion closer to the Sun than any other short-period comet except Comet Mellish, which has been seen only once, and one of the Comets Machholz. During history it must therefore have suffered many episodes of heating and depletion of its ices, and the debris from its many activations has pro-

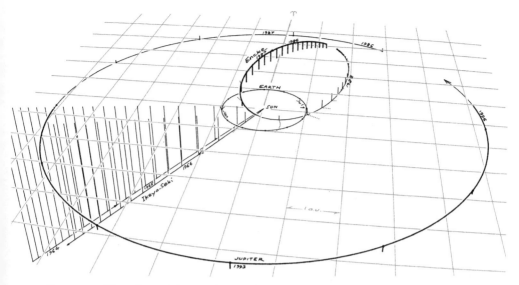

Diagram of the orbits of Comet Ikeya-Seki (long-period comet) and Comet Encke (short-period comet).

duced the great complex of diffused meteor streams that create the Taurid meteor showers of June and November and perhaps the Piscid meteor showers of June and October. Other bodies that have been suggested to be fragments of Comet Encke include the mysterious Tunguska object which exploded over Siberia in 1908, and the giant object which may have caused the formation of the lunar crater Giordano Bruno in an event that was witnessed on the evening of June 25, 1178, by a group of monks in Canterbury, England. Both of these proposals have come under fire. The latter rests on the report of the Canterbury chronicler in a superstitious age; on the rough correspondence of the crater's location with that of the distortion of the Moon's crescent; and on the finding, from laser ranging measurements of the Moon's rotation, that the Moon seems to be vibrating ever so slightly, like a rung bell, from an event that could have happened roughly a thousand years ago. Unfortunately, the low altitude of the Moon at the time of the monks' observation makes it more likely that any splitting of the Moon's cusp they saw was due to turbulence in Earth's atmosphere.

If the monks did see the formation of Giordano Bruno or some other lunar impact, the object that caused the event may not have been derived from Comet Encke. The idea that the object was a piece of Encke is part of astrophysicist Kenneth Brecher's theory of "the Canterbury

Swarm." According to Brecher, the Tunguska and Canterbury events, and also the June 22 and 26, 1975, meteoroid impacts on the Moon detected by seismographs left by the Apollo astronauts, were all caused by members of a swarm of large bodies derived from Comet Encke and following a similar orbit. Brecher posits that the swarm was caused by a breakup of Comet Encke or its parent before 1178 and is now spread across a volume roughly 13 million kilometers in diameter, in which there is a total mass of about 10^{15} kilograms in the form of numerous bodies with diameters as large as a kilometer. The next time Earth and the Moon will pass through this swarm will be in 2042, says Brecher. Brian Marsden has maintained that more than a similarity of dates of the various events (the monks' sighting, the Tunguska explosion, and meteoroid-caused moonquakes) is needed to establish the credibility of the Canterbury Swarm. Robert Young has suggested that if the event seen at Canterbury is more likely to have been an atmospheric distortion of the low Moon (which he considers not improbable), then the flock of bodies really ought to be called "the Encke Swarm." Despite his skepticism, Marsden does feel that swarms of debris in Encke's orbit are a possibility. The similarity of the orbit to that of 2212 Hephaistos suggests that this asteroid could also be a fragment of Encke.

There is also much controversy about how old Comet Encke is and how much longer it will remain an active comet. According to at least one study, meteoritic evidence argues that the comet has been around for at least 1500 orbits, or about 5000 years, but attempts to find early sightings of a brighter Encke have not been successful. The comet had a 3-degree tail in 1805 and may have just reached third magnitude in 1829, but it has had no better than about a 2-degree tail and fifth-magnitude brightness within the past few decades, which indicates that a 2- to 3-magnitude dimming per century may have occurred. This trend was once used to suggest that Encke might "die" (by exhausting its exposed ice) by the end of the twentieth century! The comet was glimpsed with the naked eye in 1980 and has continued to display its familiar fan-shaped coma, which indicates a partly (but surely not wholly) mantled nucleus.

The most important step in understanding Encke may have come in 1979, when Fred Whipple and Zdenek Sekanina used their modeling techniques to determine its spin axis orientation and rotation period, which in turn provided decisive clues in interpreting the comet's appearance and behavior. They found that Encke probably rotated in a period of 6 hours, 33 minutes in 1900 but that this spin has been slowing by perhaps 21 minutes per century—and even more in the nineteenth century. They found an equally dramatic change since 1786 in the spin axis orientation: the nucleus essentially flipped over during the past two centuries—and all because of the comet's famous nongravitational forces. Whipple and Sekanina attribute the flip to the torque from asymmetrically placed emission jets at each return, the cause of the rapid precession of Encke's axis. Their model successfully predicts the orientation of the emission fan at the nineteenth century returns. They conclude that because the northern hemisphere of Encke was directed more toward the Sun for several hundred orbital revolutions before 1900, the southern hemisphere may have become thickly covered with a dust mantle, which may now cause a post-perihelion dimming of the comet when that hemisphere comes under the Sun's stare around perihelion at each return. This effect explains the decrease in Encke's brightness since the earlier observed returns without implying that the comet is rapidly fading toward extinction. Whipple and Sekanina have even estimated that the mass of the comet is less than 10^{16} grams but that the current mass loss per revolution is only 0.09%, so Encke ought to remain active and visible for at least centuries more.

Comet Encke was the second comet (after Halley's) successfully predicted to return and was the third short-period comet established (after Halley's and Lexell's).

GIACOBINI-ZINNER   In September 1995, when it was glimmering prettily with a good ion tail in amateur telescopes, and when it was located not so very far from the approaching Halley's Comet in the sky, Giacobini-Zinner became the first comet ever to have a spacecraft fly by it—actually through the tail, very close behind the nucleus. The spacecraft, the International Cometary Explorer (ICE), was a rerouted, rechristened probe. It had originally been built to study the solar wind and so did not carry a camera to photograph Comet G-Z. And the interesting measurements of the comet that were made (confirming leading theories about ion tails) were overshadowed by the findings of the fleet that went around and through Halley the following spring. But G-Z has long been one of the most interesting of comets, and we have known some remarkable things about it for quite a while.

For instance, it is the source of the usually weak to nonexistent Draconid (sometimes called Giacobinid) meteor shower that can be intensified to storm strength in years when the comet passes near the October 9 point of our orbit a few days or weeks before or after Earth. This link between comet and meteor shower was first established in 1926, but in 1933 there occurred a Draconid storm of up to 6000 meteors per hour. Then, in 1946, Earth arrived just 15 days after the comet had passed, and up to 10,000 Draconids per hour were seen even with a full Moon in the sky (this meteor display was the first great one detected with radar—which had been invented in World War 2). Draconid displays were noted in other years the comet passed, but they were less strong because Earth arrived farther before or after the comet's passage through the intersection of their orbits. In 1985, some parts of the world saw Draconid rates up to several hundred per hour, the Earth having arrived only 29 days after the comet. The next chance to see Giacobini-Zinner will come in 1998, a good apparition that may also provide an opportunity to spot some Draconids.

An amazing 1985 paper by Zdenek Sekanina argues that the nucleus of Comet G-Z must be pancake-shaped, measuring about 2.5 kilometers by 0.3 kilometer and rotating once every 1.6 hours! A dust mantle could exist only in this object's polar regions—90% of the surface must be kept free of dust by centrifugal force. This, Sekanina argues, is why no sunward emission fan is ever seen in observations of G-Z. He believes that this nucleus is probably a slab that broke off a larger nucleus and is now, its condensation having further accelerated its rotation, itself on the point of splitting up.

HELFENZRIEDER   Despite being second-magnitude with a third-magnitude false nucleus and a 7-degree tail in 1766, this object was calculated to have a very short period, 4.35 years. And has never been seen again. The brightness may have been abnormally great at this return, and the comet may have been perturbed to an orbit that keeps it far from Sun and Earth. Another possibility is that it was perturbed to a short-period orbit so different that it was later discovered by someone else and now bears another name. The details and even the existence of such a perturbation cannot be proved because of the year or more of uncertainty in the originally calculated orbital period.

HERSCHEL-RIGOLLET   This is the comet of longest period that has definitely been observed at two returns (the Kreutz sungrazer comets have apparently done so much splitting in the past millenium or two that it seems inappropriate to refer to apparitions of the various pieces as returns of the same comet). Caroline Herschel discovered it in 1788, and it was not until 1939 that it returned and was found first by Roger Rigollet. The comet became as bright as seventh magnitude at both returns. Brian

Marsden found the period was 162 years in 1788 and 155 years in 1939—almost double that of Halley.

HOLMES   Discovered in 1892, this comet displayed a possible split nucleus, exhibited an odd detached section of tail, and underwent a brightness flare from magnitude 9 or 10 to 5 or brighter. At the next few returns on its 7-year orbit, the comet was never brighter than 13 and 15 in magnitude. Then Jupiter perturbed it out of range until it was found, 58 years later, in 1964—since which time it has never been brighter than 18 or 19 in magnitude, several magnitudes fainter still than would have been expected for it at its increased distance. The dramatic activity of 1892 seems to have been a fling that brought the comet very much nearer to extinction.

LEXELL   Messier discovered this comet only 0.21 a.u. from Earth on June 14, 1770, but it was a train seemingly headed right for Earth. It grew to a patch 2.4 degrees wide (about 5 times the Moon's apparent diameter) and about second-magnitude as it came to within 0.0151 a.u.—2.26 million kilometers, or just 354 Earth radii from our planet (about 6 times the lunar distance)—on July 1. No other comet is certainly known to have passed so close to us. After Pingre and Lambert had failed to compute its orbit, Anders Johann Lexell in 1779 showed that the orbit must be an elliptical one of 5.6 years—the first demonstrated case after Halley's that a comet was on a closed orbit.

But Comet Lexell was never seen again. The near-miss of Earth had been a shot delivered at us by Jupiter, which (as Lexell also discovered) perturbed the comet away from us again some years after our close encounter. The pre-apparition perturbation occurred in 1767, when Comet Lexell passed 0.02 a.u. from Jupiter, but in 1779 the comet passed just 0.0015 a.u. from Jupiter (almost as close as Comet Brooks 2 did in 1886). This second confrontation sent the comet into a 260-year orbit with perihelion beyond Jupiter and perhaps ultimately into a hyperbolic orbit that will take it out of the solar system.

MELLISH   This is a short-period comet with a perihelion distance of only 0.19 a.u. For a brief while, as it emerged from the Sun's glare at its discovery apparition in 1917, it was visible from the Southern Hemisphere as an object of between first and second magnitude with a visual tail length of as much as 20 degrees! The orbital period is about 145 years, so it should next arrive in 2062, the year after Halley's next return.

OTERMA   This was the one comet other than Schwassmann-Wachmann 1 with a nearly circular orbit (eccentricity = 0.14, less than that of two of the planets)—for a while. Less bright than S-W 1, Oterma was visible only at times from the 1942 return to that of 1958, before and after which its orbit was not a near-circle within Jupiter's (perihelion distance = 3.4 a.u.) but a more elongated ellipse entirely outside Jupiter's (the new perihelion distance = 5.4 a.u.).

PONS-WINNECKE   Recovered by F. A. Winnecke 39 years after Pons's discovery, it has been observed at more returns than any comet except Encke and Halley—but it may not be seen at many more. The comet seems to have lost quite a bit of its intrinsic brightness. The real problem for observers, however, is that since before its recovery, it has been locked into a 2:1 resonance with Jupiter. What this means is that its orbital period is about half Jupiter's, and so about every 12 years (one Jupiter orbital period), the giant planet and the comet are close enough for the comet to be tugged farther out. Up until 1939, this trend was favorable for us, bringing the perihelion to between 1.04 a.u. and 1.10 a.u. at the returns of 1921, 1927, and 1939, when Pons-Winnecke came as close as 0.04 a.u. from Earth—close enough for astronomers to see it as about third-magnitude and almost to see its true nucleus. Now the perihelion distance is more than 1.25 a.u., and the comet has not been brighter than 14 in magnitude in the past 40 years.

SCHWASSMANN-WACHMANN 1   This was the first of three periodic comets found on photographs by A. Schwassmann and A. A. Wachmann of Hamburg Observatory. Neither man ever found any long-period comets. This particular comet was (probably) the largest periodic known until Chiron was discovered and its cometary activity detected.

Schwassmann-Wachmann 1 is interesting enough even if it isn't the biggest periodic comet. Its orbit is becoming even more circular and stable than it already is, with perihelion about 0.5 a.u. outside Jupiter's orbit and aphelion at 6.73 a.u. from the Sun. This is an *annual comet*—that is, a comet observable all year except when we circle to the far side of our orbit and get the Sun between us and it. Thus opposition is more important for observers of SW-1 than perihelion. But unlike the few other annual comets, SW-1 undergoes strange flares that cause its usually star-like form to brighten and then expand into a larger disk or ring of coma as it fades. Although solar activity has been suggested as a cause for these outbursts, they seem completely irregular, with no correlation to the Sun's 11-year cycle of activity. The comet is normally about 18 or 19 in magnitude, which suggests that its nucleus may be bigger than any short-period comet's except Chiron's and perhaps half as large as those of the giant

Comet Sarabat of 1729 and the Sungrazer Parent. Whipple found a rotation period of 119 hours (just under 5 days) for SW-1; this is quite long. The amazing flares can increase the comet's brightness in days or hours by six or even nine magnitudes and has boosted it all the way up to magnitude 10 on several occasions. The best outburst of SW-1 in recent years occurred in early 1996, when the comet flared up to about magnitude 11.

SCHWASSMANN-WACHMANN 3   Speaking of outbursts: no one would have thought the seemingly modest Schwassmann-Wachmann 3 capable of major flares in brightness. But the autumn of 1995 proved otherwise. Before then, this comet was best known for playing tag with Earth's orbit; its perihelion was just outside our orbit at discovery in 1930 and now is just inside. The comet was predicted to brighten only to magnitude 12 in 1995, but in early October I heard it was flaring, and I quickly checked Charles Morris's Comet Observation Home Page. I was able to gather that Comet SW-3 was not as low in evening twilight as I had thought, so I raced down to a good site and saw for myself that Australian observer Terry Lovejoy's assessment that this comet looked like a "mini-Halley" made a lot of sense. The comet showed a strong parabolic outline in its head and a very intense first few degrees of tail. And like other observers, I estimated its brightness on this peak night as about magnitude 5.7, at least several hundred times brighter than predicted. The comet was glimpsed with the naked eye at its more favorable altitude in the Southern Hemisphere—this from an allegedly weak comet well over 1 a.u. away. I thought it would fade quickly and that the Internet had saved the day for me by alerting me to the possibility of an observation before the comet faded out. But SW-3 had amazing lasting power and pushed farther away from evening twilight while still unnaturally bright. It had a dramatic fine tail and posed in several noted star scenes. (For instance, Alan Hale was stunned one night to see it almost exactly where he had discovered Hale-Bopp a few months earlier.) The reason for the great flare in brightness and for the intense head and first few degrees of tail? Its nucleus split into at least four pieces around the September 1995 perihelion! The question now is whether the comet exhausted all or most of its volatile materials in the great outburst. Will SW-3 ever shine again?

SHOEMAKER-LEVY 9   This comet's story is so well known, and such thorough accounts have been given by its co-discoverer David Levy, that we need only sketch its biography here. But so event-filled was this comet's demise and so far-reaching its repercussions that even a sketch stretches several pages!

It was on the afternoon of March 25, 1993, that Carolyn Shoemaker was scanning over photographs and suddenly blurted, "I don't know what this is. It looks like . . . like a squashed comet." It was an intense bar of glow more than an arc-minute long, and better photographs soon showed that within it were not just 5, not just 14, not just 17, but 21 little bright spots, each with its own dust tail. She had found a shattered comet nucleus with each piece possessing its own coma and tail and all strung together in a line so that they looked "like a string of pearls." The strange comet was not far from Jupiter in the sky and in space, and Marsden soon determined that it had probably been broken into all these pieces by a close encounter with Jupiter on July 8, 1992, a day when Shoemaker-Levy 9 passed only about 31,000 kilometers from the colorful Jovian cloudtops. The shattered comet was now *captured* by Jupiter. It was in an orbit that would carry it no farther than about 50 million kilometers from the giant planet. But a further detail was more impressive than all the others: Not only had Shoemaker-Levy 9 been captured by Jupiter, but it was going to come back to Jupiter in just over a year and collide with the planet!

For the first time in history, humankind would witness a major collision of celestial bodies: the pieces of what was thought to have originally been a fairly large comet were going to strike Jupiter. The fragments would all hit at a fairly low southern latitude on Jupiter on the side facing away from Earth. Only the after-effect on Jupiter's clouds ought to be directly visible from Earth as soon as the impact spot from each strike rotated quickly around into view. Several unmanned craft in far space would be able to try photographing the strikes from vantage points radically different from that of Earth, even getting a direct (if very distant) view.

What would we see? There were a few predictions of tremendous effects (the naked-eye point of Jupiter briefly becoming 25 times (3.5 magnitudes) brighter at the brightest impact, thus greatly outshining Venus; the creation of a long-lasting atmospheric disturbance resembling the Great Red Spot)—that did not occur. But for the most part, the events and their after-effects were more impressive than expected. Some of the explosion fireballs soared high enough and bright enough to peek above Jupiter's edge. The largest of the pieces released millions of megatons of energy, maybe only one order of magnitude less than that of the K-T event (see Chapter 5). Each piece could have endangered human civilization and perhaps human existence if it had visited Earth.

Amateur astronomers all over the world, even those with quite small telescopes, got to see the marks left on the clouds of Jupiter by many of the strikes. These turned out to be eerie dark clouds that in some cases exceeded Earth in size and became the most prominent localized features on the planet's cloudy face—darker than the most prominent of Jupiter's darkish belts and almost as easy to see. All were at approximately the same latitude, but as the pieces of comet came in, one after another, during the course of almost a week (July 16–21, 1994), the planet rotated under them such that they formed a broken line of "bruises" most of the way around Jupiter. Each impact spot was named after the fragment that caused it, the fragments themselves having been designated with letters of the alphabet in order of their place in the several hundred kilometer-long line. This made it possible for the spots to be regarded individually, each taking on a personality of its own. Spots G and L were especially dark, huge, structured, and rather near each other so that as they elongated, they started looking like heavy eyebrows on the face of the planet (as seen inverted in most astronomical telescopes). I remember how captivating it was to look for the featured spots of each evening and watch their progress around the rapidly rotating planet. Some nights were particularly amazing. One night, spot G rotated into view at the edge and then suddenly seemed to be giving birth to a bright bump. After a moment of sheer astonishment, I realized this must be one of the Galilean moons coming out from behind the planet. But understanding the event as something less monumental than an eruption of bright material from the impact spot made the sight hardly less amazing or beautiful—an elegant jewel hovering just a whisper above the sprawling spooky monster dark spot.

The show went on for weeks—weeks of strenuous but rewarding observation of all the details discernible in all the spots available, each night in a different parade of eeriness across Jupiter's face. After a month and a half, only the strongest spots were still vivid, and even they were stretching out and graying and Jupiter was also starting to set too soon after the Sun. But even after the planet re-emerged from conjunction with the Sun, 6 months and more after the impacts, there were still some traces of the dark material. What was it? It was almost certainly produced by the interaction of comet material with gases in Jupiter's atmosphere, in the sudden furnace of the friction created when a chunk of dusty ice enters at several hundred thousand miles per hour. But how far each fragment penetrated before being consumed, how big they were, (the biggest, the G, fragment 2 or 3 km across or only a few hundred meters wide?), exactly what the comet fragments and upper atmosphere of Jupiter are composed of—these questions had not really been answered (to the best of my knowledge) even by early 1996. Perhaps the Galileo spacecraft's studies at Jupiter will help clarify what happened to the 21 shards of periodic comet Shoemaker-Levy 9. In any case, we are lucky to have witnessed an event whose like probably occurs only once in a thousand years even to so gravitationally compelling a world as Jupiter.

SWIFT-TUTTLE   This is the parent comet of what is often the strongest of annual meteor showers, the Perseids. It is also one of the few short-period comets that is about as intrinsically bright as Halley's. Those two facts alone would make it one of the most interesting of comets, but the events following its long-hoped-for recovery in 1992 have been incredibly exciting. The most sensational moment came in October 1992, when the media got wind of Brian Marsden's prediction that Swift-Tuttle had a very small but real chance of colliding with Earth at its next return in 2126!

But the story of Comet Swift-Tuttle—under that name, at least—begins in 1862. On the night of July 15 of that year, Lewis Swift, an amateur astronomer in Marathon, New York, was out with his telescope looking for Comet Schmidt, a fairly bright new comet he had read about in the newspaper. Soon he found a fuzzy patch of light in the north sky which he took to be the comet, fainter than expected. But three nights later, Horace Tuttle, at Harvard Observatory, saw the same object and realized it was not Comet Schmidt. After Tuttle announced the discovery, Swift hastened to make his claim, which, fortunately for him, was accepted.

Your first reaction might be to think poor Tuttle unfairly got second billing in this comet's name.

The inner coma of Comet Swift-Tuttle in 1862.

Horace Tuttle had already discovered several comets and would co-discover at least one more very important one—Comet Tempel-Tuttle, the parent of the Leonid meteor showers and storms. But Tuttle, after what some called heroic service in the Civil War, was dismissed from the Navy years later, when it was discovered that he had embezzled a small fortune. After rebounding to some later scientific and presumably financial success, Horace Tuttle's fate was to die in 1923 with only $70 to his name and to be buried in an unmarked grave (the research of U.S. Naval Observatory astronomer Richard Schmidt recently brought some of these details to light).

Lewis Swift lived to be 93 and went on to discover many more comets after Swift-Tuttle (his first). At least thirteen comets immortalize his name, eight of them his name alone.

But back to the 1862 appearance of Comet Swift-Tuttle. After its July discovery, the comet brightened, exhibiting for telescopic observers remarkable jets and envelopes of sunlit dust in its head (these features were sketched by many observers, but in best detail by George Bond and A. Winnecke). In late August, as Americans fought the Second Battle of Bull Run, the head shone as bright as second magnitude and had a tail spanning 25–30 degrees. Within about 2 weeks, the comet passed less than 9° from the north celestial pole, through perihelion (0.96 a.u. from the Sun), and through its closest approach to Earth (0.34 a.u.).

Over the next few years, several calculators came up with an orbital period of between 120 and 125 years for the comet. In 1866, Schiaparelli announced his discovery of a link between the orbits of Swift-Tuttle and the Perseid meteor stream. In 1889, F. Hayn computed a "definitive" orbit with a period of 119.64 years for Swift-Tuttle, which suggested that the next return would come in 1982.

In the early 1980s, anticipation of the interesting comet's return increased when, sure enough, the Perseid meteor shower's peak rates picked up dramatically. Rates exceeding 200 Perseids per hour were observed. Surely this meant the parent comet was near? When the rates returned to normal in the mid-1980s, some astronomers concluded that Swift-Tuttle had somehow slipped by unseen (perhaps delivering an inactive performance on the far side of the Sun from us).

But Brian Marsden suspected otherwise. He had long been troubled by the discordant positions given in the final observations from Capetown in 1862 and by the fact that one could not find earlier returns of this intrinsically bright comet if one assumed a 120-year orbit. Marsden was intrigued by the possibility that a comet found in China in 1737 by Father Ignatius Kegler, a Jesuit priest, might be an earlier return of Swift-Tuttle. If so, the current orbital period of the comet should be about 130 years, and 1992 could be the year of return.

In 1991, a brief but intense Perseid display was seen in some parts of the world: about 350 meteors in an hour. In 1992, despite a bright Moon and widespread cloudiness in the favored area of Earth (turned toward Perseus at the right hour), there were observations of numerous meteors bright enough to light up the clouds—like little moving moons, one account said. This outburst happened even closer to the time when Earth passed through the orbital plane of the comet. Marsden had refined his prediction to a perihelion passage on December 11, 1992, but he was starting to lose his confidence.

It was restored when, on September 26, 1992, a Japanese amateur astronomer found a comet that was headed to perihelion just one day off from Marsden's latest prediction, and the rest of its orbit was the appropriate orbit. It was Swift-Tuttle!

This time the comet was coming nowhere near as close as in 1862, but it still became visible to the naked eye, passed near the great globular cluster M13 (which it then outshone), and exhibited marvelously vigorous jet activity as it had in 1862 (seen this time from much farther away, but with far superior instruments). But observation of the comet fell into the background when a media frenzy kicked up after Marsden made a unique announcement on one of the IAU (International Astronomical Union) Circulars, publications normally only heeded by astronomers. The return of Swift-Tuttle had enabled Marsden both to confirm that 1737's Comet Kegler had indeed been an earlier appearance of Swift-Tuttle and to calculate the circumstances for the next return. Marsden announced that his calculations showed a very small but distinct possibility that Comet Swift-Tuttle would collide with Earth at its next visit in 2126!

There was uncertainty, perhaps largely because of nongravitational forces (the rocket effect from powerful emission jets). Marsden therefore urged astronomers to make accurate observations of the comet for as long as possible on its way out. What were the chances of collision? Only 1 in 10,000. But of course the mere fact that collision was possibile at some specific time—even 134 years in the future, farther ahead than most people ever look—was enough to make it electrifying news.

With plenty of new observations, Marsden was later able to rule out a collision in 2126. The comet will not even come as close as 1996's great

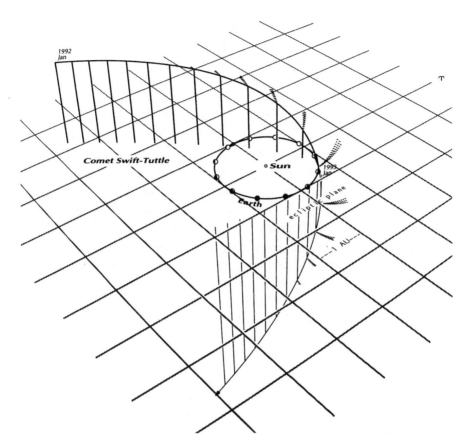

Orbit of Comet Swift-Tuttle.

Comet Hyakutake did, though it should certainly put on a very impressive show (the return after 2126 will also be a close, good display). And there is a closer approach scheduled for the year 3044. We should bear in mind that Comet Swift-Tuttle is by far the largest object we know of whose orbit intersects that of Earth—perhaps the only one that could cause a cataclysm as great as the K-T event 65 million years ago.* In fact, David Levy argues that the dinosaurs are much more likely to have been killed by a comet than by an asteroid, because no known asteroid whose orbit intersects Earth's now is anywhere near large enough, and the situation 65 million years ago is unlikely to have been much different. Of course, only one exception would have sufficed, but Levy's point is well taken.

---

*A new study, however, suggests that in the next million years the 40 Km × 14 Km asteroid Eros has a 40 percent chance of *becoming capable of* hitting Earth!

Of course, smaller objects than Swift-Tuttle might be capable of causing an impact winter not much less severe than that inflicted upon Earth by the K-T object. Thus we shouldn't be complacent. We should increase our efforts to survey the heavens for all asteroids and periodic comets that might pose a danger—and as a matter of fact, the 1992 Swift-Tuttle scare may have helped push U.S. government efforts and funding in that direction.

If we knew a comet or asteroid was headed for a collision with Earth, could we prevent the catastrophe? To blow a small celestial object to bits with nuclear weapons might only cause us to be hit by a swarm of smaller objects and dust whose effects could be even worse. The best bet would be to explode bombs near to the object to deflect it ever so slightly. But how fragile is a particular comet nucleus? And could we judge how close and big the explosions could be without breaking sizable slabs off a nucleus and multiplying the danger? The key to deflecting even a small comet or asteroid is to deliver the deflecting nudge as long before it nears Earth as possible. If you made the doomsday object deviate even the tiniest fraction of a degree from its path many months or years in advance, that angle could translate into thousands or millions of miles of "miss" by the time it finally reached Earth.

Should we keep a nuclear arsenal ready for this purpose? Should we get a little practice at deflecting asteroids and comets? In his book *Pale Blue Dot*, Carl Sagan explores the issues and finds great danger in these activities. Besides the chance of fragmenting a comet or asteroid into a worse threat, there is the possibility of a test of a harmless object mistakenly sending it toward Earth or of a psychotic dictator deliberately ordering an object nudged toward Earth.

There is more to say about Swift-Tuttle, and some of the findings are being published even as this book is being finished in early 1996:

- Various studies of the 1992 return confirm that Zdenek Sekanina's 1981 paper was almost exactly right in its calculations of the Swift-Tuttle nucleus's rotation period and location of active areas. The comet's nucleus apparently does rotate in close to 67 hours, and the active areas Sekanina identified as firing in the 1862 return seem to have been active in the same locations again in 1992. That the same areas become active at two successive returns implies a semipermanent and fairly stable crust.
- In 1994, Donald Yeomans, Kevin Yau, and Paul Weissmann confirmed other researchers' calculations identifying Comet Swift-Tuttle

as the comet seen in China in A.D. 188 and 69 B.C. They noted that although the comet should have been a conspicuous second-magnitude object in 447 B.C. and 574 B.C. there are no known records of its being sighted in those years. In order to fit the observations from the ancient and modern returns of Swift-Tuttle, Yeomans and colleagues had to do two things. One was to leave out the series of observations from Capetown at the end of the 1862 apparition (the same ones that troubled Brian Marsden). They also had to assume no significant non-gravitational forces, and because Swift-Tuttle's jets are vigorous, this meant assuming a nucleus so massive that the jets couldn't affect its path. Brian Marsden has argued that there might have been a recurrence of the "Cape effect" in 1993 observations, when the comet was moving away, and that scientists should not ignore the Cape observations, which are internally consistent. He feels that "we have not properly modeled the forces on Swift-Tuttle. . . ." But the assumption of a giant nucleus for this comet is supported by evidence assembled by other researchers. Marina N. Fomenkova and colleagues studied a variety of the 1992 observations (including a higher temperature in the coma than expected) and concluded that the nucleus of Swift-Tuttle is a huge 30 kilometers (plus or minus 6 kilometers) wide and 34 times more massive than Halley's nucleus!

• Bradley E. Schaefer argues that records that have led researchers to believe a supernova (the brightest kind of exploding star) occurred in A.D. 185 are actually a distorted account in Chinese annals of both the sighting of a nova (a less powerful exploding star) in 185 itself and the return of Swift-Tuttle 3 years later. If Schaefer is right and this second brightest of identified supernovae in our galaxy didn't happen, it shakes up some important cosmological work about the age of the universe that was being based on this supernova.

And what about the possibility of collisions between Earth and Swift-Tuttle in the far future? When Marsden announced the 10,000-to-1 odds against an impact in 2126, a friend of mine noted that if Swift-Tuttle was a long-lived object, and if its orbit did not deviate greatly, then a collision would eventually become not just possible but likely. Swift-Tuttle probably can survive for at least a million years (I speculate), maybe even many times longer if it is as big as we now suspect. And a study by John Chambers points out that Swift-Tuttle's mean orbital period is exactly 11 times that of Jupiter—a resonance that renders the comet's orbit extremely stable, making close encounters with our world possible for tens of thousands of years into the future. But certainly we can relax for the moment.

Even so, the last I heard about observations of Swift-Tuttle was that in 1994, it still burned at magnitude 22 in the constellation Vela. And for the very last word on Swift-Tuttle, what could be more dramatic than this assessment, which appeared in *Sky & Telescope?*

> Given its large size and potential impact speed of 60 km per second, Comet Swift-Tuttle is the single most dangerous object known to humankind.

TAYLOR   This comet was discovered in late 1915 and reached about magnitude 8.7 in the first week of January 1916. On February 10, Barnard found a secondary nucleus, which remained visible until late March. Although the primary was observed until late May, it was then 3 magnitudes fainter than predicted. The comet's period was about $6^1/_2$ years, but efforts to recover it at subsequent predicted returns (even rather favorable ones) failed. This led to speculation that Comet Taylor had suffered a fate similar to that of Biela's Comet. But calculations in 1974 by V. V. Emel'yanenko for both the A and B nuclei led in 1977 to the recovery (by Charles T. Kowal) of the B (secondary) nucleus. The B nucleus was only 16 in magnitude and was near the predicted position, but the A nucleus was nowhere to be found!

TEMPEL-TUTTLE   The parent comet of the Leonid meteor shower is trailed by dense (as such things go) swarms of meteoroids that have produced the most intense meteor storms in history.

This comet may have been routed into approximately its present orbit by the planet Uranus in A.D.126, over 1600 years before Uranus was discovered (the first record of a Leonid shower seems to be the one in A.D.137). The great Leonid displays since then have occurred with the comet relatively near perihelion, near aphelion, and even in various other places in between. But in the past two centuries, the really awesome Leonid storms have come after the comet's perihelion. For instance, in 1965 Tempel-Tuttle was at perihelion on April 30 and in November of *the following year* the stupendous sight of Leonids as thick as 150,000 per hour (with brief spates even more intense) was visible from the southwestern United States and a few other places. (I myself caught the opening phase in New Jersey, but the Sun rose over my horizon before the incredible peak torrents occurred.)

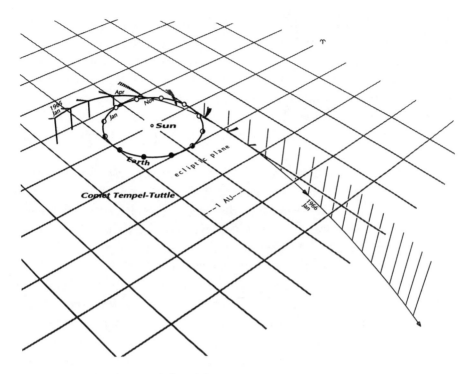

Orbit of Comet Tempel-Tuttle.

The comet itself was discovered by Tempel and Tuttle at the end of 1865. Its connection with the Leonid shower (which was magnifcent in November 1866) was first proposed at the end of 1866, when Oppolzer calculated the comet's orbital period and many astronomers pointed out the similarity of the orbit to that of the Leonid stream. The period then was 33.52 years, and in 1965 it was 32.92. But for many centuries, it has averaged close to $33^1/_3$ years, making the comet and its meteoroid swarm (which has sometimes hit and sometimes missed) a thrice-a-century visitor. (Similarly, Comet Encke is an almost perfectly thrice-a-decade visitor, with like apparitions occurring every 10 years.)

After the connection of the comet with the Leonids was made, John Russell Hind pointed out similarities between Tempel-Tuttle and the comets of 868 and 1366. The 1366 apparition proved unquestionably to be an earlier appearance of Tempel-Tuttle and a very interesting one, because the comet became about third magnitude as a result of passing just 0.02 a.u. from Earth. This is either the second- or third-closest known passing and is somewhat closer than our 1983 encounter with IRAS-

Araki-Alcock. Tempel-Tuttle is the one of just two short-period comets other than Halley that has been certainly identified at a return before the invention of the telescope. After the 1965 return, J. Schubart was able to identify the fourth magnitude 1699II comet observed by Gottfried Kirch as an early return of Tempel-Tuttle. Thus the 1366, 1699, 1866, and 1965 returns were observed, and the next return should occur in 1998. In both that year and 1999, the Earth is very likely to pass through the great Leonid swarm again.

TUTTLE-GIACOBINI-KRESAK   Originally found by Tuttle in 1858, this comet was lost until Giacobini came upon it in 1907 and then lost again until Kresak noted a tenth-magnitude object in 1951. At this third appearance, the comet was at last observed well enough to have its 5.5-year period determined and its identity with the earlier apparitions established. The comet was at another typical magnitude of about 13 on May 20, 1973. But 6 and 7 days later there were independent "discoveries" of an eighth-magnitude and a fifth-magnitude comet that turned out to be Tuttle-Giacobini-Kresak. The comet reached a maximum magnitude of 4 and then took only 5 days to fade back to tenth magnitude. In early July a similar flare occurred, this time documented as taking about 3 days to brighten by about 10 magnitudes (that is, to become 10,000 times brighter!)—the greatest flare ever recorded in a comet. At the time of the first flare, the comet had been around perihelion at a distance of 1.15 a.u. from the Sun, but the spectacular behavior was not seen at an earlier return (or at the 1978 apparition) and remains unexplained.

WESTPHAL   The discovery apparition of Comet Westphal in 1852 ended with the comet fainter than predicted, but 61 years later, Westphal's next return was remarkable for a dramatic fading and diffusing as it approached its perihelion distance of 1.25 a.u. Westphal had been expected to remain at about magnitude 8 for the 2 months before perihelion, but it instead dimmed in that timespan to about 16 or 17 magnitude, when it became too faint and diffuse to detect—a few days before perihelion! Attempts to recover it at the next predicted return in 1975–1976 all failed, and the comet is believed to have disintegrated or exhausted itself right before observers' eyes at that final return in 1913.

WILSON-HARRINGTON   See Chapter 4.

Chapter 8

THROUGH THE AGES WITH HALLEY'S COMET

Each return of the great visitor Comet Halley is a vision-flash, a revelation of the next frame in the movie of humankind and Earth. We are inside the movie; it is for us not a fiction but life itself. The watcher is Halley's Comet. Its once-a-lifetime returns suggest very easily the idea of this object as a witness, passing not only to remind, inspire, and warn but also to see us, as if appointed by duty or its own concern.

No doubt we inject our own spirit into this ice-hearted cloud of dust and gas. Let us nevertheless look through its bright eyes as though they were our own. We'll see what people and events were astir at each return of Halley, and we'll discover what lore and other reactions the comet would have seen itself sparking. And of course we'll also describe what the visitor comet itself looked like, how it behaved at each return.

As you read this history scored to the beat of Comet Halley, remember that some of the dates of early rulers' births, deaths, and reigns are uncertain and that I've chosen swiftly what seemed to me the most reliable figures. At least the returns of the comet from 1404 B.C. to A.D. 2134 can be precisely and confidently dated. In addition to these, however, I have added two hypothetical returns: a very close visit in the remote past and an even closer one in the remote future.

We know that sometime about 25,000 to 27,000 years ago—or maybe many "libration cycles" earlier—there came the closest of Comet Halley's passes of Earth. There may be a limit to how close the comet could have

157

come since achieving anything like its present orbit. But I'll ignore that because even if Halley never did come as close in the past as the encounter I imagine, it still could, sometime in its immense future. Because of the librating nodes of Halley's orbit, even a collision of Earth and Halley's Comet is presumably possible, however unlikely, some day (but no day within the next many thousands of years).

In the Remote Past

In distant prehistoric times, perhaps during an Ice Age, we imagine Comet Halley coming to within about a million miles of Earth (that is, about 4 times the Earth–Moon distance, and not greatly closer than the small Lexell's Comet came just over 200 years ago).

The total magnitude and size of Halley at such a visit are difficult to estimate because of the coma's diffuseness, but we can make some guesses. The magnitude might be something on the order of −6, with a head perhaps more than 5 degrees wide. The naked eye might be able to see jets emerging from the perinuclear cloud, and binoculars would show incredible internal structure.

But forget binoculars. These are prehistoric times—there is a herd of wooly mammoths shifting uneasily a few hundred feet away from us on the brittle starlit snow. We see a glow along the eastern horizon and think the crescent Moon is rising hours before dawn, but we climb higher up the hill to look from a less obstructed vantage point.

What we see is something shaped roughly like a crescent Moon but many times larger, surrounded by a glow much larger still! The radiance is virtually as great as a crescent Moon's, but it is spread out over so much larger an area that stars—including a first-magnitude one—are fairly prominent in the outer portions of the hazy extended glow. Just within the crescent of the head is what looks like a star almost as bright as Jupiter! The "star" is yellow like Jupiter, but shades of pale gold and blue are visible in parts of the surrounding coma. In the course of an hour or two, the overall cloud of the head, large enough to fill much of the Big Dipper's bowl (which is distorted somewhat from its modern configuration because we are in the distant past), has moved dramatically across the starry background, rolling past with incredible speed. As dawn comes and even as the Sun rises, the central "star" and some of the crescent remain dimly visible in the blue sky. And yet the views seen this night and morning, whose richness and wonder surpass the products of the most vivid imagination, are many times less strange and magnificent than what we will behold on another night, many, many millennia in the future.

Earliest History

But first we enter history, at 1404 B.C., and take not only our views from Earth but also assume the perspective of the comet itself.

The Middle Kingdom of Egypt is near its zenith after the reigns of Queen Hatshepsut and Thutmose III. Amenhotep III rules, and it is only a decade before the birth of the great poet and heretic, the revolutionary monotheist who will call himself Ikhnaton. Before the next return of Comet Halley, Ikhnaton will have been overthrown and the reign of the famous Tutenkhamon will have come and gone.

At this great south polar passage of Halley's Comet in **1404 B.C.,** the comet would have seen a partly lit blue and white globe with Antarctica in its center—the Earth would actually have been visible as a globe from the comet with unaided human vision. The brightness of Earth would be roughly that of a thick crescent Moon in our own sky, though only about one-third the diameter. And only about one Little Dipper-length away from Earth (Earth's position would be not far from the Little Dipper's bowl, near the Draco–Ursa Minor border) would be a fat yellow star or tiny globe perhaps as bright as Venus: Earth's Moon.

The next three visits of Comet Halley were not so spectacular, but they saw Assyria growing stronger and Egypt in the 67-year reign of Rameses II (1300–1223 B.C.); Halley's **1266 B.C.** return was almost exactly in the middle of that reign. The return of **1198 B.C.** may have seen Moses and the Hebrews still seeking the Promised Land and the trouble leading up to the Trojan War perhaps already coming to a head (the dates of the Exodus and Trojan War—even, among many scholars, the historicity of the events themselves—are of course conjectural). The comet in 1198 B.C. was closest to Earth after perihelion but no nearer than at the A.D. 1986 return, and this closest approach was just 2 days after inferior conjunction with the Sun. The comet was then still well above the plane of Earth's orbit (its "descending node" was out near the orbit of Mars).

Along with the **1059 B.C.** visit (which happened around the time of the fall of China's Shang dynasty), the **986–985 B.C.** return is most in-

Table 5.   **Returns of Halley's Comet**

	Year and date of perhelion passage		Elapsed years since previous return	Orbital period in years (osculating)	Perihelion distance (a.u.)
(1)	1404 B.C.	Oct. 15	.00	71.86	.6209761
(2)	1334 B.C.	Aug. 25	69.86	79.82	.6265826
(3)	1266 B.C.	Sep. 5	68.03	78.15	.6341965
(4)	1198 B.C.	May 11	67.68	78.89	.6289955
(5)	1129 B.C.	Apr. 3	68.90	70.52	.6217410
(6)	1059 B.C.	Dec. 3	70.67	78.68	.6118822
(7)	986 B.C.	Dec. 2	73.00	74.53	.6020087
(8)	911 B.C.	May 20	74.46	75.06	.5983184
(9)	836 B.C.	May 9	74.97	74.97	.5987312
(10)	763 B.C.	Aug. 5	73.24	74.87	.6016488
(11)	690 B.C.	Jan. 22	72.47	74.35	.5996958
(12)	616 B.C.	July 28	74.52	75.70	.5925521
(13)	540 B.C.	May 10	75.79	75.73	.5917786
(14)	466 B.C.	July 10	74.19	76.15	.5902897
(15)	391 B.C.	Sep. 14	75.16	76.12	.5880409
(16)	315 B.C.	Sep. 8	75.99	76.17	.5874293
(17)	240 B.C.	May 25	74.71	76.75	.5853647
(18)	164 B.C.	Nov. 12	76.47	76.88	.5845470
(19)	87 B.C.	Aug. 6	76.73	77.12	.5856047
(20)	12 B.C.	Oct. 10	75.18	76.33	.5871999
(21)	66	Jan. 25	76.29	76.55	.5851046
(22)	141	Mar. 22	75.15	77.28	.5831377
(23)	218	May 17	77.16	77.37	.5814660
(24)	295	Apr. 20	76.93	79.13	.5759148
(25)	374	Feb. 16	78.83	78.76	.5771940
(26)	451	June 28	77.36	79.29	.5737438
(27)	530	Sep. 27	79.25	78.90	.5755915
(28)	607	Mar. 15	76.47	77.47	.5808315
(29)	684	Oct. 2	77.55	77.68	.5795841
(30)	760	May 20	75.63	77.00	.5818368
(31)	837	Feb. 28	76.78	76.90	.5823182
(32)	912	July 18	75.39	77.45	.5801559
(33)	989	Sep. 5	77.14	77.14	.5819144
(34)	1066	Mar. 20	76.54	79.26	.5744956
(35)	1145	Apr. 18	79.03	79.08	.5747921
(36)	1288	Sep. 28	77.45	79.12	.5742108
(37)	1301	Oct. 25	79.08	79.14	.5727097
(38)	1378	Nov. 10	77.04	77.76	.5762013
(39)	1456	June 9	77.58	77.10	.5797014

(continued)

Table 5.   (continued)

(40)	1531	Aug. 26	75.21	76.50	.5811975
(41)	1607	Oct. 27	76.14	76.06	.5836150
(42)	1682	Sep. 15	74.89	77.40	.5826084
(43)	1759	Mar. 13	76.49	76.89	.5944466
(44)	1835	Nov. 16	76.68	76.27	.5863423
(45)	1910	Apr. 20	74.42	76.00	.5871888
(46)	1986	Feb. 9	75.81	76.00	.5871045
(47)	2061	July 28	75.47	74.77	.5728000
(48)	2134	Mar. 27	72.66	74.21	.5932149

teresting for being possibly the dimmest (of course, flares of the comet's brightness at a particular return could always alter this). It has been calculated that King David reigned from 1010 to 974 B.C. If Halley's Comet got no brighter than second magnitude when near the Sun and was fainter at the pre-perihelion passage of Earth, then perhaps neither David nor anyone else paid much attention.

The **911 B.C.** and **836 B.C.** visits were similar to each other, with the comet's closest approach near perihelion on similar dates, but no record of its having been seen (though surely it was) has come down to us from those days of the divided kingdom of the Hebrews and the time when Homer may have walked on earth. However, the years when Homer recited the stories of Ilium and Odysseus (perhaps seven returns of Halley after the real Trojan War) may have been about the time the ascending node of Halley's orbit passed Earth's orbit. Perhaps the first great Orionid storms of meteors flew in Homer's lifetime. When Halley's Comet came in **763 B.C.**, it pierced up through the plane of Earth's orbit just 0.014 a.u. (about $5^1/_2$ lunar distances) outside of that orbit.

Ten years after this last date is the traditional date for the founding of Rome. At the next return, in **690 B.C.**, the comet came much closer (0.213 a.u.) than at any visit since 1266 B.C., but the **616 B.C.** visit was another distant perigee (closest approach to Earth) around the time of perihelion. The comet at these returns saw some of the greatest prophets of Israel, the beginning of the Chaldean empire, and Zarathustra walking on Earth. The first of the great Greek scientists, Thales of Miletus, was born in about 624 B.C. and lived almost until the Halley visit of the next century.

In **540 B.C.**, another rather typical perihelion-oriented apparition, Comet Halley was witness to the greatest assortment of important religious leaders ever to live at the same time: the very young Confucius, the young Buddha, and Lao-Tze in old age. The Jews were in their last days of exile in Babylon, and it is believed that "Second Isaiah" wrote his great words about 540–538 B.C. A mystical but scientific thinker also in his prime at this visit was Pythagoras.

The next return of Halley's Comet, **466** B.C., marked the start of the Golden Age of Greece. In that year, the Athenian navy defeated the Persians in a decisive battle 24 years after Marathon. For a long time, researchers believed that a famous comet of 467 B.C. (which ancient scientists erroneously connected with a meteorite fall) was a return of Halley's Comet, but the Yeomans-Kiang calculations of Halley's past visits disproved this.

About 8 years before the apparition of **391** B.C., Socrates was sentenced to death and drank the hemlock. But Plato might have seen Halley's Comet (he is credited with writing a little poem about the joy of stargazing).

About 8 years before the return of **315** B.C., Alexander the Great died—with no worlds left to conquer but Halley's Comet yet unseen.

The Ancient Recorded Visits

Now we come to the recorded visits. I provide a title for each one (some of them invented by Guy Ottewell) as an assistance to memory (and a touch of appropriate color).

240 B.C. FIRST RECORDED VISIT   The next return of Halley's Comet saw an important step in the military growth of the next giant to rise, Rome. But it is not in Roman but in Chinese and Mesopotamian sources that the earliest records of the comet are found. "During the 7th year of Ch'in Shih-Huang-Ti a broom-star first appeared at the north and during the 5th month it was seen at the west. [Later] it was again seen at the west."

So reads the first known record of Halley's Comet. The comet should have been moderately bright and should have behaved much as the passage says. Who was Ch'in Shih-Huang Ti? When he grew up, he was the self-proclaimed emperor of China (whose reign began in 221 B.C.). He is best known for his Burning of the Books and for joining smaller walls to make the Great Wall. As Guy Ottewell relates, "He had a bedroom for each night of the year so that death should not find him and deprive the world of him. But die he did, in 210, and his meteoric dynasty only three years later."

164 B.C. MOST RECENTLY RECOVERED VISIT   The Romans had de-feated Carthage for the first time in the First Punic War, and Judas Mac-cabeus was in the midst of retaking Jerusalem from the Seleucid descen-dants of Alexander, to restore it to Jewish control. This was a missing return of Halley's Comet until April 1985, when a paper by F.R. Stephen-son, K.K.C. Yau, and H. Hunger reported three clay tablets found in the British Museum that appear to be a Babylonian record of the apparition.

This return was of an unusual type in which Halley arrives from be-yond Earth and appears in the midnight sky, moving into the morning sky. No similar apparition occurred until 1759.

87 B.C. OCTAVIAN TROUBLES   It is now believed that this return of Halley was alluded to in Chinese, Babylonian, and Roman records. It should have been only moderately bright (zero magnitude at best), but it is probably this return that Pliny later cited as an example of a "terrible star, announcing no small shedding of blood in the consulate of Octa-vian" (an Octavian before the one who became Augustus Caesar). Mar-ius had seized power in Rome the year before.

Yeomans lists this as one of Halley's many apparitions that qualify as "great." Its maximum magnitude outside of twilight was 2.

12 B.C. AGRIPPA'S DEATH PRESAGED   Although authorities have re-peatedly disproved the claim that Halley's Comet could have been the Star of Bethlehem, Magi must surely have studied the 12 B.C. return of Halley. It should have been very impressive, with the comet passing just 0.16 a.u. from Earth—though before perihelion, so the comet would have been less developed. Halley passed in front of Earth on its way in toward the Sun, reaching a maximum magnitude outside of twilight of about 1. The comet swept its tail across the Big Dipper around September 10. The later Roman historian Dio Cassius apparently mentions this return of the comet, in the year "before the death of Agrippa," Augustus's right-hand man. The Roman Empire lay mostly in peace, but unrest was brewing among the Jews.

In the period before the next visit, Jesus of Nazareth lived out his life on earth without ever seeing Halley's Comet.

66 A.D. SWORD OVER JERUSALEM   This approach was not so close as the previous return, but it was after perihelion and so is assumed to have had a maximum magnitude of 1 outside of twilight.

This apparition of Halley is the very famous one that "hung in the form of a sword over Jerusalem." This phrase is from the writings of the Jewish historian Joshephus and is taken to refer to the destruction of

Jerusalem, but A.D. 66 was only the start of the war brought about by the Jewish uprising. The Roman victory and ravaging of the city did not occur until A.D. 70 (the last Jewish stronghold, Masada, fell in A.D. 73).

Is it possible that the people of this time noticed that a bright comet was returning every 75 years or so? The track of Halley across the sky is not necessarily very similar from visit to visit, and sometimes it is extremely different. And most centuries feature several other comets at least as bright as Halley. On the other hand, we should not underestimate the intelligence of "primitive" or ancient peoples. And so it is with real interest that we look at the Talmudic tale of the first-century sages Rabban Gamaliel and Rabbi Joshua.

The two were starting off on a long sea voyage, and Rabbi Joshua carried not just bread but also flour because he thought the trip might take longer than usual. Why? Because he believed a bright star that appears every 70 years could be due again and would confuse the sailors! In 1910, Mlle. G. Renaudot argued that this "star" was Halley's Comet and that its periodicity was known to the ancient Hebrews. Admittedly, the figure of 70 years is very compelling. But it might merely represent the "three score years and ten" of a full lifetime, or it might be derived from numerical mysticism and therefore have nothing to do with any real phenomenon. To surmise that Halley's Comet was the same comet returning again and again would almost certainly require mathematical knowledge that the ancient Hebrews did not possess. Unless more evidence can be advanced, the star of Rabban Gamaliel and Rabbi Joshua must be regarded as only a delightful coincidence.

141. ACROSS THE LEAPS OF THE GAZELLE   This was a close (0.17-a.u.) and bright pass of Halley's Comet after perihelion. The comet raced behind Earth on its way out, with a magnitude of perhaps -1 outside of twilight. Strangely, no mention of it has been found in Roman records. It came 3 years too late to be blamed for the death of Emperor Hadrian. The Chinese said it spanned about 9 degrees long and was bluish white when first spotted on March 27. The comet passed closest in late April, when its tail swept north across the three pairs of stars in southern Ursa Major that later became known as The Three Leaps of the Gazelle.

I've read a secondhand report that the comet at this return was blamed for a plague that supposedly spread from China to much of the civilized world, but I have not found primary source information on this.

218. THUMBS DOWN ON MACRINUS   The capture and killing of the cowardly Roman emperor Macrinus were associated with this return of

Halley's Comet. It should not have been a particularly bright one, but the Chinese record of it seems to suggest a tail that spanned 120 degrees (surely not true?). Its greatest magnitude may have been about 0 outside of twilight.

295. IN THE REIGN OF DIOCLETIAN   This was an average return, recorded in China but not in Rome, where a rare period of relative stability (as opposed to death and destruction that could be related to a comet) prevailed under the emperorship of Diocletian. The comet was 0 magnitude at its brightest outside of twilight.

In 313 the Edict of Milan helped advance religious tolerance, and in 325 the Council of Bishops at Nicaea established the Nicene Creed concerning the Trinity. Constantine ruled the empire, declared it Christian, and moved its capital to Byzantium, which was renamed Constantinople.

374. THE SECOND CLOSEST VISIT ON RECORD   This was the second closest of Halley's recorded returns and should have been very impressive. The comet passed 0.09 a.u. from Earth and may have outshone the brightest star for a week. It was not recorded in Europe, however—perhaps because this apparition was visible entirely in the morning sky. The comet dived down just in front of Earth.

Four years later, the Visigoths won a great victory over the Romans at Adrianople. Alaric captured Rome briefly in 410, just 19 years after paganism had been forbidden by Theodosius.

451. ATTILA'S BANE   Comet Halley suddenly came brightly from the morning back to the evening sky as Attila the Hun suffered his first and very important (though hardly crushing) defeat. But Attila was far less effective ever after and died of a drunken nosebleed in 453 (an event associated with a different comet, which appeared in that year).

This was a distant but bright and readily visible apparition of Halley. The comet was closest to Earth just a few days after perihelion and perhaps about 0 in magnitude outside of twilight.

530. LAMPADIAS   Halley's Comet at this return was apparently called Lampadias—which means "lamp-like" or "torch" (the name of one type of comet)—by Byzantine sources. The comet passed only moderately close to Earth, and before perihelion, so it was only first or second magnitude out of twilight. This year, as in 443 and 446, the heavens glittered with thousands of Halley's Eta Aquarid meteors. In 585, there was a display of the Orionid meteors in which "hundreds of meteors scattered in all directions." These are the annual meteor showers associated with

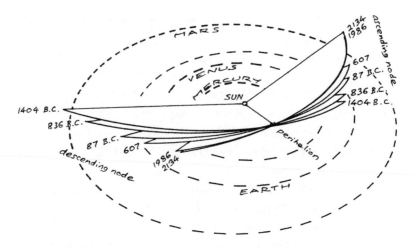

Diagram showing the libration of Halley's orbital nodes. About 800 B.C. the ascending node crossed Earth's orbit; about 87 B.C. the nodes were symmetrically positioned; about A.D. 600 the descending node crossed the Earth's orbit.

the inbound (Orionids) and outbound (Eta Aquarids) parts of Halley's orbit. Between the 530 and 607 returns, the descending node of Halley's Comet, the place where it dives down through Earth's orbital plane, passed right through the orbit itself, so it is not surprising that there were some impressive showings of Halley-derived meteors.

It was a few decades after Justinian's reign that the descending node passed through Earth's orbit. Justinian ruled over most of the former Roman Empire from 527 to 565, its last era of greatness. This was $7^1/_2$ Halley periods after the founding of the Empire, 10 Halley periods after the First Punic War, $13^1/_2$ Halley periods since the establishment of the Roman Republic, and 17 Halley periods after the legendary founding of Rome by Romulus and Remus. Thus Rome from its founding to its final glory under Justinian lasted almost the entire time between the two Halley nodal crossings; another crossing should not occur for over 20,000 years.

If King Arthur ever lived, this is the apparition he is most likely to have seen.

607. THE SECOND BRIGHTEST VISIT ON RECORD   This was probably the second-brightest recorded return of Halley. The descending node of the comet's orbit was apparently right on Earth's orbit in about 600. With the descending node so close to Earth's orbit in 607, there was the potential for this to have been the closest Halley visit for about 26,000 years. In 1985, Guy Ottewell calculated that if Halley's Comet had arrived just 5 days earlier than it actually did in 607, it would have passed about two-thirds of the Moon's distance from us.

As it was, in 607 Halley passed 0.09 a.u. from Earth, crossing behind us. And because inferior conjunction of Halley with the Sun was just 1 day before closest approach and just 5 days before Halley passed through descending node, the tail must have come very close.

Halley rushed through almost a constellation a day at close approach. Its maximum brightness outside of twilight should have been about −2. And yet only in China has record of its passage been found. This is surprising, although Europe and the Middle East were in even greater turmoil than usual as the remains of the Roman Empire thrashed about. However, this brilliant apparition must have been seen by Muhammad 3 years before he received his call.

The followers of Islam expanded their empire immensely in the next Halley period. Just 7 years before the 684 apparition, the Arabs were driven from the siege of Constantinople with help of "Greek fire," whose composition remains a mystery.

The Medieval Recorded Visits of the Comet

684. WOODCUT IN THE NUREMBERG CHRONICLES   In 1493 there was printed a history that has become known as the *Nuremberg Chronicles*. On the page where text concerning the year 684 appears, there is a crude woodcut of a comet—Halley's Comet of that year, it is believed. The chronicle claims the comet brought 3 months of storm and rain, withering of crops, an eclipse of the Sun and Moon, and a plague.

This return of Halley is the first mentioned in Japanese annals. A great cultural flowering was going on in Japan. The Yamato rulers had recently begun to style themselves emperors, in the manner of the Chinese, and in 710 would found their first imperial capital at Nara.

At this visit, Halley passed fairly close in front of Earth. But it was on its way in to perihelion and so should not have been as bright or had as well-developed a dust tail as at post-perihelion close approaches. Its maximum magnitude outside of twilight was 1 or 2.

760. A BEAM IN BYZANTIUM   Eight years before Pepin's son Charles—Charlemagne—began his grand reign, there was a rather typical Halley return, the comet passing north of the Sun between a period of morning and another of evening visibility. The best magnitude may have been 0. Byzantine writers mention a bright comet, in the shape of a beam, appearing for 10 days in the east and then for 21 days in the west. This was in May and June, and it is remembered that the Sun was eclipsed on August 15.

837. THE CLOSEST AND BRIGHTEST VISIT ON RECORD   Of all the returns of Halley whose comet-Earth circumstances can be characterized accurately—those from 1404 B.C. to A.D. 2134—only the 1404 B.C. visit was the tiniest bit closer than the one in A.D. 837. But no record of the 1404 B.C. visit has been found, and that time the comet squeaked past Earth *before* perihelion. Thus the 837 return, with a post-perihelion close approach, should have been the brightest in the known span of over 3500 years. The comet should have shone at −3 or −4 (as bright as Venus) at this return. It passed just 0.03 a.u. from Earth—5 times closer than it did in 1910, over 3 times closer than Hyakutake in 1996, and closer than any *great* comet in the historic record. (The only *great* comet to come anywhere near as close was the comet of 1132, which came to within 0.04 a.u. and reached −1 out of twilight.) On its closest approach, on April 10, 837, Halley's Comet was only 13 times the Moon's average distance away. The comet moved 60 degrees across the constellations (one-third of the way across the sky) in this single day.

The tail was awesome. Starting a few days before perihelion, it had several branches and already spanned about 60 degrees. By April 11 the perspective changed and branching disappeared, but the tail lengthened, and on April 13 and 14 it spanned over 90 degrees. Charlemagne's son,

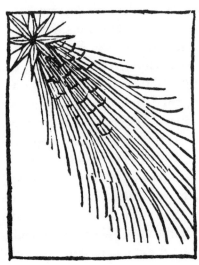

The A.D. 684 apparition of Halley's Comet as depicted in a woodcut from the Nuremberg Chronicles.

King Louis the Pious, thought the comet was a sure omen of his death, but he did not die until 840.

Between the Halley returns of 837 and 912, Alfred the Great (871–901) ruled in England, but after his death the Viking invasions worsened and England was for a time ruled by the Danish dynasty of Canute. About 874, Iceland (already home to some Irish monks) was first settled by Norsemen. The Vikings were being Christianized, and in 912, as Halley's Comet looked on rather brightly, Rollo converted to that religion. He and the Norsemen were rewarded with the section of France's coast that became known as Normandy—the land from which the most successful invasion of England of all time was to take place 2 Halley periods later, with Halley's Comet again in the sky.

912. JAPANESE RESCUE   This was a reasonably bright (magnitude 0) return of Halley of the common kind (like 451) in which the comet is closest to Earth—and not very close—near perihelion. But the records that could refer to Halley are largely vague or confused. The most convincing are those in the Japanese *Dainihonshi*. Were it not for the positions given there, it would be harder to argue that this return of Halley was recorded.

989. A SECOND NONDESCRIPT RETURN   This return was fairly unremarkable—less bright than the previous one but more observable. Maximum magnitude out of twilight was only 1 or 2. It was one of Earth's pre-perihelion encounters with Halley, and indeed a Chinese source describes the comet as blue-white (gas without much dust yet). If this color was noticed by naked-eye observers, I wonder—judging from observations of Comet Hyakutake in 1996—if perhaps the comet was not a little brighter than calculated, at any rate first magnitude.

Just 2 years before 989, Hugh Capet had taken over the throne of France from the descendants of Charlemagne. In the year of Halley's return itself, Prince Vladimir of Kiev became a Christian, and Russian chronicles mention at least five of the next seven visits of the comet. About 3 years before this return of Halley, Eirik the Red colonized Greenland and Bjarni Herjolfsson probably sighted America. Not until about the year 1000 did Leif Eiriksson land on the American shore.

1066. ISTI MIRANT STELLA   After two modest returns that seem to have stirred little interest, Halley's Comet came back for one of its two or three most impressive visits ever, just in time for the most important historical event it ever became associated with: the Norman conquest of England.

The comet became a portent of doom for the multiply harried Saxon king of England, Harold. His defeat was chronicled in the 77-yard-long, 20-inch-wide Bayeux Tapestry, which was made in the decade after 1066 and is preserved today in a museum near the cathedral of Bayeux in Normandy, France. The most famous section of the tapestry shows Harold imagining a fleet of invading ships as an advisor informs him of more bad news and a crowd of Saxons points up at a comet with the words *Isti Mirant Stella* beside it: "These marvel at the star."

What a display Halley put on in the spring when events were leading up to the climactic happenings on Earth! It passed 0.10 a.u. from Earth after perihelion, going behind Earth. Its calculated magnitude was -2 or -1, but there are indications that it underwent an unusually strong post-perihelion surge and may have brightened to as much as -4, rivaling or surpassing the return of 837. Several witnesses compared it to the Moon (perhaps more in reference to size than to brightness). One cleric thought the comet "looked like the eclipsed moon, its tail rose like smoke halfway up to the zenith." But some of the records can be interpreted as suggesting a tail that spanned over 90 degrees. The tail at one time was apparently visible all night long, swinging through remarkable changes of angle due to perspective as it passed. There's no doubt that it fanned out into three long rays at one stage (hence the nineteenth-century poet Tennyson has a character in his play *Harold* speak of "Yon grimly-glaring, treble-brandish'd scourge" and "these three rods of blood-red fire up yonder").

1145. EADWINE'S PSALTER   At the 1145 return, the comet, or a watcher on the comet, would have seen that the Moslems had just retaken Edessa and that the Second Crusade was being organized. The comet's bright and well-placed apparition was observed for 81 days. There was a post-perihelion close approach, and the comet should have been about 0 in magnitude, but Kronk offers evidence that the comet may have been observed as bright as -3 and displayed a 60-degree tail. The Eadwine (or Canterbury) Psalter (psalm book) dates from around 1145, and it appears that the monk Eadwine decided to decorate one of its pages with a lovely stylistic (and captioned) illustration of a comet— presumed to be Comet Halley.

1222. DAYLIGHT HALLEY?   At Halley's return in 1222, the University of Paris was 22 years old, the University of Cambridge 13, and Kublai Khan 8. The cathedral at Chartres had been under construction for a little more than 30 years, the one at Rheims for about 12. The Magna Carta had been in force for 7 years. In 1222, Frederick II's reign as Holy Roman

Depiction of 1145 return of Halley's Comet by Eadwine in the Canterbury Psalter.

Emperor began just as the Mongols, under Jenghiz Khan, made their first appearance in Europe. The death in 1223 of Philip Augustus of France, who had seized Anjou and Normandy from the English, was said to have been announced by the comet.

This should have been one of the dimmer returns, with a pre-perihelion encounter that was not close. But a Korean source says the comet was seen during the day on September 9. Could this be a mistranslation or a mistake? A Japanese source says that between 7 and 9 P.M. the previous evening, a broom-star appeared with a center as large as a half-moon; its color was white but it had red rays, and its length was more than 17 degrees. Perhaps the red color was from haze at twilight. But possibly Halley underwent a tremendously great and brief pre-perihelion brightness flare? The question remains debatable.

From Giotto to Edmond Halley

1301. GIOTTO'S CHRISTMAS STAR   This was also a pre-perihelion encounter with Halley, but it was a close one—the comet passed just 0.18

a.u. from Earth. Halley was perhaps only of magnitude 1 or 2, but it displayed a tail of 50 degrees or maybe even 70 degrees. The Florentine artist Giotto—"the first flower in the spring of the Renaissance"— painted a Nativity scene in the Arena Chapel at Padua perhaps 4 or 5 years after the 1301 return. And there, above the people and animals, is a remarkably naturalistic and beautiful comet—almost certainly Comet Halley—serving as the Star of Bethlehem! (See color photo insert.)

The period until the next visit of Halley was in many ways a bleak one in European history. The Hundred Years' War began in 1337, the very warlike Ottoman Turks rose in power, and the Black Death occurred in 1348–1349.

Other things happened. Robert II was king of Scotland (1371–1390) and founder of the Stuart dynasty. Around 1376, the Wycliffe Bible was published. Thanks to Philip the Fair of France, from 1305 until 1377 (almost exactly the period between the Halley returns), the Popes were all French and all resided at Avignon. But starting in 1377 (and continuing until 1417), there were as many as three claimants to the papacy at a time; the period is known as the Great Schism. Thus Europe was in even greater turmoil than usual when Halley's Comet soared north over the Earth in 1378, early in the writing career of Geoffrey Chaucer.

1378. POLE VAULT   Only once in the 3500 years of Halley returns that are accurately calculated has the comet passed very near the north celestial pole. This remarkable encounter occurred in 1378, with the comet inbound. It was probably about first magnitude—and only 0.12 a.u. from Earth, the second-closest return in the more than a thousand years between 1066 and 2134. Its maximum northern declination was 82 degrees 33 minutes (in 1378 coordinates) on October 2. It passed right through the bowl of the Big Dipper the next day.

This return is mentioned in both Asian and European sources, but they say less about it than we would think it deserved.

1456. THE NON-EXCOMMUNICATED COMET   The 1456 visit of Halley came at a key time in history. Two years earlier, Johannes Gutenberg had invented movable type, and the War of the Roses had begun a year before. But the attention of Europe was on the attempt to win back Constantinople from the power of the Ottoman Turks, who had first invaded Europe (Gallipoli) about a hundred years earlier. "God save us from the Comet and the Turk!" was a popular cry. But scholars have exploded the myth that Pope Calixtus III excommunicated Halley's Comet. In the end, it was the Turks who fared worst. They were besieging Belgrade when, on the night of June 5, a sentry woke the camp of the Turkish army with the warning that there was in the sky an apparition "with a long tail like a dragon's." The comet's tail then spanned about 60 degrees; its magnitude was about 0. On July 22 the Turks were defeated by the Hungarian hero Janos Hunyadi, who died of plague just 4 weeks later. One of Hunyadi's important generals for the past few years had been a certain Vlad Tepes, or Vlad the Impaler. He did not join Hunyadi on this final campaign. But Vlad must have been an important figure in these wars, and he must have seen the mighty comet with interest (at least) as he heard of the Turks' defeat and then of his mentor and rival Hunyadi's death. What is so notable about this Prince Vlad? Only that he is the real-life basis for a legend that has fascinated many, in our own time: He was the historical Dracula.

The great observer of this return was Paolo Toscanelli, a physician and cosmographer whose idea of reaching the East by sailing west eventually inspired Columbus. In the next inter-Halley period, the great explorers set sail for America and India and even circumnavigated the world (a feat first performed by the crew of Ferdinand Magellan in 1519–1522). Michelangelo wrought his great works (the ceiling of the Sistine Chapel was painted between 1508 and 1512), living with seemingly boundless creative energy from 1475 to 1564. Thus he had his chance to see Halley's Comet (in 1531), which Raphael did not and which Leonardo could have done only if he were aware of it as an infant in 1456. The Spanish Inquisition was established in 1478. Martin Luther nailed the 95 Theses to the church door at Wittenberg in 1517.

1531. APIAN'S HALLEY   When Halley appeared in the sky in 1531, the Aztecs had already been slaughtered, and their culture destroyed, by Cortes (in the few years after Montezuma beheld the first bad omen—possibly a comet—in 1517). But another genocide was beginning as Halley shone in 1531: Francisco Pizarro began the conquest of the Incas and their rich, vast civilization.

The prime observer at this return was Peter Apian. His positional measurements of the comet would eventually be of great value to Edmond Halley. Apian studied the tail and announced that it always pointed away from the Sun. This was a pre-perihelion encounter with Earth, and not particularly bright (probably about first magnitude). Most of the important figures of the Reformation commented on the comet. Luther thought it boded ill and said that "a comet is a star that runs and won't keep still at its post like a planet, a whoreson among planets . . . like a heretic, who thinks he alone knows it all." Willy Ley quotes, as a typical example from comet listings of the sixteenth century, the following: "Anno 1531, 1532, 1533 comets were seen and at that time Satan hatched heretics." There were also treatises relating an earthquake in Germany to Halley's 1531 return and announcing a rainbow that at last put an end to the comet-caused disorders.

1607. KEPLER'S HALLEY   The 1607 return was the last before Halley himself saw the comet. It was another pre-perihelion encounter, but this time much closer to Earth and less close to the Sun at perigee. Most of the magnitude estimates of the (now more numerous) astronomers of the time suggest that Halley was in the first to second magnitude range for at least a few weeks. But the tail seems to have been surprisingly bright for pre-perihelion: Kepler was able to see 10 degrees of it on October 5 despite the full Moon. Kepler first noticed the comet at the end of a fireworks display he was watching in Prague on September 25 (September 15 by the Julian calendar then still in use).

As far as I know, there is no record that the settlers at Jamestown saw Halley's 1607 apparition but surely it *was* noticed by Pocahontas and John Smith? Another group of prospective colonists farther up the American coast very likely did record Comet Halley. Peter Broughton writes that George Popham's band, which landed near the mouth of the Kennebec River in Maine (where they started building Fort George) in the late summer of 1607, reported a "blasing starre" in the northeast on the morning of September 15 (old style). There can be scarcely any doubt that it was Comet Halley.

Meanwhile, back in Europe, a German work of the time blamed this return of Halley's for a plague in Saxony, Thurinigia, and other places; for a 7-year famine; and of course for the attacks of the Turks.

1682. HALLEY'S HALLEY   The visit of "his" comet that Edmond Halley saw was a rather typical one, in which the comet was fairly bright (0–1) but never came very close (perigee occurred a few weeks before perihelion). The comet was observed by all the famous astronomers of the time, and careful observations were made on a number of nights by Halley himself. The tail spanned as much as 30 degrees, and its position was sketched on several nights by Isaac Newton.

Among the wider public, this visit was somewhat overwhelmed by the vast lore and consternation over the Great Comet of 1680, and later writers have repeatedly confused the two. There apparently was on August 26, 1682, however, a comet egg ("egg with a starry design on its shell") laid by a hen in Marburg (see p. 223) under the supposed influence of Halley's Comet—as there had been a comet egg laid a year and a half earlier under the supposed influence of the Great Comet of 1680. This old tradition was carried on 3 Halley periods later when Mme. Haydee Bouyard of Gard reported that one of her chickens had laid, on May 17, 1910, an egg with an image of Halley's Comet on its shell.

In New England, the Puritans noted comets with superstitious awe. The famous Puritan clergyman Increase Mather (father of the even more famous Cotton Mather) published what is probably the first book in English on comets, *Kometographia; or, discourse concerning comets*. . . . Mather attempted to list all comets in history known up to this time (1683), and he mentions what are probably the Halley returns of 66, 684, 837, 912, 1066, 1145, 1301, 1456, 1531, 1607, and of course 1682. Surely it was the comets of 1680 and 1682 that provoked Mather's effort, and indeed the book contained "two sermons occasioned by the late blasing stars." Apparently Mather supported the idea that Comet Halley was God's warning of some forthcoming disaster but sternly rejected the claims of the astrologers that they could forecast specifically which mis-

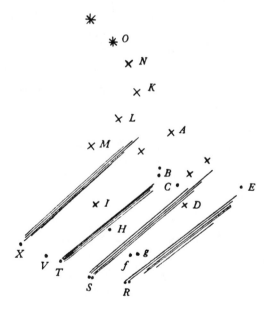

Newton's sketch of the positions of Halley's Comet and its tail from August 29 to September 1 (Gregorian calendar) 1682. (The comet's positions are marked by R, S, T, and X and the other dots and crosses are the stars of Ursa Major, including those of the Big Dipper at top.)

Christiaan Huygens's sketch showing the parabolic outline of Halley's Comet at its 1682 return.

fortunes would occur. Increase Mather's son Cotton wrote an article about the comet in a 1683 almanac.

Back in Europe, Christiaan Huygens, another multiplex genius in some ways like Edmond Halley, observed Halley's Comet and sketched it in a letter to his brother, though he seemed more interested in the conjunction of Jupiter, Saturn, and Mars visible at the time. One Christopher Ness referred to the comet and the conjunction in a 1683 work entitled *A Strange and Wonderful Trinity; or, a Triplicity of Stupendous Prodigies, Consisting of a Wonderful Eclipse, as well as of a Wonderful Comet, and of a Wonderful Conjunction, now in its Second Return.* The second return is of the conjunction (Kepler had seen a similar configuration almost exactly one Halley period earlier), not a second return of the comet! The idea that this comet had made many returns was Edmond Halley's to develop.

The First Predicted Returns

In 1705, Halley published his prediction that the comet would come back in 1758 (which he later revised to late 1758 or early 1759). A variety of opinions about the prediction existed, and references to it were regularly made as the years went by and the appointed time grew closer.

The satiric wit Jonathan Swift clearly alludes to Halley's prediction in *Gulliver's Travels*, published in 1726. In Swift's description of the flying island of Laputa, we come across the following passage about the astronomers of that country:

> They have observed ninety-three different comets and settled their periods with great exactness. If this be true (and they affirm it with great confidence), it is much to be wished that their observations were made public, whereby the theory of comets, which at present is very lame and defective, might be brought into perfection with other parts of astronomy.

This passage occurs right after the famous one in which Swift by chance not only predicted the existence of two Martian moons but also came fairly close in his figures for their unexpectedly short revolution periods around the planet.

If this reference to comets and their calculators seems no more than gently mocking of Halley and his colleagues, then Gulliver's visit to the Academy of Lagado, a satire of the Royal Society of London, is much harsher. Consider also the genuine philosophy but the obvious criticism of Newton and his followers in another line from *Gulliver's Travels* (pointed out by Will and Ariel Durant): "New systems of nature were but new fashions, which could vary in every age; and even those who pretend to demonstrate them from mathematical principles would flourish but a short period of time."

Another acerbic genius, Voltaire, lived somewhat later than Swift and corresponded with the brilliant mathematician and Halley calculator Alexis-Claude Clairaut. Voltaire mocked the astronomers in 1758 as never daring to go to bed for fear they might miss the return of the comet. Of Halley's Comet itself I know of nothing that Voltaire had to say. But Voltaire did have a comment about Edmond Halley's great 1698 sea voyage. Compared to Halley's expedition, Voltaire suggested, "The voyage of the Argonauts was but the crossing of a bark from one side of the river to the other."

In 1757, Jacques Gautier, an anti-Newtonian, predicted that Halley's Comet would not come and argued that comets were meteorological phenomena. On the other hand, John Wesley (1730–1791), the founder of the Methodist denomination of Christianity, apparently wrote that Halley's Comet would come in 1758 as Halley predicted, but with special disastrous results. Wesley's pamphlet "Serious Thoughts Occasioned by the Late Earthquake at Lisbon" expressed the opinion that the comet would set Earth on fire!

The First Modern Visits

1759. FIRST PREDICTED RETURN   Excitement in the astronomy world increased as the predicted time of the comet's return neared. Clairaut enlisted the help of fellow human calculators Joseph de Lalande and Madame Nicole Lepaute to produce a more precise forecast of where and when to look for the comet. The three calculated from morning to night, often during meals, for 6 months and came up with a forecast that still proved a month off from the actual perihelion date. (It couldn't be helped: the then-accepted values for the masses of Jupiter and Saturn were in error.) But it was not the zealous skywatcher and future comet discoverer Charles Messier who was the first to see the comet. Weeks before he found it, Halley's Comet was spotted by self-educated farmer Johann Georg Palitzsch, on Christmas Day of 1758.

What kind of return was this? The comet passed only 0.1222 a.u. from Earth, and after perihelion, so it should have been quite impressive. The only problem for most of the world's observers was that the comet was diving far under Earth at closest approach: Messier saw it on March 31–April 1 as a first-magnitude object with a 25-degree tail but low in morning twilight. Up to this point, I understand why the apparition does not appear on Yeomans's list of "great" comets. But by April 25, when it was its farthest south—declination −70.79 degrees, much too far south to be seen from Europe—it must have escaped from twilight for Southern Hemisphere observers, and calculations suggest it should have brightened to something like −1.

King Kamehameha I, the ruler of Hawaii, was reputed to have been born on a night when "a strange star with a tail of white fire" appeared in the western sky, and it is possible that this was Halley's Comet at the 1758–1759 return.

Of all the American colonists of 1759, Benjamin Franklin would seem the most likely to observe and write about Halley's Comet. At the time, however, the 53-year-old Franklin was away from home, on the other side of the Atlantic. His biographers describe this time as a rather mysterious one in which little is known about his actions or motives. According to Ruth Freitag, Franklin did eventually make written reference to the comet, though.

1835. SIR JOHN AND THE SURGE   This visit was watched by all the great observers of the time but was perhaps best attended by Bessel, who drew lastingly valuable sketches of the inner coma and jets, and by Sir

John Herschel, who observed, wrote about, and sketched a certain amazing display of the comet as it pulled away.

The closest encounter with Halley at this visit came before perihelion, but it was fairly close (0.19 a.u.) and very well placed. The comet had recently crossed over our orbit and so, in mid-October 1835, though it was at conjunction with the Sun (i.e., at the same "right ascension" as the Sun in the sky), it appeared 60 degrees north of the Sun and thus was visible all night crossing the northern sky. Its declination was +64 degrees, not nearly so far north as in the "pole-vault" of 1378 but beautifully visible. The tail grew to 30 degrees. Then, as the comet plunged south, it came into view for Sir John Herschel in South Africa in late October. It was lost from view 11 days after its November 11 perihelion. But as 1836 opened, the fading comet was in Scorpius, becoming well positioned for observation by Southern Hemisphere astronomers such as Herschel and T. Maclear. It is a good thing they did not give up observing it too soon.

In the last week of January 1836, Halley's Comet was over 1.5 a.u. from both Earth and Sun (out near the orbit of Mars), but Herschel and Maclear watched it undergo a tremendous outburst, brightening by perhaps 2 to 3 magnitudes in as many days and sending out an enormous halo. The tail was no longer visible even in the telescope, but the halo expanded outward dramatically, reaching a diameter of at least a million miles in about 3 weeks. Herschel waxed poetic about its appearance in the telescope, marveling at its delicacy and definition and comparing it to "a transparent gauze or alabaster vase illuminated from within." The comet, he wrote, was keeping him up *"all* night and *every* night . . . it is altogether the most beautiful thing I ever saw in a telescope." He expended so much energy and passion that he said he both "hoped and feared" that he would be able to see it long after everyone else.

Zdenek Sekanina, in 1983, suggested that this mighty surge happened not just at the 1835–1836 return but also in 1066 and 1145 (Bortle and Morris suggested 607 also), always between 63 and 77 days after perihelion, and presumably from heating effects on the same area of the nucleus which at that point has the Sun high enough in its sky to come alive. This surge may have occurred at many Halley visits (at some, the comet was hidden in the solar glare at the key point)—but it did not, alas, happen in 1986, when our view of it would have been excellent and it would have brightened a dim apparition.

The surge kept Halley's Comet visible to the naked eye until late March 1836. Herschel may have been the last to see the comet in the telescope. He wrote to his Aunt Caroline (who had, remember, discovered comets of her own more than three decades earlier) that he had "followed him [the comet] till about May 20. . . . To say the truth I am glad he is gone."

Just weeks after Sir John's last telescopic observations of the comet, H.M.S. *Beagle* came to shore in South Africa and Charles Darwin got an opportunity to dine with Herschel. It seems scarcely possible that Halley's Comet could have gone unmentioned! Darwin in his journal called this meeting "the most memorable which, for a long period, I have had the good fortune to enjoy." Over 40 years later Darwin recalled that Herschel "never talked much but every word which he uttered was worth listening to." Herschel's words about Halley's Comet may not have been what Darwin found most interesting, however, because Herschel had been thinking about the origin and replacement of species—he had written a long letter to Lyell on the subject just 4 months earlier—and is likely to have helped focus or reinforce some of Darwin's thinking on the subject.

Two larger-than-life events in American history that occurred in 1836 have both apparently been attributed to Halley's Comet. One of them really was about to occur while the comet was still visible to the naked eye. The siege of the Alamo began in late February 1836, and it is certainly possible that the troops massing in Texas—among them Jim Bowie and Davey Crockett—noticed this oddly glowing ball shining mysteriously in the sky. On the other hand, the funeral of Supreme Court Justice John Marshall did not occur until the summer, long after even Herschel lost sight of the comet. The ill omen at the funeral was the cracking of the Liberty Bell. The bell had first been cracked in the time of the American Revolution, and many years after the Marshall funeral it was rung again (for Washington's birthday) and it cracked so badly that it has been silent ever since. Whether there is any truth to the claim that some people associated the cracking of the bell at Marshall's funeral with the comet they had last seen with their naked eye many months before, I do not know. But this claim has been made.

1910

1910. THROUGH THE TAIL   The 1910 visit of Comet Halley is (at least from our biased standpoint) its most famous return in history. The comet came close (0.15 a.u.) and after perihelion, and it got bright (theoretically, up to -1.5, though perhaps in reality dimmer). But the factors that made this return both sensational and sensationalistic were the nature of society and science at this stage in history and one thing else: the probable passage of Earth through the comet's tail.

The now mature art of photography enabled the comet to be photographed and first seen in large telescopes as soon as late summer of 1909. Public anticipation was stirred even more by a herald of Halley: the Great January Comet of 1910, brief but brilliant, long and structured of tail (see Chapter 11). Halley was first detected without optical aid on February 11 by the sharp-eyed Max Wolf. On April 17, 3 days before perihelion, Michel Giacobini was able to see the comet easily in a 4-inch telescope after sunrise—a daylight sighting of Halley. Mark Twain, who had been born 14 days after Halley's last perihelion passage, had joked he would go out with the comet just as he had come in with it—and die he did, 1 day after perihelion. And then it was time for the comet to approach and sweep its tail across us.

In early May, Halley was in the pre-dawn sky near Venus, also joined by the crescent Moon. King Edward VII of England died. There was some speculation about the connection of the comet with the death of the king. But now there rose up far more than speculation about the approaching tail—there rose up outright panic. The public had learned that poisonous cyanogen gas was a component of comets. Astronomers knew then, as we do now, that the amount of gas entering Earth's atmosphere from a comet tail passage would be mind-bogglingly little—and presumably further diffused when and if it ever reached the lower atmosphere. The Earth had probably passed through comet tails a few times in history—almost certainly through that of Tebbutt's Great Comet of 1861 (see Chapter 11). There was no danger. But it was difficult convincing the public of that.

It was as though a long finger of death were floating across the heavens to touch Earth.

The tail was certainly long. The observations of E.E. Barnard and others indicate that the tail briefly exceeded 120 degrees, perhaps even reaching 150 degrees—though never was all of this maximum length visible at the same time (compare this with the case of the Great Comet of 1861). Did Earth pass through the tail of Halley's Comet on May 19? It almost certainly went through the *gas tail* of Halley. There's no doubt that the comet nucleus transited (crossed in front of) the disk of the Sun that

Barnard's drawings of Halley's tail on the mornings of May 18 and 19, 1910. The Great Square of Pegasus in near the center of the drawings; the brightest star at upper right is Altair.

day—though it was too small to be seen in silhouette. There may have been some strange sky-glows associated with the dust tail passage—most defendably a strong display of nacreous, or mother-of-pearl, clouds that was probably really a display of the even higher (altitude 50 miles) noctilucent clouds composed partly of ice, partly of micrometeoroidal dust. How close did the *dust tail* of Halley come? Zdenek Sekanina calculates that the leading edge of the dust tail probably just missed Earth, passing 240,000 miles—closer than the Moon often is!—from our planet at May 20.32 Universal Time.

The public reactions to the tail's passage took on an air of the fantastic. The world in 1910 had just enough communication and had had just enough taste of science to lavishly misinterpret what it saw and heard of the comet's close approach. There is enough lore from the 1910 visit to fill a large book. The lore includes speculation connecting the comet

Observing Halley's Comet from a balloon. Drawing by Henri Lanois, from Graphic, *May 28, 1910.*

with Mark Twain and King Edward VII of England, what Pope Pius X thought of the comet, the observations of Mary Proctor ("the lonely watcher in the tower"), and some of the bizarre ideas of the group called the Koreshans. I can give only a brief sampling of some of the less elaborate lore here. (For more, see *Mankind's Comet* and Ruth Freitag's bibliography—both of which are listed in the bibliography at the back of this book.)

Some of the other ideas about Halley's Comet publicized in 1910 were fully as odd as the Koreshans' talk of "the breaking in of zones of cruosic energy generated at the colure." Jean B. Marchand refused to accept that the comet which appeared that spring was Halley's Comet—the real one, he maintained, would not arrive until September. Edwin F. Nulty argued that comets' tails consisted of sunlight focused and concentrated by the head, which acted as a lens and that consequently, a path of fiery destruction would be traced across Earth wherever the focal point went as the comet passed between Earth and Sun.

The comet provided a final impetus for some people to go mad (some claimed they were going to follow the comet wherever it went—literally) and for others to commit suicide. A sheep rancher in California tried to crucify himself and was badly injured. A Hungarian landowner named Adam Toma did commit suicide, saying he preferred death by his own hand "to being killed by a star." But the story that a cult in Oklahoma was stopped just short of sacrificing a virgin to the comet was apparently made up and appeared only in some Eastern U.S. newspapers.

The comet also seems to have brought about reformations and confessions. In Newark, New Jersey, Luigi Ciefice was in jail for attempting to blackmail the great opera tenor Enrico Caruso. When he heard dire predictions about the approach of the comet's tail, Ciefice supposedly panicked, confessed to a murder, and revealed the location of the body.

There was Halley panic all around the world: in parts of Japan and in Chicago, Puerto Rico, and the Philippines, on a trolley car in New York City, in the south of the United States and Russia. Hundreds of miners in Wilkes-Barre, Pennsylvania, refused to enter the mines on May 18, the supposed day of the tail passage. In York, Pennsylvania, Lee Spangler was a self-proclaimed prophet who boasted that he had predicted the San Francisco earthquake and the assassination of President McKinley; he now announced that Halley's Comet was the Star of Bethlehem and that the 1910 apparition a sign of the Second Coming. The comet, he claimed, would stand still in the sky, and then the head would collide with Earth. In Neemah, Wisconsin, many farmers took down lightning rods lest they attract dangerous substances from the comet.

One of the most famous items of 1910 Halley lore was the sale of "comet pills" guaranteed to protect people from the deadly effects of the tail's gases. Much has been said about this racket, but where was it actually practiced? Barnard in 1914 mentioned that such pills had been manufactured and sold among blacks in the south. The pills were reputed to come from Haiti. But the idea must have been picked up by other swindlers. Ruth Freitag mentions one seller of comet pills who was arrested in Texas but, because no one seemed to care about his misdeeds, was let go. In Chicago, people without comet pills or gas masks stopped up the cracks around their doors and windows with rags and newspapers while the Earth was supposed to be passing through the comet's tail.

Some people regarded the comet's close approach with no alarm but rather with great interest or jollity. Brown University students cheered the comet when it appeared, and in New Bedford young boys ran through the streets announcing the comet's visibility at dusk. While Halley was still in the morning sky, the people of one California town arranged for a siren to be sounded as a signal that the sky was clear and the comet had been sighted.

In addition to comet motifs on coins and vests and postcards and dishes and ties and even socks, Halley's Comet was also the subject of many pieces of music (at least two "rags"), a Ziegfeld Follies show, and a large amount of doggerel.

After the scare, Halley's Comet emerged into the evening sky in late May to be seen most conveniently. It raced across constellations, its tail dwindling to about 30 degrees by the time it was in full darkness. The bright Moon interfered somewhat with observation at first, but on the night of May 23–24, the Moon was totally eclipsed for all of America. (A similar event will occur on the night of March 23–24, 1997, when the Moon will be almost totally eclipsed and Comet Hale-Bopp will just have completed its entrance into the evening sky at peak splendor!)

Halley soon faded below naked-eye visibility and then was hidden in the solar glare again for a few months. It was last seen in a telescope by Barnard on May 23, 1911, when it was more than 5 a.u. from the Sun. The last photograph obtained was taken by Heber Curtis on June 15, 1911.

Between the 20th-Century Visits

The flowering of Halley's Comet lore, however, did not cease after 1910. Science fiction became a popular genre, represented by increasing numbers of writers, in pulp magazines of the 1930s and 1940s. One of the more famous science fiction writers of that period, Edmond Hamilton, wrote "The Comet Kings," a story that first appeared in 1942 but was reprinted in paperback in 1969: "Trapped in the depths of Halley's Comet, the futuremen battle fourth-dimensional monsters in a titanic struggle to save the system's solar energy!" In the 1950s and 1960s, movies and TV shows so bad they were good were common, and some of the TV shows, whose original transmissions have now reached as far out in space as the star Vega, are still occasionally re-run. One of these shows was the *Time Tunnel* TV series, which featured an episode in which the heroes, Tony and Doug, found themselves back in 1910 with Halley's Comet threatening. (In another episode, Tony and Doug solved their problems in the Trojan War by the "clever" idea of having machine guns sent back to them.)

And then there were the nonfictional but (unintentionally) fanciful attempts to relate ancient returns of Halley's Comet to almost every significant event of the prehistoric and even legendary past. The most tireless of investigators in this vein was Michal Kamienski. Kamienski was born in 1879 and served as director of the Naval Observatory at Vladivostok from 1914 to 1920, then of the Warsaw University observatory from 1923 to 1944. He made a valuable study of periodic Comet Wolf and its nongravitational forces, but after his retirement at age 65 in 1944, Kamienski pursued his attempts to calculate ancient apparitions of Halley's Comet. It took the considerable talents and efforts of scientists such as Yeomans and Kiang, aided by newly discovered or translated records, to calculate in the 1970s and 1980s the returns of Halley's Comet from 1404 B.C. to A.D. 2134. But in the 1950s and 1960s, Kamienski believed that he was making accurate calculations over much longer periods of time. The title of one of his papers was "Orientational chronological table of modern and ancient perihelion passes of Halley's comet, 1910 A.D.–9541 B.C." In 1971, at over 90 years of age, Kamienski wrote a paper discussing Halley's Comet in the vision of Jeremiah, in the time of King David, at the fall of Troy, in the year of Abraham's birth, and (first but not least) at the time of Noah's Flood. Kamienski died in 1973.

No doubt we all occasionally think about the astonishing rate of technological change that has occurred in our lives and in the twentieth century as a whole. But if you stop to consider how much we changed from 1910 to 1986 in comparison to any other Halley period—or to any other three Halley periods put together—you will gain a fresh apprehension (and perhaps *apprehension* in more than its neutral sense) of this trend.

Of special relevance to seeing Halley's Comet is this century's dramatic increase in medical knowledge and the resulting dramatic increase in average life expectancy. Many times more people than ever before saw Halley at two returns; that is, they saw Halley in 1910 and survived to see it again in 1986. Even so, as I write this book in 1996, it is sobering to think how the numbers of the "1910 Club" must have dwindled with the passage of yet another 10 years. And now, as in the early 1980s when the comet approached, it is sad to think of some of the people who remembered the 1910 visit, longed to see Halley again, and failed to survive to do so in 1985–1986.

Among those who died before the longed-for comet returned, two stand out in my mind.

One was the great amateur astronomer, sky-lover, and comet discoverer Leslie Peltier. In his gentle and lovely book *Starlight Nights*, Peltier described his memories of both the Great January Comet of 1910 and Halley, which he thought was more impressive than the January comet.

The other man I wish had lived to see Halley a second time was the paleontologist, naturalist, and gifted writer Loren Eiseley. Here is the passage about Halley's Comet from the opening section of his 1970 book *The Invisible Pyramid*:

Like hundreds of other little boys of the new century, I was held up in my father's arms under the cottonwoods of a cold and leafless spring to see the hurtling emissary of the void. My father told me something then that is one of my earliest and most cherished memories.

"If you live to be an old man," he said carefully, fixing my eyes on the midnight spectacle, "you will see it again. It will come back in seventy-five years. Remember," he whispered in my ear, "I will be gone, but you will see it. All that time it will be traveling in the dark, but somewhere, far out there"—he swept a hand toward the blue horizons of the plains—"it will turn back. It is running glittering through millions of miles."

I tightened my hold on my father's neck and stared uncomprehendingly at the heavens. Once more he spoke against my ear and for us two alone. "Remember, all you have to do is be careful and wait. You will be seventy-eight or seventy-nine years old. I think you will live to see it—for me," he whispered a little sadly with the foreknowledge that was part of his nature.

"Yes, Papa," I said dutifully, having little or no grasp of seventy-five years or millions of miles on the floorless pathways of space. Nevertheless I was destined to recall the incident all my life. It was out of love for a sad man who clung to me as I to him that, young though I was, I remembered. There are long years still to pass, and I am already breathing like a tired runner, but the voice still sounds in my ears and I know with the sureness of maturity that the great wild satellite has reversed its course and is speeding on its homeward journey toward the sun.

The Fallacies, Frisbees, and Fanfare of 1985–1986

1986. SPACECRAFT VISIT HALLEY   You may well have witnessed what I can still call "our" Halley return. But at the time of this book's publication, there are already a few readers too young to have seen this return of Halley who will live to enjoy it in 2061.

I have written much about the 1985–1986 return elsewhere. I reviewed the entire return in a chapter of *The Starry Room* (see the bibliography). I discussed my encounter with the long, dim, phantom tail briefly in Chapter 1 of the book you are holding. The nucleus of Halley, which the spacecraft Giotto and Vega 1 and 2 especially revealed, is described in Chapter 3. The fact that the great visitor comet was itself visited—by our robot probes—is probably the single most significant aspect of the 1985–1986 return.

But there are an assortment of other things to note quickly. The return was the hardest of all to see, more because of light pollution than the comet's dimness (though it reached only second-magnitude and was low for Northern Hemisphere viewers when bright). On the other hand, this return offered two different "close" approaches, making the comet potentially visible to the naked eye (and in small telescopes) for longer than any past visit. The landmark events of Halley's 1986 return, including first imaging, first and last telescopic sighting, first and last naked-eye sightings, and its most recent—but surely not last!—imaging appear in Table 6.

The 1985–1986 return produced a whole new batch of Halley lore. The pot of lore continues to be stirred, and, like folk tales in an oral culture, the stories about Comet Halley and other comets, true and untrue, become compounded with one another, as well as being elaborated and enriched. The same misinformation about Halley gets corrected at every return (for instance, that it was the comet that appeared after Julius Caesar's death) but shows up again with fresh errors and inventions. No fallacies are so good as those spiced with a touch of truth. A popular article in 1985 I read is an excellent example: It claimed that Halley's Comet was blamed for the Black Death (surely untrue—the plague occurred about 30 years before the 1378 visit), that it was credited with defeating the Spanish Armada (obvious confusion with the Great Comet of 1577, itself more than a decade too early), and that "Charlemagne was terrified of the celestial wanderer." (It was in fact his son who was scared by the comet at its return many years after Charlemagne's death.)

To attempt a thorough review of society's reaction to the 1985–1986 visit of the comet would be difficult, not least of all because I feel that somehow, even a decade later, I don't have the proper distance from it. But I can offer a brief, representative assortment of the cultural responses and an overall evaluation of the public's reaction.

Table 6. **Major events surrounding Halley's 1986 return**

Year and date	Distance from Sun (a.u.)	Distance from Earth (a.u.)	Event
1982 Oct 16	11.1	10.9	Recovery on CCD image by David C. Jewitt and Danielson, at magnitude 24.2.
1983 Oct 10	8.8	8.7	First spectrum recorded.
1984 Sep 22	6.1	6.2	First amateur photograph, by Tsutomu Seki, with Halley at magnitude 20 1/2.
1984 Sep 25	6.1	6.2	First coma detected (in CCD imaging).
1985 Jan 23	5.1	4.3	First visual sighting, by Steve O'Meara with 24-inch and comet at estimated magnitude of 19.6.
1985 Jul 12	3.3	4.2	First photographs showing ion tail.
1985 Nov 8	1.8	0.9	First naked-eye sighting, by Charles Morris and Steve Edberg.
1985 Nov 27	1.5	0.6	Pre-perihelion "close" approach to Earth.
1985 Dec 9	1.4	0.7	Two 3°–4° ion tails.
1986 Jan 9–11	0.9	1.3	Major ion tail disconnection event.
1986 Feb 9	0.6	1.6	Perihelion.
1986 Feb 22	0.7	1.4	Multiple dust tails and anti-tail noted.
1986 Mar 6–9	0.8	1.1	VEGA 1, VEGA 2, and Suisei flybys of Comet Halley.
1986 Mar 9–11	0.8	1.1	Major ion tail disconnection event.
1986 Mar 14	0.9	1.0	Giotto flyby of Comet Halley.
1986 Mar 20–22	1.0	0.8	Major ion tail disconnection event.
1986 Apr 11	1.3	0.4	Closest approach to Earth, 0.42 a.u.
1986 Apr 11–12	1.3	0.4	Major ion tail disconnection event.
1986 Late Apr–Early May			Longest observed dust tail (30° to 50°–60°?)
1986 May			Antitail noted.
1986 May 29–30	2.0	1.7	Final naked-eye sightings, by Charles Morris and Rich Keen.
1986 Jun 14	2.3	2.2	Last photo showing ion tail.
1987 Feb 2	4.8	4.1	Last spectrum showing gas emission.
1987 Apr 1	5.4	4.6	Last CCD image showing faint tail.

(continued)

Table 6.   **(continued)**

1988 Feb 23	8	7	Final visual sighting, at about 17thmagnitude, by David Levy with 61-inch.
1989 Jan 1–9	10.1	9.5	Asymmetric coma noted by Richard West around nucleus region of magnitude 23.5.
1990 Feb	12.5	11.6	No coma visible in Richard West CCD image showing Halley at magnitude 24.4.
1991 Feb–Mar	~14.3	~13.5	Amazing outburst produces coma, raises brightness from presumed 25th to recorded 19th magnitude.
1994 Jan 11	18.8	18.0	CCD image of Halley at magnitude 26.5 produced by Olivier Hainaut and colleagues. Halley imaged over halfway out, imaged almost as far out as Uranus, but surely not for the last time on the way out.

In the end, there were roughly 90 books published about Halley's Comet in 1984 and 1985 (a few came out before this too). Even before 1985, at least three dozen different Halley's Comet T-shirts were for sale. By the summer of 1985, telescope manufacturers were claiming a doubling and tripling of usual sales (in 1910, telescope sales were reported to have been greater in a few months than in all the years since the American Civil War combined). One U.S. company managed to copyright the phrase "Official Halley's Comet" for use on merchandise ranging from blazers to mugs and keychains. Commemorative coins and medallions, tie pins, dishes, and postcards were marketed. One company offered as memorabilia certificates granting buyers stock in the comet. And products that did not exist in 1910 received the Halley stamp—Frisbees, for instance.

I don't know how many comet parties there were in 1985–1986. But in early 1985, at a party at the Franklin Institute in Philadelphia, a 25-foot model of the comet hung over the heads of the attendees of a $200-a-plate charity dinner.

By the end of 1984, virtually every kind of popular magazine and newspaper had already had something to say about the comet. *People* magazine's preview for 1985 mentioned prominently, along with the usual bevy of movie and TV celebrities, a certain distinguished seventeenth-

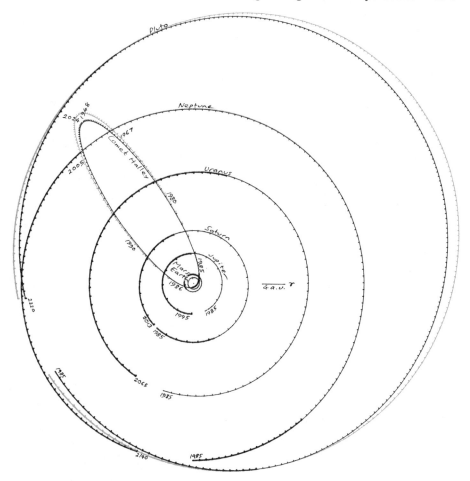

The entire orbit of Halley's Comet among the planets.

century scientist and his comet, which would both be "in" (in fashion, in style, in conversation) for the year. Even by late 1984, a tabloid was reporting one of the first of the outrageous rumors: A Russian astronomer, the newspaper said, had discovered that not Halley's Comet but an asteroid accompanying it named Ivan the Terrible would smash into Earth or, even worse, would hit the Moon and, billiards style, knock the Moon into our planet. But as it turned out, there were no serious scares associated with the comet. (Perhaps everyone was too busy buying and selling merchandise.) Editorial pages featured Halley's Comet as early as 1984. A Swiftian proposal from England to "nuke Halley's Comet" as target practice was picked up by various U.S. papers as an allegedly very funny and clever piece.

Idealistic ventures also hitched their wagons to the comet. These included Joe Laufer's "Halley's Comet Watch" newsletter and organization and certainly Harriet Witt Miller's "Kids for the Comet" column.

My own idealistic venture was the "Dark Skies for Comet Halley" campaign and newsletter, which introduced the issue of light pollution to great numbers of people. The day after Thanksgiving in 1985, *The New York Times* mentioned me on its front page! This was in conjunction with Dark Skies for Comet Halley and the role my campaign apparently had played in influencing New York City Mayor Ed Koch to propose turning off unnecessary lights in the city to facilitate seeing the comet. Koch's chief science advisor later told me that the reasons why Koch subsequently backed off from these plans were purely political. What Koch did instead, on one night in January 1986, was to have certain parts of certain city parks darkened completely (a virtually useless gesture, considering the miles of surrounding lights sending up skyglow) and given increased police protection. TV's David Letterman quipped that this darkening-the-parks plan made about as much sense as nickel beer night at Sing-Sing, and I agree. The only good that came of it was that people did gather at the designated locations, and many did through telescopes get a glimpse of the comet which they might otherwise never have gotten.

The largest crowd I know of that gathered specifically to see Halley's Comet was the throng of 40,000 at Jones Beach, New York, on the evening of January 11, 1986. In reading the comments of people at events like this, and in many personal conversations I had, I got the impression that most people were not bitterly disappointed—for the simple reason that the press had prepared them for this not being one of the closer or better returns (actually, light pollution was surely more of a problem than the comet's distance or lowness in the sky for the Northern Hemisphere). I got the impression that most of those who saw Halley were delighted to have had a look, however dim, at a legend.

Halley's Future

2061. First In-Person Visit?   No one can predict what political events on Earth between now and 2061 will do to increase or decrease the possibility that humans will make their first in-person visit near or in Halley's Comet at this next return. But such a visit is certainly a possibil-

The Great March Comet of 1843 as seen from Paris on March 19, 1843.

De Cheseaux's Comet of 1744.

Donati's Comet over Notre Dame in Paris on October 4, 1858.

Donati's Comet over the Conciergerie in Paris on October 5, 1858. Lithograph possibly by Mary Evans.

Coggia's Comet of 1874, as seen from the Pont Neuf (engraving by Charles La Plante).

Artist's conception of (in order of heads, left to right): 1882, 1907, 1811, 1861, 1819, 1835 (Halley), 1858, 1843, 1910 (January), 1874.

Halley's Comet with the Moon and Venus before dawn in May 1910.

Panels from the Bayeux tapestry, showing representation of Halley's Comet at its 1066 return.

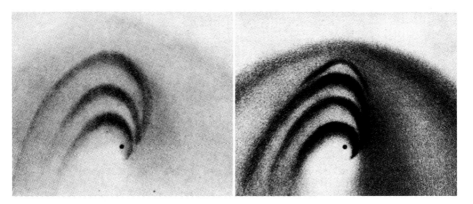

These lithographs are based on J. F. J. Schmidt's sketches of hoods in the coma of the Great Comet of 1861 at 14:59 and 15:32 Athens time. Note that a new hood has appeared in the second sketch.

The Great Comet of 1882, photographed by Sir David Gill.

The tail of Comet Ikeya-Seki and the huge diffuse glow of the zodiacal light (middle of photo) photographed by Dennis Milon in the Catalina Mountains near Tucson, Arizona, at 5:07 a.m. MST on October 26, 1965.

Close-up of head and near-tail region of Comet Bennett, by Helen and Richard Lines at 11:20 UT on April 5, 1970, at Mayer, Arizona.

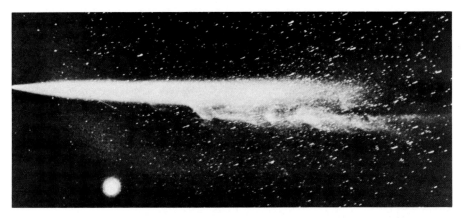

Halley's Comet with Venus, photographed on May 13, 1910, at Lick Observatory. The comet shows a 45-degree-long tail and, at bottom end of tail, a section of disconnected old gas tail flying away.

The dust tail and an unusual curved gas tail of Comet Bennett, photographed by Dennis di Cicco on April 3, 1970.

Comet de Vico, recovered for the first time in almost 150 years, was photographed by Bob Ross in October 1995.

Comet West as photographed in March 1976 by Dennis de Cicco.

Comet Halley, photographed in 1986 by Dennis de Cicco.

Comet Hyakutake shows rich tail structure in this photograph taken by Johnny Horne in 1996.

Comet Hyakutake and the Big Dipper. Glen D. Gould, with the assistance of Bunny Laden, combined a photo of part of the Big Dipper he'd taken years earlier with one he'd taken when Hyakutake's tail was passing through the other part of the Big Dipper on the night of March 24–25, 1996.

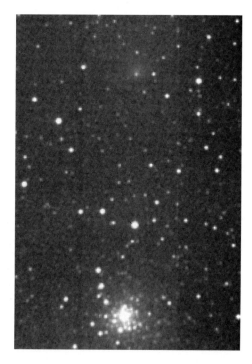

Hale-Bopp and the star cluster M70 were recorded in this CCD image by Gordon Garradd soon after discovery in July 1995.

The head and brightest tail of Comet Hyakutake were photographed in bright moonlight by Paul Ostwald in 1966.

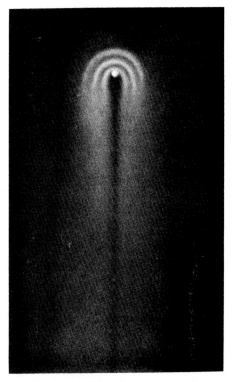

Hoods in Coggia's Comet drawn by E. L. Trouvelot on July 13, 1874.

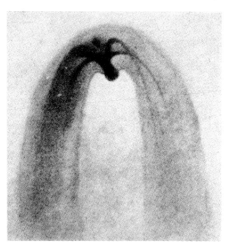

Photograph at left of Comet Daniel taken on August 5, 1907, shows ion rays in its tail. Drawing at right of same comet on same night shows jets in coma.

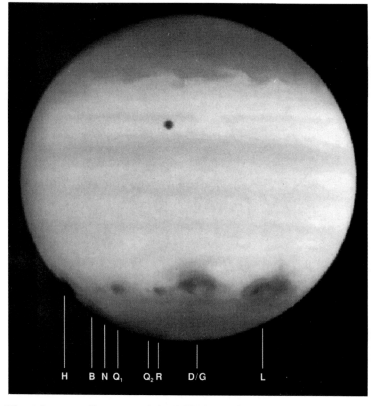

Ultraviolet image of Jupiter recorded at 7:55 UT on July 21, 1994, shows shadow of a moon and many labeled dark spots caused by the impacts of pieces of Comet Shoemaker-Levy 9. Some of the labeled spots are too faint to make out in this reproduction of a Hubble Space Telescope image.

Comet Arend-Roland, photographed at 3:13 UT on May 5, 1957.

Comet Bennett in March 1970.

Comet Tago-Sato-Kosaka, photographed at 1:26 UT on December 28, 1969.

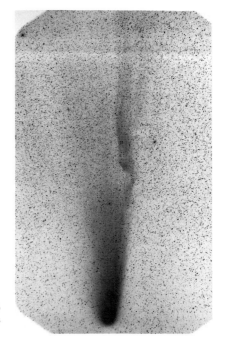

This negative taken by William Liller in 1986 shows a disconnection of the gas tail of Halley's Comet.

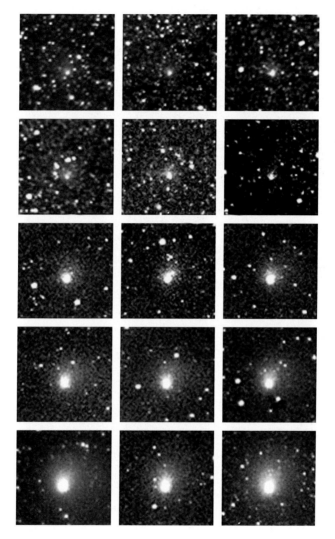

CCD images of Comet Hale-Bopp by William Liller showing the approach and brightening of the comet.

Halley's Comet, the galaxy Centaurus A, and (at bottom) the great globular cluster Omega Centauri (first recorded by Edmond Halley) are seen in this black-and-white version of a color image by William Liller from 1986.

ity. A problem (or maybe just an inconvenience to the spaceflight technology of 65 years from now?) will be the distance of this return's "close" approach: 0.477 a.u. But this perigee occurs when the comet is near perihelion (as in many past returns, especially the one this most resembles, Attila's of 451), so the comet should be much brighter than in 1986, probably about 0 in magnitude.

2134. The return of Halley's Comet in 2134 is the last one that can be calculated precisely, because the comet will come close enough to Earth to have its orbit slightly changed, as in 1404 B.C. The emphasis here is on *slightly*. Our old friend will not leave us, never to return!

But Halley's Comet will swoop 0.09 a.u. from Earth on May 7, its tail having already gone immense and tilted into Scorpius and Sagittarius before being lost from sight in mid-northern latitudes. On that date, the comet may shine as bright as Sirius and will pass about 8 degrees from the south celestial pole—its deepest plunge in our skies since the earliest precisely calculable return, over 3500 years before.

A.D. Unknown

There is one final scene for us to visit.

Imagine that we are alive thousands of years in the future and are visiting one of the vast natural areas or parks on the peaceful planet that Earth has become. We are expecting the closest passage ever of Halley's Comet. It will come to within half the distance of the Moon from Earth! (Remember, it had the *potential* to come almost this close as recently as A.D. 607.)

We have not done our homework and are unsure, but we hope that viewers at our location will get to see the head of the comet pass directly between us and the waning Moon, a few days past last quarter. But we have miscalculated: Many hours before moonrise, we are seeing ever more of the coma's glow rising, becoming brighter and brighter. At last, the inner coma itself comes above the east horizon and the view really is like that through a telescope—but swollen across much of the heavens!—with an apparent nucleus as big as a full Moon and indescribably luminous. The visible glow of the coma now stretches across most of the eastern sky.

The night is illuminated as brightly as by a gibbous Moon, but no moonlight has ever had this tint or eeriness or omnipresence—the light is too diffuse for even the brightest inner coma to cast shadows. The tails have become foreshortened but are still immense jet-like structures amid the coma and almost as bright as the apparent nucleus and its surrounding sheath of dust. What we cannot understand is a broad region of yellow glow that we notice is moving away from the apparent nucleus. Then understanding dawns: The nucleus has passed between Earth and the Moon already, but the broad yellow glow is the near side of a dust halo which is expanding rapidly outward. The bottom and brightest edge of it (the line of sight along which we look through the greatest thickness) is now well above the horizon and is followed by the thick crescent Moon, which is not so bright as the comet's brightest structures.

Over the next hour, we watch in astonishment as the nucleus moves farther away from the Moon in the sky but the forward edge of the expanding halo at last catches the Moon from our point of view, and we are seeing Luna through a most strange glow.

Before the ends of this halo actually reach Earth and Moon in space, the coma has traveled onward too far—or has it? We also wonder if someone somewhere sometime has seen the shadow of the Earth as a dark red splotch searching through the luminous blue and gold fog. As daylight comes—we almost feared, even with all our scientific knowledge, that it would not—a huge patch of brightest coma remains visible like a cumulus in the blue, and we wonder what it would be like if that wave of dust had made it close enough to Earth to be captured by our atmosphere. Yet is it not likely that the Earth–Moon system has captured a lot of dust and gas as the great coma brushed by? Where are the meteors? Perhaps tomorrow the heavens will be filled as never before in all history with meteors as numerous as stars, their flashes like strange lightning all about the receding but still brilliant cloud as it rolls away. What eerie glows in Earth's atmosphere or beyond might be seen in weeks to come in the form of silver, gold, and blue noctilucent clouds, a zodiacal light tower, or enhanced cloud-satellites of Earth?

COMETS WILD

Chapter 9
WHAT MAKES
A COMET GREAT?

Tame comets versus wild comets: That's how Guy Ottewell suggested we think about the distinction between short-period and long-period comets. "Wild comets are flakes of the virgin material of the solar system; tame comets are stripped-down and used-up," he wrote. "Wild comets are young and prodigal; tame comets are old and doomed."

The rest of this book is devoted to the long-period comets—the comets fierce and swift, wild and free. Of course, the stolid lecturer interrupts—not all long-period comets are bright or boisterous, and not all short-periods are dim or quiescent. And as for terms like *wild*, *fierce*, and *free*. . . .

We know. We know the objections and qualifications, but it is important to convey compellingly the difference between most long-period and most short-period comets.

Wild? Fierce? Free? Perhaps a less objectionable term is *great*.

According to the criteria for cometary greatness proposed by Donald Yeomans, there are no "great" apparitions of any short-period comet on record except for some—actually most—of Comet Halley's. (Another Guy Ottewell way of explaining what is special about Comet Halley: "Halley's is the wildest of the tame comets.")

Thus there is only one "great" (usually, though not in 1835 or 1986) short-period comet on Yeomans's list. How many "great" long-period comets? 47.

Before we start our two-chapter journey through many dozens of history's most memorable comets—some of them "great"—we should in this chapter consider briefly what people in general and comet experts like Yeomans in particular have meant when they have applied that term to a comet.

Ideas of Greatness

We might first look at the historic record and propose to accept as "great" only those comets which came to be known by a title which included the word: examples would include "the Great Comet of 1811" and "the Great Southern Comet." But we immediately discover that this scheme is unsuitable. There are many irrelevant cultural and circumstantial details that have entered into whether a comet came to have "Great" in its title. Even in past centuries, if one individual clearly discovered a comet or became firmly associated with a comet for other reason, the comet might become known by that individual's name: Donati's Comet was certainly a great comet, and Halley's Comet was at many of its returns, but they have not been known as the "Great Comet" of the years they appeared. Now in the 20th century, with the official scientific procedure to name comets after no more than their first three independent discoverers, and the improvement of communication and widespread vigilance for new comets, there is much less chance of a comet first being sighted by crowds or for other reason coming to be popularly known with a title like "Great Comet of . . . [whatever year]."

What do comet experts and enthusiasts say a great comet is? An article in the July 1996 issue of *Sky & Telescope* presents a selection of such statements in the context of discussing whether Comet Hyakutake qualified as a "great" comet. Some of the experts polled felt that a bright dust tail was a necessity, that only a comet bright enough to be seen impressively from large cities could qualify (according to them, it has become much harder in the twentieth century, with its increasing light pollution, for a comet to qualify as great).

There is much food for thought in this article, but I can speak in greater detail about the ideas that I myself have had on this issue.

Over a decade ago, I tried my own hand at making up a list of great comets. I specified the period from 1485 to 1985 and attempted to choose the ten greatest comets to appear in those 500 years. Halley's I excluded, but I acknowledged that the 1910 apparition would have made my list. The comets I chose were

The Great Comet of 1577
1618 II
The Great Comet of 1680
DeCheseaux's Comet (1744)
Messier's Comet of 1769
The Great Comet of 1811
The Great (March) Comet of 1843
Donati's Comet (1858)
The Great Comet of 1861 (Tebbutt's)
The Great (September) Comet of 1882

Some veteran comet enthusiasts might scoff: 1965's great Ikeya-Seki not on the list? And where does the bright, broadside, striated, split-nucleused, high-altitude Comet West get ranked?

But, if forced to make a choice, I would still stand by this list—though perhaps it would be safer to limit myself to the 400 years starting with 1577. We have too little information about the comets of 1532, 1533, and 1556, which suggests why rating the top ten comets in all of history would be even more ill-advised. But I would have to explain my criteria and then admit that some of them are not strictly numerical and that the final ranking ends up having a subjective element. What were my criteria? Not just brightness, tail length, height in a dark sky, and duration of visibility, but also whether most of the world's population could have seen it (comets were penalized if they were visible only from the Southern Hemisphere) and any other interesting features (special tail displays, large apparent coma size, split nucleus, and so on). Ikeya-Seki was not seen easily enough or long enough in the Northern Hemisphere. Some of Comet West's best features were much exceeded by the also many-branched tail of March-dawn deCheseaux's Comet. (I was also afraid of being biased toward the comets I had personally seen and loved.)

All in all, this list is an interesting effort (I think), but it uses too many subjective judgments and the appropriateness of some of the criteria is highly debatable. The idea of selecting precisely ten comets makes far less sense than just listing all—however few or many—that qualify as great according to certain more objective requirements.

This is what Donald K. Yeomans did in his paper "Cometary Apparitions: Great Comets in History."

Yeomans's "Great" Comets

Yeomans kindly sent me a preprint of this paper in early 1990, after I had talked with him as I was preparing a magazine article on that year's Comet Austin (we all hoped that Comet Austin would become a great comet by anybody's standards, but it didn't).

Yeomans's table of great comets is reproduced (with his generous permission) as Appendix 2 of this book. You may wish to browse it for a few minutes right now.

The table lists 72 cometary apparitions, plus a new one I've added, which clearly now deserves inclusion: 1996's Comet Hyakutake. Of the 73 apparitions, 25 are returns of Comet Halley.

All kinds of interesting facts can be gleaned from Yeomans's table.

But the first things to discuss are his criteria for greatness. He does not detail them and even comments, "Although applying the appellation 'great comet' to a particular cometary apparition is necessarily a subjective process, Table 1 is an attempt to list the most impressive cometary apparitions from 373 B.C. through 1976 A.D." The most important single figure by which a comet is judged here, however, is interesting because it attempts to combine others: the maximum apparent brightness when seen in a dark sky (that is, after the end of evening twilight or before the start of morning twilight). It is vital to account for a comet's angular closeness to the Sun, because even a very bright comet will obviously not produce a widely observed spectacle if it appears too deep in the solar glare.

Yeomans's other major criterion for cometary greatness appears to be the duration of a comet's naked-eye visibility. This may create a slight bias toward modern comets, because many of the world's best naked-eye observers have been alerted to look for them. How long a comet is visible to the naked eye is surely a worthy criterion in itself. But I think it complements the maximum brightness in a dark sky very well, because the two act as checks to prevent automatic inclusion of some comets all of us would agree were not great (visually great, that is; here we are not counting factors of purely intellectual interest). This combination of criteria excludes any comet that was bright for just a few days—the naked-eye observation interval is not long enough—and any comet which though at least marginally visible without optical aid for a long time, never became a first-magnitude object outside of twilight (thus neither IRAS-Araki-Alcock nor the 1985–1986 apparition of Halley, fascinating as they were in some ways, qualifies as a visually great comet).

It is interesting to note that first magnitude is not quite the cutoff point for the Yeomans comets. A number of these comets are listed as being of magnitude 1 to 2 and two (the 87 b.c. apparition of Halley and the 1965 apparition of Ikeya-Seki) are listed as having maximum magnitude of just 2 outside of twilight. But the wisdom of having a bit of flexibility at the limits is clearly shown here: Although Ikeya-Seki's lesser brightness away from the Sun (and its being visible to less of the world's population as a result of its southerliness) kept me from putting it in my top ten monumental tail and tremendous brightness in twilight and broad daylight make it unthinkable to eliminate this comet from a list of history's 73 greatest cometary apparitions. By the way, not limiting the number that can qualify for the list removes the need to weigh latitude as part of a competition. (It might be nice, however, to have in Yeomans's table some measure of the southerliness of an apparition. Perhaps how far south a comet was at maximum brightness outside of twilight should be included.)

Yeomans's table is not just a curiosity; it is useful. It's a great means of identifying important comets far back into history. It also gives a sense of how often large comets achieve certain levels of brightness or closeness to the Sun or Earth.

Proof of its value recently came when brightness and closeness predictions were announced for Comet Hyakutake. While many comet enthusiasts were scrambling to figure out when the last comet this bright had come this close to Earth, I simply turned to the Yeomans list. I immediately discovered that the answer was the year 1556. In another few moments, I could identify still earlier comets of this sort, see what the best combinations of close approach to Earth and Sun for large active comets were compared to what Hyakutake would do, and more.

Three Ingredients for Greatness

In the paper accompanying his table of great comets, Yeomans summarizes three major properties that can contribute to a comet's becoming visually impressive from Earth. Most of the great comets have at least two of these going for them.

The two properties that most commonly make comets spectacular are closeness to the Sun and closeness to Earth. A comet can be great without too much of one of these as long as it excels tremendously in the other. For instance, none of the well-established apparitions of large Kreutz sungrazers brought these comets closer than 0.84 a.u. from Earth. But their close passes of the Sun made their true brightness so dazzling that they were spectacular even from relatively far away. On the other hand, the Great Comet of 1861 never came closer than 0.82 a.u. from the Sun. But then it moved to within 0.13 a.u. of Earth, became visible for many hours in a dark sky, showed a large developed dust tail from a broadside view, and even swept its tail right across Earth. Yeomans does not list maximum tail length and thus probably does not take it into consideration as a major factor (this is probably wise, because the angular length of a tail can be made immensely longer if there happens to have been one well-placed, eagle-eyed observer detecting the faintest extension of a tail on even one night). Comets West and Donati both offered wonderful broadside views of their well-developed tails—but neither was deficient in intrinsic brightness or in coming fairly close to Earth and Sun either. If a comet is too far from Earth or Sun, its apparent brightness will not be great enough for us to see much of its tail, no matter what angle or fascinating details it boasts.

Almost every object on Yeomans's list has a perihelion (closest to Sun) *and* perigee (closest to Earth) distance of no more than about 1.0 a.u., and usually both considerably less. At only 2 of the 73 apparitions (the comets of 1807 and 1811) was a comet's perigee somewhat more than 1.0 a.u., and at only 1 apparition was the perihelion somewhat more than 1.0 a.u. (the remarkable comet of 442 had a perihelion way out at 1.53 a.u.!).

There *is* a third way a comet can become great, even if its distance from both Sun and Earth are fairly large (both 1 a.u. or somewhat more). A comet can become visually impressive from Earth if it is intrinsically an extremely large and active comet. This is very rare. The only "great" comet in all of history that has both a perihelion and a perigee distance of more than 1.0 a.u. (1.04 and 1.22, respectively) is the Great Comet of 1811. This comet was visible at zero, first, and second magnitudes for impressive amounts of time. But what shows up on Yeomans's table is how long it was a naked-eye object. Only 7 of the 73 apparitions on the Yeomans list featured naked-eye visibility of 100 days or more. Only 2 featured naked-eye visibility of 120 days or more. The second-longest spell of naked-eye visibility was that for the mighty sungrazer, the Great

Comet of 1882, at 135 days. But the Great Comet of 1811 was detectable with the naked eye for 260 days—almost twice as long as the nearest contender among the great comets!

Of course, here again we find that no one criterion can give us a perfect measure of greatness. The timespan (from first to last naked-eye sighting) over which Halley's Comet was at least marginally visible to the naked eye in 1985–1986 was about $6^1/_2$ months, yet for the vast majority of that time it was shining feebly. As we noted earlier, it never reached better than about second magnitude and even then was dimmed in horizon murk for most of the world's population in the sleepy hour before dawn.

The Great Comet of 1811 was not at all like that. It spent many weeks as a high and brilliant object with a large head and a hefty tail. And now, for just the second time in the historical record, there is a good chance we will get another comet very much like it in these key respects. I mean Comet Hale-Bopp, which has only to live up to the average predictions for it to resemble the Great Comet of 1811 in important ways.

Chapter 10

THE COMETS BEFORE 1700

This chapter and the next survey the range of what comets can do and mean to us through a sampling of the long-period comets of history. The survey is far from complete, of course, and my selection criteria are far from objective. But I've tried to include the most visually interesting comets of which we have sufficient record. I've noted all the long-period comets that satisfy the conditions of greatness established by Donald Yeomans (see Chapter 9; the italicized word *great* in this list is used only for these comets). I've left Halley out (we've already treated it separately) but have included a few of the brightest and most interesting returns of other periodic comets.

Many comets I've included here for their interest rather than for any visual spectacle they may have produced. Some were chosen merely because the contemporary description of them was so colorful or because they were closely connected with the death of a famous person or some other historical incident. I've even included a smattering of the most outrageous mystery objects, some of which may have been comets but others of which may be spun of even thinner substance than comets themselves—pure imagination.

This chapter deals with comet appearances up until 1700—just before Halley published his famous prediction and just before telescopic discoveries of comets began to become common. My source for much of the information on these objects has been Donald Yeomans's catalogue of naked-eye comets before 1700, which appears in his book *Comets—A Chronological History of Observation, Science, Myth and Folklore* (see the bibliography). I have supplemented that work with a small number of other, sometimes odd or hard-to-find sources. For the comets from 1700 to 1995 in the next chapter, I have relied heavily on Gary Kronk's *Comets: A Descriptive Catalog* (see the bibliography and the listing of web sites, including Kronk's, at which he offers a few fascinating samples of the new multi-volume "cometography" he is preparing). But there are many other sources of information for these later comets, too.

Here, then, is a selection of the most fascinating comets of history for your delectation and wonderment.

ELEVENTH CENTURY B.C.   (maybe around 1059 B.C.). The first comet for which there are some details and chance of datability: "When King Wu-Wang waged a punitive war against King Chou, a broom star appeared with the handle of the broom star pointed east."

613 B.C.   Chinese records relate that a broom star comet entered the constellation of the Great Bear. Yeomans says, "This is probably the first comet for which a verifiable record exists."

467 B.C.   A tailed comet was recorded in both China and Greece. This has often been claimed to have been a return of Halley's Comet, but it was not. This is the comet that Plutarch mentions as having appeared prior to the falling of a giant meteorite at Aegospotami in Greece.

373–372 B.C.   This is the first comet in history which Yeomans thinks we have enough information about to call *great*. The most exciting speculation about this object is that it was a Kreutz sungrazer and that it may have split to produce the great comets seen in 1843 and (just possibly) 1680. This comet could itself have been one of the two largest pieces of the original "parent of the sungrazers," which Marsden thinks may have last visited the inner solar system as one body sometime between 18,000 and 8000 B.C. (The other piece would be the comet that was seen to split in A.D. 1106, whose major fragments Marsden suggests came back as the Great Comet of 1882 and 1965's Ikeya-Seki.)

What are the basic facts about the 373–372 B.C. comet? In winter "a comet was seen in the west at the time of the great earthquake and tidal wave at Achaea, Greece." Aristotle said the comet first appeared in the west near the Sun and later had a tail that extended one-third of the way across the sky—about 60 degrees. If Aristotle was speaking from personal experience, he was remembering seeing it when he was 12 years old. Pingre was the first to argue that this comet of Aristotle's was the same one which Seneca mentioned Ephorus as describing. Seneca scorned Ephorus for claiming that this comet was seen to split into two divisions, but if the comet was a Kreutz sungrazer, and possibly the progenitor of two later great sungrazers, then the claim of the comet's splitting may be correct.

Too little is known about where and when the comet appeared to determine an orbit for it.

147 B.C. This is the second *great* comet on Yeomans's list and the first for which an orbit has been determined. It passed perihelion on June 28 (0.43 a.u. from the Sun) and should have been closest to Earth—only 0.15 a.u.—on August 4. But the record that survives has it being first reported by the Chinese on August 6. On this latter date it was in the southwest, below Scorpius in the evening sky. Then on August 8 it was north of Scorpius with a tail stretching 90 degrees. Seneca refers to a comet occurring after the death of Demetrius, king of Syria, and a little before the Greek Achaean war in 146 B.C.—a comet that was as big as the Sun, reddish like fire, and "bright enough to dissipate the darkness."

44 B.C. In May-June, observers in China and Korea saw a reddish-yellow comet spanning about 12 degrees in the northwest. In a few days, it was near Orion with a 15-degree tail pointing toward the northeast. This may be "the Julian comet," the one that several Roman authors said was seen in the north for 3 to 7 days during the games Octavian held in honor of the assassinated Julius Caesar.

A.D. 79 A tailed comet was seen, first in the east and then in the north, from Korea; it disappeared after 20 days. This is the best candidate to be the comet the Roman emperor Vespasian joked about when he said, "This hairy star is an omen for the king of the Persians," referring to the fact that the Persian kings wore their hair long and that he himself was balding.

178 A comet with a reddish tail spanning 70 to 90 degrees, visible for 80 days in China. Its orbit is undetermined, but this comet clearly deserves its place on Yeomans's list of *great* comets.

191 A white comet with a tail over 100 degrees, recorded in China and Korea. It appeared south of the "sidereal divisions" of Spica (Alpha Virginis) and Kappa Virginis in October. A *great* comet.

240 This comet, recorded in China, first appeared on November 10, the day it was at perihelion 0.37 a.u. from the Sun. It measured 30 degrees and was in Scorpius before going through Sagittarius. It "trespassed" against Venus, and on December 19 "trespassed" against Aquarius. The closest approach to Earth was about 1.0 a.u. on November 19. It was visible for 40 days. A *great* comet.

262   This comet was either less than a degree long or 50 degrees long. Yeomans follows Ho Ping-Yu's translation on this point. But the fact that it is said to have been visible for 45 days seems to suggest either that the longer tail figure is right or that something else is garbled or missing in this account.

287   A comet with a tail spanning 100 degrees—Ho translates "hundreds of degrees" long—was visible for only 10 days in Sagittarius.

302   In May or June, Chinese and Korean records say, "a broom star comet appeared in the day." That at least is the translation of Ho Ping-Yu, whichYeomans follows, and which Seargent uses to establish this as the third comet on his list of daylight comets. But in George F. Chambers's turn-of-the- (last) -century catalogue (as Seargent himself notes) the translation is "comet was visible in the morning"—perhaps meaning pre-dawn visibility?

336   A "broom star comet" appeared in Andromeda in the western evening sky in February. This is probably "the hairy star of unusual size" that was said to have presaged the death of the Roman emperor Constantine in May 337.

390   Appearing on August 7 in Gemini, this *great* comet passed only 0.10 a.u. from Earth on August 18. It was recorded in China and Korea—and also in Rome, where it was said that a sign appeared in the sky hanging like a column and blazing for 30 days. On September 5 it was at perihelion 0.92 a.u. from the Sun. "On September 8 it entered the region of northern Ursa Major" and its tail was white and over 100 degrees long.

Comet of 384, in a woodcut from the sixteenth century.

On September 18 "it entered the north polar region and went out of sight." Yeomans gives its maximum magnitude outside of twilight as −1.

400   This comet is the only *great* one to have perihelion and perigee (closest approach) both as small as Hyakutake's—in fact, its close approach to Earth was a bit closer: 0.08 a.u. The perigee occurred on March 31, only 6 days later in the year than Hyakutake's in 1996. Maximum magnitude outside of twilight occurred on March 19 and, again like Hyakutake, was about 0. But the rest of the facts about this comet are quite different. It reached perihelion 0.21 a.u. from the Sun on February 25—*before* perigee, not after as Hyakutake did. It was reported in China, Korea, and Rome. Our first record of it is on March 19 (when brightest); it was then located in the morning sky between Andromeda and Pisces with a 45-degree tail extending into Cassiopeia. It moved to Ursa Major and next, between April 10 and May 9, into Leo (evening sky). It was called sword-like, and Socrates Scholasticus apparently thought it the most terrible comet on record.

418   There were either one or two (or three?) comets in this year, but whatever the number, one of them sounds as though it was truly tremendous. The Chinese recorded a "bushy star" comet appearing in Ursa Major on June 24. But Pingre quotes Roman sources that speak of a comet—or something that sounds comet-like—seen in Europe during the (total?) eclipse of the Sun on July 19 and also speak of a comet that was seen from midsummer until the end of autumn. Starting on September 15, we have Chinese records of a comet in Leo whose "rays" extended more than 100 degrees to northern Ursa Major.

442   This is the *great* comet with by far the largest perihelion distance—1.53 a.u. Only 2 of the 47 other long-period comets on Yeomans's list have perihelion at greater than 1.00 a.u.: the great comets of 1664 (1.03 a.u.) and 1811 (1.04 a.u.). Yeomans estimates this comet's maximum magnitude out of twilight as just 1–2, but the key reason for its place on his list is no doubt its more than 100 days of visibility. This was the rare case of a comet in the middle of the night far beyond Earth's orbit (out near that of Mars, in fact) that was bright. It was first reported on November 10 and was closest to Earth—0.58 a.u.—on December 7 (8 days before perihelion passage). This comet's path took it from Ursa Major to Auriga, Taurus, and Eridanus.

453   In February-March, both Chinese and Roman sources report that a comet was seen in the west. This could be the comet said to have preceded Attila the Hun's death in 453 (it was the 451 return of Halley, remember, that was associated with Attila's defeat at Chalons in France).

520   Another possible daylight comet whose daylight visibility rests on whether a Chinese word (or phrase) means "day" or "dawn" (before sunrise).

565   The second comet of the year, this *great* one was visible for about 100 days as it went from Ursa Major to southern Pegasus (where it was over 15 degrees long) and onward to disappear in northern Aquarius. It passed perihelion (0.82 a.u.) on July 15, was first reported on July 22, and was closest to Earth (0.54 a.u.) on September 13. Estimated maximum magnitude of 0–1 at perigee.

568   Only 3 years after the previous one, another *great* comet came, this one passing only 0.09 a.u. from Earth on September 25. It was first reported in Libra on July 28, unless that was a nova which occurred just before the comet! The comet went through perihelion (0.87 a.u.) on August 27 and by September 3 was near Antares. Two days after closest approach to Earth, it "trespassed" Delphinus and entered Pegasus. The comet ended up in western Aries on November 5. Estimated maximum brightness outside of twilight: 0 at closest approach.

615   Yeomans follows Ho in saying about this comet: "It measured $1/2$ degree, looked black, and pointed and scintillated as it moved toward the northwest for several days until it reached Ursa Major." But Chambers follows Gaubil in attributing to this comet a tail that spanned 50 to 60 degrees and saying that its "extremity had an undulatory motion."

770   This *great* comet was first noted on May 26, went through perihelion (0.58 a.u.) on June 5, and passed 0.30 a.u. from Earth on July 10 (when its magnitude was 1–2). It was seen in the north, and its tail extended about 75 degrees. On June 19, it moved eastward approaching northern Auriga. On July 9 (near closest approach), it was in northern Canes Venatici. It was last seen on July 25.

813   From Constantinople on August 4 there was seen a comet (?) that resembled two moons joined together. The two parts separated and, taking different forms, looked like a man without a head. Pingre accepted this as the description of a comet. Yeomans comments, "This is a questionable appearance at best."

817   On February 17, a tailed comet appeared in Taurus. It was 3 degrees long and after only 3 days came near Orion and disappeared. But "On Feb. 5, at the second hour of the night, a monstrous comet was seen in Sagittarius"—this account is from Vita Ludovici Pii in Bouquet's Collection, vi, but not mentioned in Yeomans.

875   In Japan, a red "broom star" comet "with pointed rays" was seen in the northeast on June 5, extended over 15 degrees in northern Auriga on June 9, and was a "bushy star" (tailless) comet on June 24. In France it was said that the death of Louis II was "announced by a burning star that showed itself on June 7 in the north."

886   The second comet of this year was reported in China on November 16. Here's what Yeomans writes: "A comet known as the *long-path* type was seen coming from the west. It was white in color, 21 degrees in length, and bent at an angle. Eventually, it fell like a meteor. . . . Possibly, this comet exhibited a tail directed towards the Sun as well as away from it."

891   This *great* comet was first seen on May 12, in Ursa Major. It went east, "swept" Arcturus, and proceeded to the region of Hercules-Aquila-Serpens-Ophiuchus. Its tail reached a maximum span of 100 degrees. It was last seen on July 5. Except for the great comet of 1106, which we think was a Kreutz sungrazer, the comet of 891 is the latest on Yeomans's list of *great* comets whose orbit remains undetermined.

892   There were two or maybe three interesting comets this year. European records mention a comet in the tail of Scorpius that lasted 80 days (a long time!) and was followed by drought in March and April. Chinese records say that in June a "white, cometlike banner was seen" that was 3 degrees long, was "shaped like hair," and "after several days . . . stretched from the midheaven to the horizon." Finally, the Chinese records say that on December 28, "a comet of the type called *celestial magnolia tree* was seen in the southwest" but that on December 31 "it turned into a cloud and faded away." (Another translation says merely that on December 31 the sky was cloudy, so it was not seen.) Hasegawa believes this final comet was in Sagittarius.

893   There is record of a comet with a tail 100 degrees long this year, but Yeomans thinks it is suspiciously similar to the great comet of 891. Several old translators say that this comet's tail was reported to have gradually lengthened to 200 degrees (!) and that "the clouds then hid it."

896   "Guest stars" can be novae (exploding stars) or comets. The following Chinese report, from November–December of this year, is reprinted here as Yeomans gives it based on Ho's translation: "Three guest stars, one large and two small, appeared in northern Aquarius. Moving eastward together, they sometimes approached one another, then separated, giving the illusion that they were fighting among themselves. After 3 days the two smaller ones disappeared while the larger one faded away in northern Aquarius."

905   The first record of this *great* comet is from May 18, when "a star resembling Venus" was seen in the northwest evening sky. "It emitted rays of 45 to 60 degrees and was blood-red in color." But on May 19 its color was compared to white silk. The perihelion had been on April 26 (0.20 a.u.), but on May 25 it passed 0.21 a.u. from Earth. At this close approach it was moving from northern Gemini to Ursa Major. In June it went down into the summer constellations and disappeared by mid-month. Estimated maximum brightness outside twilight: 0.

962   This comet is listed as *great* by Yeomans. It was visible for 64 days, and got as bright as first magnitude. Yeomans says it was first seen in western Pegasus on January 28, then on Ferbuary 19 moved southwest and entered Libra, and then went out of sight in the region near Alphard (Alpha Hydrae) on April 2.

1014   On February 25, this comet passed just 0.04 a.u. away from Earth, one of the few closest encounters on record. It was then in Auriga, so it should have been prominently visible. But existing reports say nothing about its being remarkable for brightness or tail length. It was seen for at least a month, with what may have been the last observation on March 7 in Perseus. This comet reached perihelion (0.56 a.u. from the Sun) on April 6.

1037   From Korea: "Five comets, each measuring 7 to 9 degrees, were seen." A bolide (exploding meteor) display, perhaps?

1097   There is radical disagreement between Yeomans and Kronk about how close this comet came to Earth. The old catalogue of Chambers has the tail many times longer than either of them—a span of 50 degrees was seen in China, he claims—and mentions that the tail was "much bifurcated."

1106   This *great* comet may have been a Kreutz sungrazer that split and whose two biggest pieces came back as the Great Comet of 1882 and Ikeya-Seki in 1965 (see Chapter 4). Pingre was the first to identify the daylight sighting of this comet. European records mention that on February 4 or 5, a star or comet "one foot and a half" (supposedly 1 to 2 degrees) from the Sun was seen for most of the day. Seargent estimates that the magnitude at this point may have been about −10: brighter than a half Moon! By February 7, people in Constantinople and Palestine saw a comet in the southwest after sunset. On February 9 and 10, observers in the Orient saw a comet "the size of the mouth of a cup" in the west. "Its rays scattered in all directions as if broken into fragments." Yeomans says it was more than 60 degrees long, Kronk says nearly 90 degrees long and 5 degrees wide. The comet remained visible for more than a month. A medieval annalist said that a meteor separated itself from the comet on February 16 and fell to Earth. Could this be a confused observation combining the sighting of a real meteor with the sighting of a fragment (perhaps the Great Comet of 1882?) splitting off from the main nucleus?

1132   This *great* comet became as bright as −1 as it went on October 7 just 0.04 a.u. from Earth—one of the closest passes in history. At that time it was in northern Aries and displayed a tail with intense rays over 45 degrees long. It moved south into eastern Pisces and then further south before disappearing. All of this happened well after perihelion (0.74 a.u. on August 30).

1240   Yeomans rates this *great* comet as having been about 0 in magnitude when out of twilight. But on February 1 "it was seen at the side of Jupiter, the same size as Venus"—could this mean as bright as Venus? Its rays stretched for 7 degrees at this time. The comet's color was reddish-white, at least when it was first seen on January 27 (6 days after its perihelion at 0.67 a.u.). It passed 0.36 a.u. from Earth on February 2.

1264   After being noted on July 21, this *great* comet extended a tail 100 degrees long and was "illuminating the heavens" from its position in northern Hydra. The comet continuted on to Hydra, Gemini, and eastern Orion. It passed closest to Earth on July 29, when it probably shone at about 0 in magnitude. It remained visible until October or November.

1274   "Three days before the death of Thomas Aquinas, a comet appeared," says an old source. But Yeomans lists no comet for this year.

1345   A comet passed just 0.05 a.u. from Earth.

1351   Another comet passed 0.05 a.u. from Earth!

1366   Periodic comet Tempel-Tuttle passed 0.03 a.u. from Earth on October 26 (see Chapter 7, on periodic comets).

1402   One of the greatest of the *great* comets, this object shone at −3 outside of twilight (it is among only five comets in the historical record that Yeomans estimates to have done so). This occurred around March 12. On March 19 it was lost in the solar glare, but then it went on to become visible in daylight for a record 8 days (March 22–29). Perihelion passage was on March 21, and despite the great brightness, the comet was neither a sungrazer (its minimum distance from the Sun was 0.38 a.u.) nor a close approacher of Earth (perigee was 0.71 a.u. from Earth on February 19).

1468   Although it came close to neither Earth or Sun, this comet shone at about magnitude 0 outside of twilight, and in late September its tail exceeded 45 degrees. Another report mentioned a 30-degree bluish tail. It was visible for over 80 days. A *great* comet.

1471 (brightest in 1472)   Just over 3 years after the previous one, another *great* comet. This one, like the comet of early 1402, was one of the greatest of all, one of the few in the historical record to shine at possibly −3 outside of twilight. Unlike the 1402 comet, however, this object benefited from a very close pass of Earth. It was first reported on Christmas Day of 1471. On its way inward toward perihelion, it was seen in Virgo with its tail pointing west in the second week of January 1472. It passed to a region near Arcturus, sweeping eastern Coma Berenices. It soon arrived in Leo, with the tail still facing west (away from the morning Sun, though now in the midnight sky). On January 23, the comet passed only 0.07 a.u. from Earth (it is one of only two cases in the historical record of a *great* comet having a closer pre-perihelion passage to Earth than Hyakutake). It was moving almost 40 degrees per day, and on January 24, Chinese observers said "its rays grew and stretched across the heavens from east to west." Korean observers reported "only" a 30-degree tail in the third week of January. Perhaps this presumably gas tail was much like Hyakutake's, with a very long extension to it visible to careful observers for only a small number of days and only in good sky conditions. In any case, on January 24 the great comet zoomed north to Ursa Major and was even sighted at midday. By February 17, it had raced over to western Aries and eastern Pisces. It was last seen in late February, and it passed perihelion (0.49 a.u. from the Sun) on March 1. Paolo Toscanelli, who had studied comet tails and noted their anti-sunward direction, was now 75 years old. He recorded this comet from January 8–28.

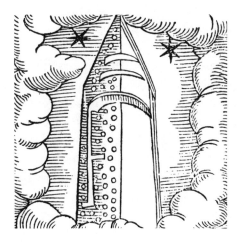

Comet of 1479 seen through a break in the clouds.

"Montezuma Transfixed by a Comet in 1519–20," from Diego Duran's Historia de las Indias de Nueva Espana, *written in 1574–81.*

1531   A *great* apparition of Halley, soon to be followed by two other great comets.

1532   The second of an unprecedented three *great* comets in three years. It passed close to neither Earth nor the Sun, but it must have been an intrinsically bright object continuously well located for observation: It was observed for 119 days (third or fourth longest among *great* comets in the historical record) and is estimated to have reached a maximum magnitude outside of twilight of −1. The observations of Peter Apian, Girolamo Fracastoro, and Johannes Vogelin permit us to determine its orbit and to know that it was observed passing from southern Gemini through Leo, Virgo, and Libra.

1533   The third *great* comet in three years. This one was visible for 80 days, and we know it was observed by Copernicus. Its path took it from northern Auriga in late June and then through Perseus and western Cassiopeia to Cygnus. Maximum magnitude outside of twilight: about 0.

1556   This *great* comet has been mentioned a lot recently as the last object so bright to have come closer to Earth than Hyakutake. Interestingly, its path in the sky was in some ways similar to that of Hyakutake, and indeed, like Hyakutake, it passed closest to Earth over a month before reaching perihelion (between 1 and 2 weeks earlier than Hyakutake in the calendar year). This object was first seen in Corvus and passing not far southwest of Spica, at the end of February. Its closest approach found

A German broadside depicting the great comet of 1556, the last comet both brighter and closer than 1996's Hyakutake (many have been brighter or closer).

it racing northeast through Bootes and Draco. At that time, it came to within 0.08 a.u. of Earth and shone at a very impressive estimated brightness of −2 (significantly closer and brighter than Hyakutake). We apparently have no record of its possessing a long tail, but it was described as a "terrifying, prodigious, extraordinary star." It rushed onward through Cassiopeia and Andromeda to eastern Pegasus. It passed perihelion, 0.49 a.u. from the Sun, on April 22 and was last seen on May 10.

1577 This *great* comet is most famous for being the first to have been proved more distant than the Moon and therefore not the atmospheric phenomenon that Aristotle had claimed all comets were (see Chapter 2). Although Tycho was among the scientists responsible for this proof and the comet is often referred to as his, he was not the first to observe it.

The comet was at perihelion on October 27, when it was only 0.18 a.u. from the Sun. The first observations were from Peru, on November 1 where the comet "shone through the clouds like the moon." This description and other facts suggested to the modern comet scientist Vsekhsviatskii that the comet's magnitude may have been about −7: perhaps 15 times brighter than Venus!

During the next few days, the comet was noticed from London, appearing with a tail 7$^1/_2$ degrees long. On November 8, Japanese observers

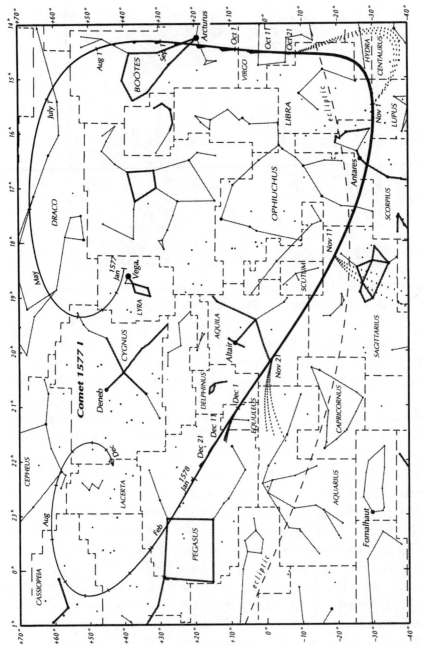

The great comet of 1577's path among the constellations.

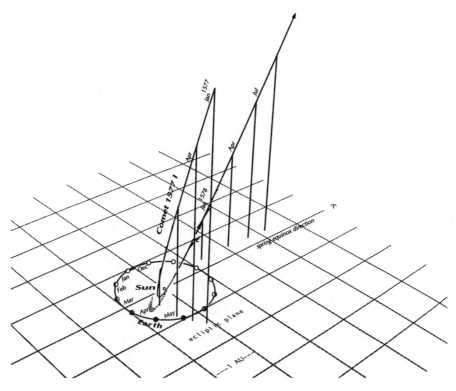

Orbit of the great comet of 1577.

saw it in the southwest after sunset with a white tail more than 60 degrees and perhaps 80 degrees long (almost halfway across the sky). They described the comet as being "bright as the moon."

On November 13, Tycho Brahe, still unaware of the comet's existence, was trying to catch fish for supper when he first noticed the comet, shortly *before* sunset. After nightfall he estimated the comet's brightness as comparable to that of Venus (−4). The gold (he called the color "Saturnlike") coma was 8 arc-minutes across (about one-quarter of the apparent width of Sun or Moon) and had an intense pseudo-nucleus. Tycho noted a curved dust tail that was slightly reddish and 22 degrees long. (This comet also apparently displayed a Type III tail at some time, but no gas tail has been established from the observations.) The next day, November 14, Chinese observers could see perhaps 50 degrees of tail, which indicates a true length of 0.8 a.u. or more—the tail would have stretched from Mercury's orbit to well beyond Earth's.

Though not the first, Tycho was probably the comet's last observer. His final glimpse came on January 26, 1578 (thus it was visible to the naked eye for almost 3 months). At that time, the great 1577 comet was 2.07 a.u. from the sun and 2.67 a.u. from Earth.

The true brightness this comet achieved was perhaps as great as any on record that was not a sungrazer, and the absolute magnitude has been estimated as about −1 or −0.5. (Comet Hale-Bopp's absolute magnitude may be similar, or a little less bright.)

The most famous lore about the Great Comet of 1577 concerns Queen Elizabeth I. Elizabeth rejected the advice of her courtiers, who argued that going out to look at the comet might bring bad luck. After getting a view of the majestic object, she supposedly said, *"Lacta est alea"* ("The die is cast.")

Tycho produced two reports on this comet: his important scientific report in Latin and another in German, in which he felt obliged (perhaps without any great conviction on his own part) to resort to astrology. Much space in the German paper is devoted to interpreting the comet's significance in human affairs. Because the comet had first been seen at sunset (the west, where Spain was) and was seen in Sagittarius (which supposedly ruled the Spaniards), Tycho predicted the Spaniards would suffer—they would lose many cattle and many of their best men. He said these effects would disappear by 1583, but an astrologer stretching this nonsense further might have claimed with hindsight that the power of the comet endured longer and in 1588 led to the historic defeat of the

Comet of 1596 being discussed by an astrologer and his client.

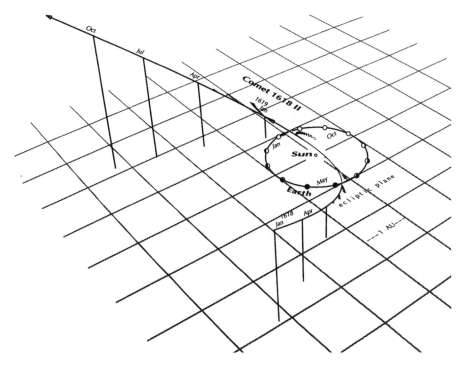

Orbit of the great comet of 1618.

Spanish Armada. That other prominent comets had occurred closer to 1588 could of course be ignored—no need to let facts get in the way.

1618 I   The first of three comets seen in this year was spotted first on August 25 in Hungary and 2 days later by Kepler near Linz, Austria. On August 28, Korean observers reported it below Ursa Major in the morning sky with a bluish-white tail more than 15 degrees long (it had been at perihelion on August 17 and nearest to Earth—0.52 a.u.—on August 20). On September 6, Kepler apparently became the first person to use a telescope for observing a comet when he turned his telescope on this object. He last observed it on September 25.

1618 II   This *great* comet passed perihelion on November 8 (0.40 a.u., about the average distance of Mercury from the Sun). But it was not observed until before dawn on November 16, when several observers noticed a tail sticking up from the horizon in Libra. In the next few weeks, as the head became visible in Libra in a darker sky, the tail extended all the way across the eastern sky to the bowl of the Big Dipper!

The 1618 II comet passed 0.36 a.u. from Earth in early December, when it was seen in broad daylight, and yet estimates in this month—

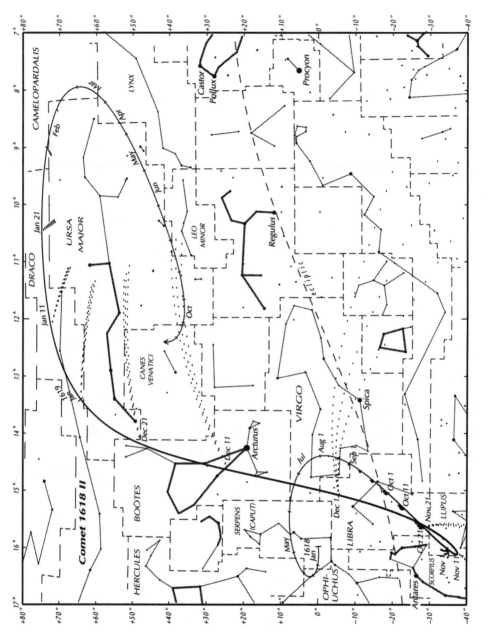

The great comet of 1618's path among the constellations.

220

later in this month?—placed the brightness at magnitude 1–3. In early December, the comet (the dust tail?) had a reddish hue. No gas tail has yet been established, but according to Christianus Longomontanus, the dust tail reached to as long as 104 degrees on December 10. The maximum length in space must have been more than 0.5 a.u.

Vsekhsviatskii says that "Cysat (with a small telescope) and Wandelin noticed the splitting of the comet's tail into numerous star-like pieces" but offers no interpretation of this strange report. In the latter part of December, the comet's head passed roughly 5 degrees from the brilliant star Arcturus, and its tail extended to the westernmost reaches of Ursa Major. John Bainbridge wrote that "the comet's hair was spread over the faire starre."

The great comet of 1618 was last viewed on January 22, 1619, by which time it was north of the Big Dipper. (By coincidence, the calendar period during which it was visible—November to late January—was almost the same as that of the 1577 comet.) Its absolute magnitude was perhaps about $4^1/_2$.

The monarch of Shakespeare's later years, Elizabeth's successor James I, was highly superstitious about comets. Both King James I and the diarist John Evelyn believed that the 1618 II comet presaged the Thirty Years' War.

1664   The first of two *great* comets in consecutive years, comets associated with the great plague and fire in London. This one was first seen in Corvus in November in the morning sky, but it moved to the evening sky starting in January. The comet was visible to the naked eye for about 75 days, but use of the telescope extended observations to March 20. "In November its grayish tail was about a degree in length and pointing to the southwest." It was at perihelion on December 5 at 1.03 a.u. from the Sun—one of only three on Yeomans's list with a perihelion distance of more than 1.0 a.u. Closest approach to Earth nevertheless came over 3 weeks later (December 29) and found the comet only 0.17 a.u. from our world. The comet was in Gemini at that time (thus in the midnight sky, just outside our orbit) and probably shone at about magnitude −1. It raced on through Taurus to Aries in just the next few days, and by January 8 its blue tail (another gas-dominated comet!) was pointing east from Aries. This comet was extensively observed by such famous scientists as Christiaan Huygens and Johannes Hevelius.

1665   The second *great* comet in as many years. Halley correctly figured its perihelion date as April 24. It passed only 0.11 a.u. from the Sun, and although Yeomans gives its maximum brightness outside of twilight as -1, Kronk suggests that because it was seen on April 20 just minutes after sunset, it may then have rivaled Venus (-4) in brightness. It was first seen on March 27 and, presumably as a result of the unfavorable angle formed by Earth, Sun and comet after perihelion, was last viewed at that April 20 sighting. The comet was closest to Earth on April 4 but fully 0.57 a.u. away. On April 6, the comet was in Pegasus, and Hevelius was able to see it with the naked eye in moonlight with a tail nearly 20 degrees long. Between April 10 and 13, this extensively observed comet was seen to have a tail 30 degrees in length.

1668   The third *great* comet in a five-year span. It is probably a member of the family of Kreutz sungrazers and, if so, probably passed less than 0.01 a.u. from the Sun on March 1. But the "official" figures for it are still those which Carl Heinrich Friedrich Kreutz himself calculated in 1901—0.07 a.u. from the Sun on February 28—not a member of his comet group. This comet was first seen on March 3 from the Cape of Good Hope and was always better placed for Southern Hemisphere observers. By the time it was out of twilight on March 8, its estimated brightness was just first to second magnitude. But on March 10, J. D. Cassini, in Italy, observed its tail stretching about 30 degrees from Cetus to the middle of Eridanus, and on March 18, Chinese observers saw the tail extend over 40 degrees in Eridanus.

1680   This *great* comet was the first comet discovered with a telescope. It was found by Gottfried Kirch while he was looking at the Moon and Mars before dawn on November 14. Just three other astronomers saw it before it passed perihelion: Dorfel, Cassini, and Hevelius (Hevelius had just experienced the devastating fire which consumed his instruments, books, and other possessions).

This comet was a sungrazer, though it has not been established as a member of the Kreutz family. Interestingly, Halley did speculate that this comet was one with the comets of 43 B.C.., A.D. 531, and A.D. 1106, the last of which was a sungrazer and may have been a member of the Kreutz group (possibly the parent of several of the great Kreutz sungrazers of the nineteenth and twentieth centuries). But Halley was wrong. It was the less impressive comet of 1682 that he was correct in connecting with certain earlier ones—the 1682 object was indeed a return of Halley's Comet. (There was, by the way, much confusing of the 1680 and 1682 comets in the years to come.)

How close to the Sun did the 1680 comet pass? It missed the Sun's surface, the photosphere, by much less than the Earth–Moon separation, only

The Marburg comet-egg of August 26, 1682. From Das Weltall, April 1, 1907.

Title page of a pamphlet about the great comet of 1680.

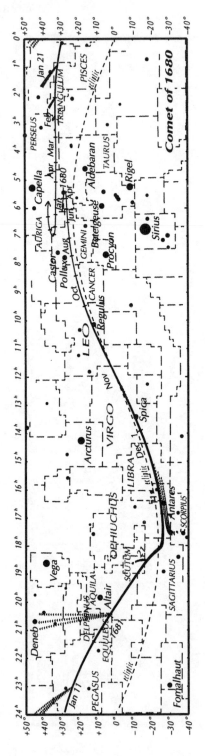

The great comet of 1680's path among the constellations.

about one-seventh of the Sun's diameter. The date of perihelion passage was December 18. Flamsteed at Greenwich saw the tail appear after sunset on December 20 but didn't see the head down low in bright twilight until December 22. The head appeared red "like a coal," and the tail, stretching almost halfway across the sky, was curved backward "like a sabre."

The great comet's tail spanned 20 degrees before it passed perihelion and was still 30 degrees long on January 23, 1681. Between these times, the maximum apparent length was 70 to 90 degrees in late December. The maximum true length of this tail may have been just over 2 a.u., exceeded or rivaled only by the great comets of 1769 and 1843.

The last naked-eye sighting of the tail (and perhaps of the comet) was on February 18. Isaac Newton, using a telescope, was the last to see it, on March 19. On this date it was about 2.2 a.u. from the sun and 2.5 a.u. from Earth. Newton was trying to learn more about cometary phenomena by observing this object intensively.

Like the 1618 comet, this one has been calculated to have had an absolute magnitude of about 4.

The Great Comet of 1680 was so impressive that it set off a particularly strong flurry of pamphlets and, like the 1577 and 1618 comets, appeared on coins and medals. But it also produced a new response. The news came from Rome that on December 2, 1680, an egg with a picture of a comet on its shell had been laid by a "virginal hen"! An Italian print of the time shows the comet egg and two other odd ones, along with a representation of the heavens and zodiac showing the comet's head at the knees of Virgo (perhaps this was why the hen was virginal). Other "comet eggs" were reported under the imagined influences of Halley's Comet in both 1682 and 1910.

1686   This *great* comet was first seen on August 12 from the Cape of Good Hope, when it was in the constellation Lepus with a 35-degree tail reported to extend up into Gemini (though how this tail direction could be correct I don't know!). On August 14, as seen from Brazil, the comet's head was just below Orion's Belt and was of magnitude 1 with an 18-degree tail. It passed 0.32 a.u. from Earth on August 16, when Jesuits in Siam estimated the tail length as 15 degrees. It was brightest out of twilight—about first to second magnitude—on about August 27. The comet passed on to Cancer with a dwindling tail and was seen in northwest Hydra on September 14. This was 2 days before it reached perihelion 0.34 a.u. from the Sun. It was last seen on September 22.

1689   A possible Kreutz sungrazer; a 68-degree, strongly curved tail while third or fourth magnitude; Southern Hemisphere visibility.

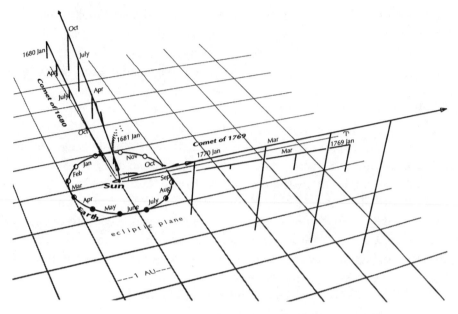

Orbit of the great comets of 1680 and 1769.

1695   A possible Kreutz sungrazer; tail 40-degrees long.

1699   Periodic Comet Tempel-Tuttle observed a second time (see Chapter 7).

A Lore Epilogue

A little had been learned about comets by 1700, but superstitions about them were deeply entrenched.

The sixteenth-century French political thinker Jean Bodin was an opponent of belief in witchcraft, but comets were a different matter. Bodin wrote that "Comets are spirits, who, having lived innumerable lives on Earth, and being at last near death, celebrate their last triumph or are brought again to the firmament as shining stars."

A few years after the great comet of 1618, Guillaume du Bartas, in *Du Bartas, his Divine Weekes and Works*, wrote (as translated into English),

> A Blazing Star
> Threatens the World with Famine, Plague, and War;
> To princes death; to Kingdoms, many crosses;

To all Estates, inevitable Losses;
To Herdmen, Rot; to Ploughmen, hapless Seasons;
To Sailors, Storms; to cities, Civil Treasons.

There were comets just before the Great Plague in England in 1665 and great Fire of London in 1666, and the widespread belief about their connection with these events was referred to as late as 1722, by Daniel Defoe in *A Journal of the Plague Year*. But comets were not seen only as portents of evil in seventeenth-century England. In 1642, a pamphlet appeared whose exceedingly long title synopsizes its subject rather thoroughly: "A Blazing Starre seen in the West at Totneis. Wherein is manifested how Ralph Ashley, a deboyst [debauched] Cavalier, attempted to ravish a young Virgin; also how at that instant a fearfull Comet appeared; likewise how he, persisting in his damnable attempt, was struck down with a flaming sword so that he died."

John Milton was only 10 years old when the 1618 II comet appeared. It stretched away from the constellation Ophiuchus, so perhaps he had another comet in mind (or, more likely, an imaginary one) when he wrote, in *Paradise Lost*,

> Satan stood
> Unterrified and like a comet burn'd
> That fires the length of Ophiuchus huge
> In th'Artick sky, and from its horrid hair
> Shakes pestilence and war.

Across the Channel, the death of the powerful French statesman and cardinal Jules Mazarin was reputed to have been announced by a comet. But Mazarin died in a year of no known comets (1662), so one from the preceding year is presumably meant—another case of delayed results from a comet. But comets were seemingly on everyone's mind in that century. An English diplomat (apparently Sir William Temple), annoyed at the arrogance of Louis XIV, wrote of the Sun King as "this Great Comet that is risen of late and expects not only to be gazed at but adored."

Even while Edmond Halley was hard at work on the study that would throw open the doors of comet science, Reverend William Whiston was elaborating and adding to an unpublished speculation by Halley that an ancient comet could have caused the Flood of Noah. Not Halley's Comet in its 1682 appearance but the great comet of 1680 was regarded, in this theory, as the possible return of the Noah comet. Now Whiston wrote *The Cause of the Flood Demonstrated* and his preposterous *A New Theory of the Earth*. In the latter, Whiston proposed the novel idea that Earth itself had once been a comet! (Another comet struck it and set it rotating.) Then in the time of Noah, the punishment of God came in the form of a disastrously close approach of the 1680 comet.

Whiston determined that the comet that caused the Flood came closest at noon, Peking time, on Monday, December 2, 2926 B.C. Unlike the earlier comet that had set the Earth spinning, this one had plenty of coma and tail vapors to shroud Earth, and those vapors (claimed Whiston) condensed on our world to cover it with waters 6 miles deep. Whiston argued that the next comparable disaster would see the world destroyed by fire, and that the same comet—the Great Comet of 1680—would again be the divine instrument of destruction. The predicted date for this holocaust, the comet's next return, was 575 years later (an orbital period Whiston borrowed from Halley's incorrect reckoning). Thus the world would be destroyed by fire by this comet in the year 2255. Voltaire wrote that Whiston was "unreasonable enough to be astonished that people laughed at him." But many did not laugh, and there was panic in 1719 when some people supposed that Mars at a close, bright opposition was in fact the comet headed for Earth.

Comets were on everyone's mind and superstitions still prevailed. But as the seventeenth century ended, Edmond Halley was pondering a certain comet, and a paper and prediction of his were about to strike the world like a lightning bolt of sanity.

Chapter 11

THE COMETS FROM 1700 TO 1995

This chapter presents a sampling of some of the most interesting comets from 1700 to 1995. Before we plunge onward with the survey of particular comets, however, I wish to continue the discussion of lore with which we ended the previous chapter, bringing it, as we will our list of comets, up to the present.

A Lore Prologue

As the eighteenth century dawned, the old fears of comets began to give way to new fears set off by misunderstandings of the astronomers' findings. Whereas comets had previously been construed as the causes or at least the heralds of earthly disasters, now there was the terror that they would be the disasters themselves by colliding with the Earth.

One of the great scares of this kind occurred in the early spring of 1773, when Joseph Jerome de Lalande, presumably inspired by the recent "close" encounter of Earth with Lexell's Comet, announced a lecture entitled "Reflection on comets which can approach the earth." An overcrowded schedule that day caused the French Academy to drop Lalande's lecture. But its cancellation ignited a rumor that the police wanted to prevent a panic: Many Parisians were convinced that the lecture included an announcement that the world would end on that May 12 because of a collision with a comet. Although Lalande had his lecture published immediately, nothing could quell the public fears until the fateful day came and passed.

In 1832, Biela's Comet was due for its first predicted return, and again it was a respected astronomer who innocently set off a panic. H. W. M. Olbers announced that on October 29 of that year, the head of Biela would pass virtually right through Earth's orbit. The fact that the Earth would then be about 90 million kilometers away was not understood by the newspapers, and the alarm was sounded. Joseph Johann von Littrow quickly issued a pamphlet to explain the real situation; miraculously, the public was convinced and the panic subsided.

Another scare was due to a comet that never even existed—but was widely expected in 1857. An English book called *Will the Great Comet Now Approaching Strike the Earth?* predicted that a comet seen in 1264 and 1556 (that 1556 comet again!) was returning and would at least inflict great heat on the world as it passed. The French caricaturist Honoré Daumier poked fun at people who were out looking for the comet just as Voltaire had the eager Halley seekers a century earlier. James Howard, Third Earl of Malmesbury, wrote in his diary in June 1857, "We are suffering under an extraordinary heat. People are really getting alarmed for if it is occasioned by the comet which is not yet visible, what must we expect when it reaches our Globe!" Some minor comets were observed in that year, but none with an orbit like that of the earlier comets and none that came close to Earth.

The next year a great (but not very close) comet did appear—Donati's Comet—and the Earl of Malmesbury wrote, "The largest comet I ever saw became visible with a very broad tail spread perpendicularly over the sky, the weather being very hot. Everyone now believes in war." Yet another truly great comet appeared 3 years after Donati's, suddenly becoming visible at practically full length and unprecedented breadth to viewers in the United States and Europe at the end of June 1861. Little imagination was needed for U.S. viewers to see this as a portent of war. Fort Sumter had been fired on back in April, and the first major battle of the American Civil War, the First Battle of Bull Run, was fought on July 21—with the comet still bright and long in the heavens.

The awards offered to comet discoverers, Barnard's house that comets built, and Barnard's dream of many comets in the sky miraculously confirmed by the Great Comet of 1882—these topics are discussed elsewhere in this book, as is the lore of Halley's Comet in 1910.

The best observations of some of the most spectacular comets take on a legendary quality (many scientists did not believe de Cheseaux's report of a fan of multiple tails on the comet of 1744, yet it was true). Such observations are treated in the rest of this chapter. After 1910, you might think that the only comet lore left would in fact be the charming or thrilling personal stories of great comet discoverers. Surely comet superstition and frenzy are things of the past, aren't they?

Perhaps not. We've already looked at the cultural and commercial impact of the 1985–1986 return of Halley's Comet and noted that there was surprisingly little real craziness or fear-mongering in advance of that visit. But there is still tremendous public interest in fringe ideas and outright nonsense about comets, and the potential for widespread panic probably remains.

When the topic of "fringe ideas" in astronomy and other fields comes up, the name still most likely to emerge is that of Immanuel Velikovsky, author of the 1950 book *Worlds in Collision*. Velikovsky's interweaving of ancient and primitive legend with some of the more fascinating mysteries of astronomy, history, archeology, and geology makes for colorful reading. But there is scarcely a trace of it that could be considered science. This is no place to rekindle the debate which a few patient scientists spent so much of their valuable time trying to set fairly to rest. Let us just note here that Velikovsky's *Worlds in Collision* attempted to establish (1) that Venus was originally a comet ejected from Jupiter in historical times, (2) that this comet's game of billiards with the Earth and Mars caused physical events which can explain major miracles of the Old Testament and many events described in the *Iliad*, and (3) that the comet Venus eventually attained its present very un-comet-like circular orbit and has fooled modern astronomy by masquerading as a planet.

Superstition and the deliberate spreading of misinformation about comets have continued up to the very present. In 1970, Comet Bennett was a "sky-broom" to Vietnamese soldiers, who thought it portended a worsening of the war, and the same comet was supposedly feared by some in Egypt to be a secret weapon of Israel. Comet Bennett was in fact the last thing the Apollo 13 photographed before the explosion occurred which put the astronauts in deadly peril.

Comet Kohoutek's approach in 1973 provoked a pleasant enough reaction in the form of coins, cartoons, and cruises, but there was also a "religious" pamphlet warning of the danger of the "Christmas monster" Kohoutek (the comet was due to reach perihelion on December 28); it was then, according to the pamphlet, to be just 40 days until collision and end of the world. Certain conscienceless hustlers proclaimed that Kohoutek would usher in a new era of peace—on their own terms, for their own silly glorification, and to their own financial benefit.

Fortunately, when the closest comet in over two centuries passed Earth in May 1983, there was not enough advance notice for the public to get alarmed and for the phony prophets—and profiteers—to come out of the woodwork. The same was true with Comet Hyakutake in 1996. What will happen in response to the long-awaited Hale-Bopp remains to be seen.

Now, on with our comets of 1700–1994.

The Comets After 1700

1702 I   Another possible Kreutz sungrazer; visible mostly in the Southern Hemisphere; 43-degree tail.

1702 II   Passed only 0.04 a.u. from Earth.

1718   Passed 0.11 a.u. from Earth; moved nearly 2 degrees an hour in Ursa Minor. Kirch estimated it as first magnitude then, and it was obviously outside of twilight, so doesn't it deserve to be one of Yeomans's great comets? No. This comet had no tail and was only 11 arc-minutes across (this measurement was apparently taken in the telescope, though, which would have made the outer coma harder to see)—and it faded remarkably quickly. (Did Kirch exaggerate its brightness or observe a brief flare that other people missed?)

1723   Passed 0.10 a.u. from Earth.

1729. SARABAT   This object was found by Sarabat, an ecclesiastic of the Jesuit order. It remained visible to the naked eye for over a month after reaching perihelion fully 4.05 a.u. out from the Sun—about twice as distant as any other comet that became visible to the naked eye and also considerably farther out than the limit where a coma from water ice should start forming. Yet a coma 1.5 arc-minutes wide was observed by Jacques Cassini. The absolute magnitude of Comet Sarabat may have been about −3, about two magnitudes brighter than that of any other comet in at least the past several centuries (10,000 to 20,000 years ago, the Parent of the Sungrazers may have had an absolute magnitude of −5—and may have come hundreds of times closer to the Sun). The nucleus of Comet Sarabat may be one of the rare giants, possibly 10 to 20 times wider than the average. Sarabat moved just 15 degrees in the sky during the entire 6 months it was observed (some comets have moved that far in less than one night).

1731 I   Seen in daylight?

1737 II. KEGLER—but now known to be an earlier return of periodic Comet Swift-Tuttle (see Chapter 7).

1743 I   Passed 0.04 a.u. from Earth. Marsden thinks this was probably a periodic comet.

1744. DE CHESEAUX   This great comet has often been referred to as de Cheseaux's Comet, even though Dirk Klinkenberg spotted it 4 days before Philippe Loys de Cheseaux, and later investigations have revealed that other observers saw it on November 29, 1743, which was 11 days earlier still. Perhaps de Cheseaux deserves the honor, though, for his observations of the comet's most spectacular feature. De Cheseaux's account was so amazing that it was doubted by the scientific community for many years until confirming observations were identified.

The comet was first spotted with the naked eye when about 2 a.u. from the sun and more than 1 a.u. from Earth. That was more than 3 months before perihelion. In those 3 months, the comet kept getting brighter and brighter. Few comets have become anywhere near so bright at such distances from Earth and Sun, and perhaps none has maintained great brilliance so long. By mid-January of 1744, the magnitude had reached about 1.

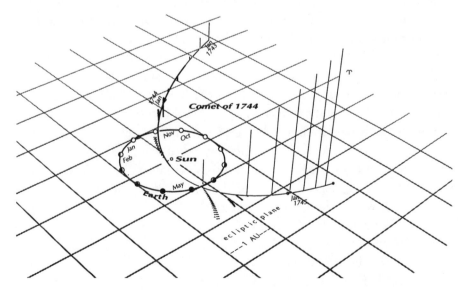

Orbit of the great comet of 1744 (de Cheseaux's).

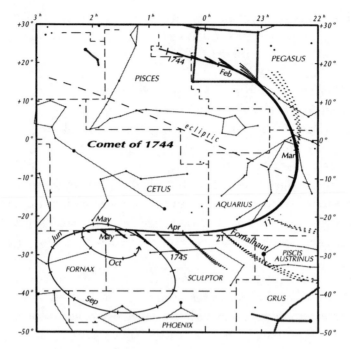

The great comet of 1744's path among the constellations.

By February 1 de Cheseaux's Comet was brighter than Sirius, by February 25 brighter than Venus, and on February 28 plainly visible in broad daylight while a full 12 degrees from the sun—at an estimated magnitude of −7. Perihelion was reached the next day at 0.22 a.u., and about 0.82 a.u. from Earth. Was the comet's full month of visibility at a magnitude brighter than Jupiter going to be its most outstanding feature? During March it was due to recede rapidly from Earth and so would surely fade.

But no one could have foreseen the remarkable tail that appeared after perihelion. Before perihelion it had grown from 7 degrees in mid-January to 15 degrees on February 1, and then on February 18 (when the comet was about magnitude −5), the comet displayed a dust tail 24 degrees long and a gas tail 7 degrees long. A hint that far more was in store came on February 24, when Heinsius saw a reddish-yellow tail sticking up from below the horizon in the dawn sky.

What de Cheseaux and other observers witnessed from March 6 to March 9 defied belief: While the comet's head was still 20 degrees below the horizon before dawn, there were between five and eleven separate tails sticking up as much as 20 degrees or more from the horizon in a broad fan over 30 degrees wide! More than a century later, Bredikhin concluded that what was seen was a gas tail and numerous branches—actual synchrones (discussed in Chapter 2)—of a great fan of dust tail presented broadside. The most nearly comparable other tail, Comet West's in 1976, extended up toward Delphinus that very same week 232 years later. West's tail was magnificent, containing as many as five branches visible to the naked eye, but the fan was less broad and the branches probably less discrete or distinct than de Cheseaux's. The full tail of de Cheseaux's comet (including that below the horizon on de Cheseaux's famous diagram) was also about $1^1/_2$ to 2 times as long as Comet West's at this juncture.

And by March 18, the tail of de Cheseaux's Comet had lengthened immensely more, reaching almost 90 degrees (halfway across the sky). The true physical length must have been more than 0.7 a.u.

The comet faded rapidly as it receded from Earth, but on April 22 it was visible without optical aid a last time—after almost 5 months of naked-eye visibility. On that date it was about 1.33 and 1.94 a.u. distant from Sun and Earth, respectively. The absolute magnitude has been estimated as about 1, but around perihelion it was perhaps -1. Along with the 1577 comet, de Cheseaux's had about the greatest intrinsic brightness known apart from the greatest sungrazers. Oddly, like the great comets of 1577, 1618, and 1680, it was first seen in November.

1747   Rivaled de Cheseaux's Comet in absolute magnitude, trailing only Sarabat's 1729 comet in this category among comets of the past few centuries. (Note, however, that the values for absolute magnitudes are very controversial and can change during an apparition. This comet and Sarabat might not have lived up to their potential if they had gotten closer to the Sun.) The comet of 1747 was barely visible to the naked eye but its perihelion distance was 2.2 a.u. and it was then on the opposite side of the Sun from us. The absolute magnitude was thus about -1 to 0. So its nucleus may be one of the largest known.

1758   Seen in twilight only 7.5 degrees from the Sun, which suggests a magnitude of about -3.

1759 III   Passed only 0.07 a.u. from Earth; moved as fast as 1 degree an hour.

1763. MESSIER   Passed 0.09 a.u. from Earth.

1766 II   The one visit of periodic comet Helfenzrieder.

1769. MESSIER   The famous Charles Messier found this *great* comet in Aries on August 8, 1769, with a telescope but could also dimly detect

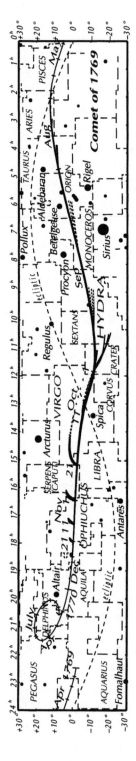

The great comet of 1769's path among the constellations.

it with the naked eye. The special wonder presented by this object in the weeks to come was the growth of a prodigiously long tail *before* perihelion. The tail grew from 6 to 15 degrees between August 15 and August 28 and then to 34 degrees on August 31, as its distance from the Sun dwindled to 1.13 a.u. and from Earth to only 0.45 a.u. At this time it had a brightness of almost first magnitude.

In the first week of September, the tail began to display some curvature (this comet was observed to have both a dust and a gas tail) and shot out to over 43 degrees as it crossed under the Earth's orbit. The head passed only 3 degrees south of Betelgeuse, which it outshone, on September 5. On September 10, the comet was closest to Earth—only about 0.35 a.u. away—and two separate observers reported a tail length unprecedented for a comet before perihelion.

Pingre was on a ship between Teneriffe and Cadiz. He says the tail was 90 degrees long but was so faint at the end that "when Venus rose above the horizon, its light shortened the tail by several degrees." This observation might be seriously doubted were it not for de la Nux's testimony of having seen a tail length of 98 degrees the same night. Even though both of these men were presumably skilled observers, these observations *have* been questioned. The reason for the skepticism is the fact that the physical tail length worked out from de la Nux's observation is 3.5 a.u.! If this figure is correct, the tail exceeds even that of the Great Comet of 1843 and is by far the longest tail known. The head was then only slightly within Earth's orbit, so the end of the tail would have stretched all but a small fraction of the way out to the distance of Jupiter's orbit! Even if the observations of Pingre and de la Nux are discounted, the other reports around this time suggest a true length of much more than 1.5 a.u.

On the date of its maximum apparent tail length, Messier's Comet was passing about 8 degrees south of the star Procyon and was roughly as bright as that star (about 0 in magnitude) but ruddy. The tail, then about 2 degrees wide, stretched right across the sword of Orion, just north of Rigel, and across the huge constellations of Eridanus and Cetus. On September 13, it was almost 60 degrees long, the first 40 degrees being very bright. The tail was also said to be surrounded by a 1-degree "atmosphere"—perhaps a beautiful sheath of fainter exterior tail?

In the next few weeks, the head dropped ever lower in morning twilight as it moved into Hydra, but plenty of tail was still visible by September 26, the day the comet was last seen before perihelion. On October 8, perihelion was reached (0.12 a.u.—quite close to the Sun), but the viewing angle for head and tail remained bad, and the object was not spotted until Messier glimpsed it with the naked eye, with difficulty, and saw only 2 degrees of tail.

By October 26, the tail was 3 degrees long and the pseudo-nucleus was clearly visible as a third-magnitude object, but the comet was 0.65 a.u. from the Sun and a huge 1.36 a.u. from Earth. The tail was reported to be as long as 6 degrees on November 3, but then bright moonlight interfered, and late in the month the comet faded to the naked-eye limit as it receded to over $1^1/_4$ a.u. from the Sun and over 2 a.u. from Earth. The last sighting with a telescope was on December 3. Absolute magnitude has been estimated as about 3.

After the fulfillment of Halley's prediction in 1758–1759, you might think the association of comets with plagues, earthquakes or the death of kings would be over. In fact, this tradition persisted even into the twentieth century, but if we discount the mild association of the King of England's death with Halley's Comet in 1910, the last outstanding examples of linking comet and ruler involved two comets connected with Napoleon. Napoleon apparently believed quite strongly that the great comet of 1769 had occurred in honor of his birth that year. Even the comet's discoverer, Charles Messier, agreed and wrote a pamphlet on the topic in 1808 (Messier's old age). But as omens, comets are two-edged swords. The Great Comet of 1811 appeared as Napoleon planned his invasion of Russia, and the emperor supposed that it presaged his glorious victory. Instead, of course, he met with disastrous failure. It seemed that the comets had turned against the mighty Napoleon, and his road now led to his inevitable defeat.

1770 I   Periodic comet Lexell comes closer to Earth than any comet certainly known in history.

1788   This comet, discovered by Caroline Herschel, was recovered in the twentieth century and became known as periodic comet Herschel-Rigollet.

1797. BOUVARD-HERSCHEL-LEE   Passed only 0.088 a.u. from Earth.

1806 I   Brightest return ever of periodic comet Biela, which passed just 0.04 a.u. from Earth.

1807   One of Yeomans's *great* comets, this is one of only two (the other is the Great Comet of 1811—just 4 years later!) on his long list that never came closer than 1 a.u. from Earth. The 1807 comet reached perigee, 1.15 a.u., on September 27 yet a few days later was of magnitude 1–2 in brightness outside of twilight. Perihelion was on September 19 (0.65 a.u.). It displayed well-separated dust and gas tails, the latter being longer and reaching a maximum span of 10 degrees.

1811. FLAUGERGUES   This *great* comet was found on March 25, 1811, by Honoré Flaugergues. It never came at all close to Earth or the Sun yet was so huge and intrinsically bright that it was an impressive sight for months. Its greatest claim to fame is that it remained visible to the naked eye much longer than any other comet in the past 500 years—or in all of history so far as we know. This is the comet which Hale-Bopp initially most seems to resemble.

The comet was visible to the naked eye when Flaugergues first saw it and was viewed as late as June 16 (5 days after conjunction with the Sun) by Humboldt in Paris. It had been well over 2 a.u. from the Sun when first observed and was heading for a perihelion just farther out than Earth's orbit (1.04 a.u.). But from mid-June until mid-August, it remained hidden in the sun's glow, emerging on August 20 and finally arriving at perihelion on September 12. On that date the comet was a full 1.6 a.u. from Earth and a dim naked-eye object, but the autumn was to see a dramatic improvement.

In early October the comet became a north circumpolar object displaying twin tails, each about 15–16 degrees long. On October 15, the all-night object had a gas tail 24 degrees long and a curved dust tail almost 7 degrees wide. The apparent nucleus was very prominent; Sir William Herschel described it as having a ruddy hue and the surrounding coma as having a bluish-green tinge (this blue-green may have resembled what I observed in Comet West, perhaps a mixture of sunlight-yellow

The memorable "Comet Wine" was credited to the influence of the Great Comet of 1811.

dust and blue gas). On October 20, the comet was nearest Earth, and though still about 1.1 a.u. away, it was nearly magnitude 0. Its head was measured as being up to 28 arc-minutes wide, similar to the apparent size of the Sun or Moon. This means that the true diameter of the visible head must have exceeded that of the Sun: at one point, 1.7 million kilometers compared to the sun's 1.4 million kilometers. (How large might the invisible hydrogen coma have been?)

Although a few comets undergoing outbursts (including Halley's in 1836) have had halos that temporarily matched or even slightly exceeded this size, no other known comet has maintained a coma so gigantic. Yet the show was still not over. As the comet receded from Earth and Sun, the tail's length in the sky increased. Between mid-October and December, it grew from 24 degrees out to 70 degrees. The maximum calculated true length was 1.3 a.u. Even on December 31, with the comet now so far away as to be hardly visible to the naked eye, a section of tail could be seen.

The Great Comet of 1811 was last visible without optical aid in mid-January 1812, about 10 months after it was first spotted with the naked eye. Even subtracting the 2 months when it was hidden by the sun's glow, this is easily the longest known period of naked-eye visibility.

Next the comet passed through conjunction with the Sun a second time and was rediscovered in telescopes on July 11 and July 31. On the latter date, Wisniewsky saw it as a faint yellowish nebulosity about $1\frac{1}{2}$ arc-minutes across but did not see the tail 10 arc-minutes long observed from Cuba on July 11. The comet was then 4.8 a.u. from the sun and 3.3 a.u. from Earth and had remained visible for a timespan not surpassed until near the end of the nineteenth century, when far larger telescopes and photography were in use.

A number of tumultuous events in the United States appeared to confirm the American newspapers' speculations that the Great Comet of 1811 was an omen of evil. On December 16, 1811, what was possibly the strongest earthquake in U.S. history occurred. The epicenter was near New Madrid, Missouri, but the quake rang church bells in Richmond, Virginia, and even (we are told) in Boston, more than 1000 kilometers away. The quake also caused sections of the Mississippi River to flow backwards. In Kentucky, John James Audubon sought shelter initially, fearing that the roar and the strange brightenings and darkenings of the sky were caused by an approaching tornado. (These phenomena associated with severe earthquakes are still not fully understood.) The Great

Comet of 1811 was dimming but still had a long tail at this time, and it had been prominent all autumn.

There was much severe weather in America that year, as well as a stormy political climate leading up to the United States' declaration of war on England on June 18, 1812. The amazingly enduring Great Comet of 1811 was last seen telescopically about a month later.

Unlike all these ill effects blamed on the widely observed comet, its most famous imagined influence was a good and lasting one: Comet Wine. The weather for the vineyards of Europe that year was ideal, and for many decades the 1811 vintage was enjoyed with a nod to the comet. One English writer in 1813 mentioned that the figs, melons, and "wall fruit" were abundant and ripe in 1811 and that human fertility had also been at its greatest: An abundance of twins were born "and a shoemaker's wife in Whitechapel produced four at a birth."

The 1811 comet came perhaps a bit late for the older generation of English Romantic poets (Wordsworth and Coleridge were past their youthful prime) and a bit early for the younger generation of Byron, Shelley, and Keats. Of course, the fiery William Blake continued to write and sketch his visions, and it is not surprising that both his verse and his visual art feature some comets.

French caricature of the panic caused by the 1811 comet.

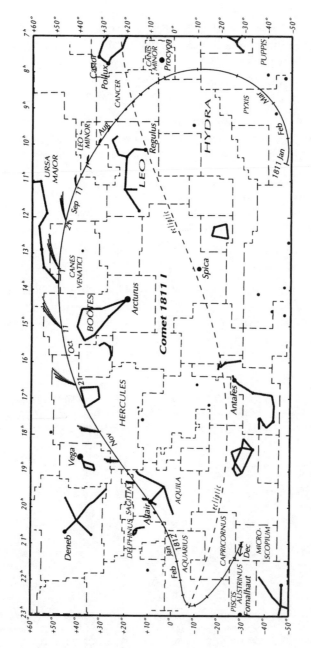

The great comet of 1811's path among the constellations.

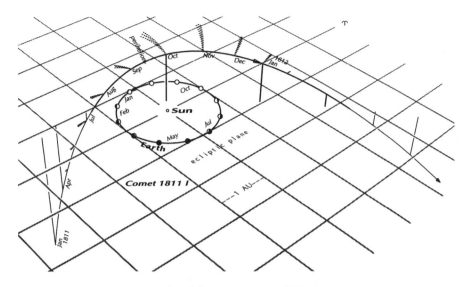

Orbit of the great comet of 1811.

Among the English Romantic poets, Shelley had the greatest interest in science (though principally as a means to the fervently hoped-for betterment of the human condition). In his poem "Epipsychidion," Shelley uses a comet as a symbol of a part of the spirit, "beautiful and fierce," whose unruliness disrupts another part but will now be brought into peace with the latter (another interpretation, not necessarily inconsistent with the first, sees the comet as Mary Shelley). But Shelley wrote elsewhere that Hell is "distributed among the comets . . . a number of floating prisons of intense and inextinguishable fire." In the same prose work, Shelley quotes Byron's line about "A wandering Hell in the eternal space," but whether he got the idea from Byron is unclear. To the poets, of course, this concept was essentially a useful imaginative device, but one wonders whether the idea was taken literally in the early nineteenth century. At any rate, Shelley was only 19 in 1811 and had not written any major poetry yet. But one of his biographers mentions that Shelley and his first bride, on their "honeymoon" in Scotland, would walk out in the evening and admire the great comet.

1819 II   Split tail; magnitude of false nucleus almost 1. This comet probably transited (passed directly in front of the Sun's disk) on June 26, and Earth may have passed through its tail sometime around then—unbeknownst to the human race, which did not discover this comet until it emerged low in the west after sunset on July 1.

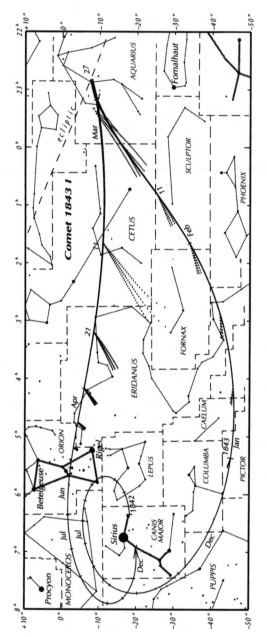

The Great Comet of 1843's path among the constellations.

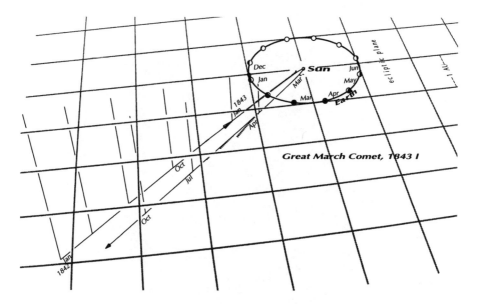

Orbit of Great Comet of 1843.

1823   Bright anti-tail.

1825 IV. PONS   Magnitude 2; 14-degree, five-branched tail.

1826 V. PONS   About magnitude $2^{1}/_{2}$ at one point; tail very long and two were seen for a while; passed only 0.03 a.u. from the Sun at perihelion on November 18, when it may have transited the Sun (just 7 years after another of these extremely rare events seems to have occurred!).

1843. THE GREAT MARCH COMET   The first of the three mightiest Kreutz sungrazers of modern times, this *great* comet was initially reported in the Southern Hemisphere at about magnitude 3 or 4 on February 5. It did not become widely known, however, until the very day of perihelion, February 27, when it was spotted little more than 1 degree from the edge of the Sun in broad daylight! It showed a dagger-like tail about 1 degree long, and Clark thought the head would have been bright enough to see in transit across the Sun's disk. For it to be so widely noted by the general public so close to the Sun, the magnitude must have been much, much brighter than the −7 often quoted. Thousands of people in various places around the world (for instance, the whole populace of Waterbury, Connecticut) beheld it near the Sun that day.

After a few more days of similar visibility, during which it was described as resembling "a flow of fire from a furnace," viewers in Bombay and on a ship in the Atlantic saw the slightly curving tail as much as 68

"Peregrinations d'une comete," by J. Grandville, from his 1844 work Un autre monde.

to 70 degrees long after sunset. Other estimates on March 4 ranged up to 90 degrees, which would indicate a tail extending 2.15 a.u. in space (this tail is usually listed as the longest on record, because the claims for the 1769 comet are often discounted).

The comet around perihelion showed little coma, and even on March 6 the apparent nucleus appeared as a planetary disk 12 arc-seconds wide with a nebulous surrounding only 45 arc-seconds across. Starting around this time there appeared the first traces of a dark lane that would later clearly bifurcate the tail. On March 13, the tail was the typical sungrazer brilliant ribbon—only 33 arc-minutes wide at 15 degrees from the head and 60 arc-minutes (1 degree) wide at a distance of 30 degrees from the head. As with Ikeya-Seki in 1965, there seem to have been very disparate magnitude estimates in this month of the "Great March Comet"; this is probably because of the comet's dissolution and the odd situation of having a tail scarcely less bright than the head.

Only on March 17 and 18 did the comet get far enough north to be universally seen in its entirety in the evening sky from most of Europe and the

United States. This was when Peters at Naples made a marvelous observation: He reported that the comet's tail was so bright that 40 to 45 degrees of it was visible straight above Vesuvius despite the presence of the full Moon in the sky! (This suggests that immense lengths of sungrazer tail could be seen even from today's largest, most light-polluted cities.)

From March 19 to March 27, Schmidt saw the tail go from 56 degrees to as much as 64 degrees (March 21) and then to 50 degrees, but it dwindled to 38 degrees by March 30. During its great nights, the tail was located below Orion and lay for a while parallel to the celestial equator. At the time of one of the most famous illustrations of it (see black and white photo insert), the comet's tail stretched from Eridanus to a bright end extending right over and beyond Sirius. As the comet headed away from Earth it faded rapidly, probably because of exhaustion from the perihelion encounter. By early April, the comet must have been lost to naked-eye view. It was last seen in the telescope on April 11, when it was probably tenth magnitude or dimmer. At that time it was 1.48 a.u. from the Sun and 1.92 from Earth.

1845 III. GREAT JUNE COMET Magnitude 0, with prominent false nucleus; tail only about 5 degrees long at best.

1847 I. HIND Seemingly not very bright before perihelion, yet observed in telescope in daylight by Hind around the time of perihelion passage.

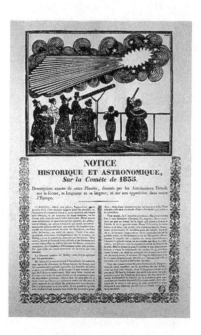

1853 comet notice.

1853 II. SCHWEIZER   Passed only 0.08 a.u. from Earth.

1853 III   Became brighter than −1 when at a small elongation from the Sun (perihelion distance, 0.31 a.u.).

1854 II   Magnitude 0; orbit similar to that of the comet of 1677.

1858 VI. DONATI   This *great* comet is often called the most beautiful ever seen.

Giovanni Battista Donati discovered it near the nose of Leo as a seventh-magnitude object on June 2. The comet did not show the first traces of tail until August 20 and did not become a naked-eye object until August 29, around which time it was far enough north to be widely observed before sunrise and after sunset. The first sight of Donati's famous tail curvature was on September 6, and at Harvard College Observatory a sudden increase in brightness was noted on September 12: The third-magnitude head contained a fifth-magnitude false nucleus, and the tail was 6 degrees long. During the next week, the magnitude improved to about 1.5, and Winnecke was the first to see one of the two straight rays which later became so prominent. On September 30, Donati's Comet reached perihelion (0.58 a.u.—almost the same as Comet Halley) and became even more impressive as it continued to approach Earth and the true length of the tail no doubt increased.

On October 2, the comet was brighter than Arcturus, and so perhaps magnitude −0.5 or −1, and the tail had grown to 25 degrees. On October 5, the false nucleus passed south of and just 20 arc-minutes from Arcturus, which shone on unimpeded through the bright tail. The comet's head was then of similar brightness to the star. By this time, however, the tail extended 35 degrees, and the entire scene became the subject of lovely illustrations, all the more wonderful because of the size and beautiful shape and structure of the tail. The broad dust tail was curved back to the west like a scimitar taller than Bootes as the comet and constellation declined in the northwest in the evening, and there were also, to the east of the broad curve, two delicate rays extending for about as far—presumably twin components of the gas tail. The tail increased in length until October 10, when the comet was closest to Earth (but still 0.5 a.u. away), and the dust tail reached 60 degrees. The true length in space was probably well over 0.50 a.u. The head or false nucleus was then about magnitude 1, a brightness it maintained through October 15, but by the end of the month the departing comet was about fourth magnitude. The last naked-eye view of it came on November 8 and the last telescopic view on March 4, 1859, when its magnitude was perhaps around 10.

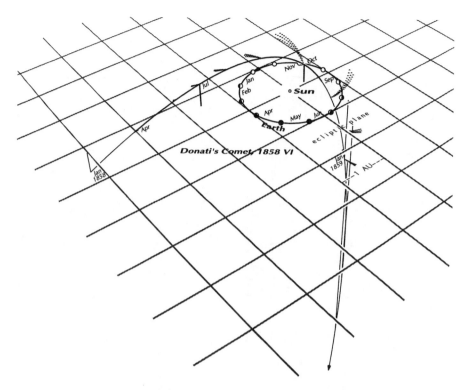

Orbit of the great comet of 1858 (Donati's).

1860 III   Many independent discoverers; magnitude 1.5; 20-degree tail.

1861 I. THATCHER   Maximum magnitude about $2^1/_2$; period about 415 years; possible parent of the yearly Lyrid meteor shower of April, which was once very strong and produced the earliest shower on record— in 687 B.C. In 1982, the Lyrids had one of their rare and astonishing revivals with rates of 75 per hour at peak and much higher for brief periods.

1861 II. TEBBUTT   One of the *great* comets. No *great* comet between the one in 1556 and Hyakutake in 1996 came as close as this one. But there were additional wonders unique to this comet. Certainly no comet in at least the past 500 years filled the sky with so broad a fan of observable tail, and perhaps there is none whose tail Earth is so likely to have passed through. And, as far as I can determine, this comet presented the angularly longest completely visible tail on record (at least in recent centuries).

Jerome L. Tebbutt of Australia discovered the comet with the naked eye on May 13, and by early June it was second magnitude with a tail 5

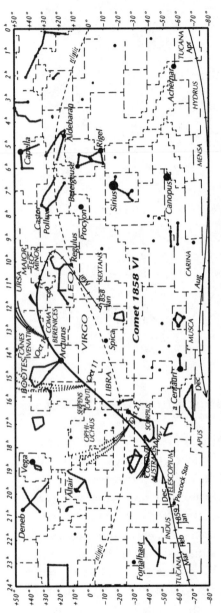

The great comet of 1858's (Donati's) path among the constellations.

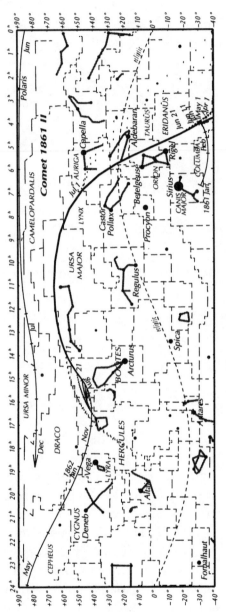

The great comet of 1861's path among the constellations.

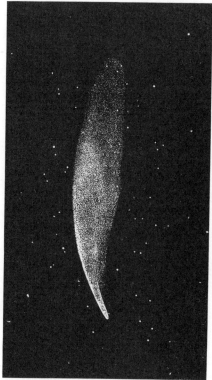

Donati's Comet on October 5, 1858.   *Donati's Comet on October 9, 1858.*

degrees long. On June 11, it would have been roughly first magnitude, and the tail was 40 degrees long, but it was lost in the Sun's glare. Perihelion occurred on June 12 (0.82 a.u.), but still no word had gotten up to people in the Northern Hemisphere about the display they could soon expect to see. The comet's orbit was inclined 86 degrees from Earth's, and the comet's distance was dwindling toward a minimum of 0.13 a.u. from us on June 30. Thus it passed unheralded at tremendous apparent speed into the view of Europe and North America.

The reaction of astronomers on these continents was absolute astonishment. Schmidt in Athens made the following report:

On Sunday the 30th of June at 8:30 p.m. a comet of enormous size appeared at the north-western horizon of Athens. The twilight behind Mt. Parnassus had not yet faded away when I was informed, and I can truthfully say no other surprise could have made so deep

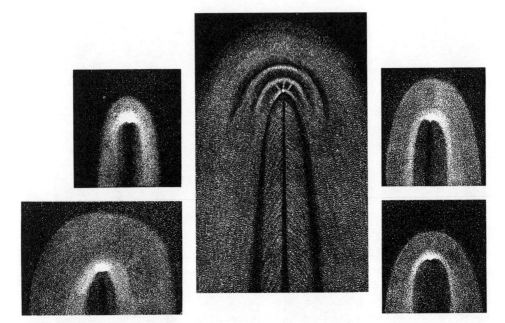

Various views of Donati's Comet.

an impression. The night before had been absolutely clear and I had not seen a trace of a comet. Now the sky was filled by this majestic figure, spreading the tail from horizon to beyond Polaris and even across Lyra. It was, to use the langauge of the past, a comet of truly fearful appearance. At 9 o'clock the head of the comet, looking as large as the moon, was next to Mt. Parnassus. The head and the very wide lower part of the tail appeared like a distant fire, and the tail seemed like windblown smoke illuminated by the fire. After the head had disappeared below the horizon and it had grown dark, one could see that the tail extended to the Milky Way in the constellation Aquila. At 11 p.m. I went to the observatory to watch [for] the reappearance of the head in the northeast. . . . At midnight and for some time after the tail stood nearly vertically above the northern horizon, its most brilliant portion and the nucleus hidden, the tail reached 30 degrees of arc beyond the zenith [AUTHOR'S NOTE: *In other words, the tail was more than 120 degrees long!!*] At 4:27 a.m. the head of the comet became visible again, following reappearance of the brightest parts of the tail which produced weak but noticeable shadows. Neither the Great Comet of March 1843 nor Donati's comet of October 1858 had been so bright. . . . I watched the rising

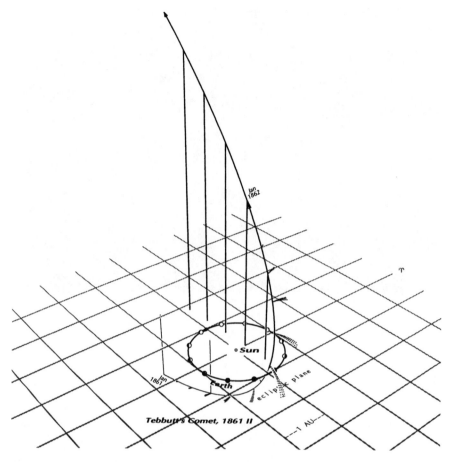

Orbit of the great comet of 1861.

of the comet's head with the naked eye; it was an incredible phenomenon that cannot be compared to anything else. The great mass of light hung like a dull smoky fire over the dark outline of the mountains. As it grew lighter the tail disappeared, I could only see about 4 degrees of arc of the tail at 5:30 a.m. But at 6:08 a.m. when Capella was the only still visible star the nucleus was still clearly luminous.

Halley's Comet in 1910 may have had an even longer tail, but never was all of it visible at one time, as the tail of this comet was for Schmidt. Many authorities list this comet's maximum brightness as 0, which is Capella's brightness. But if Schmidt, a keen-eyed and reliable observer,

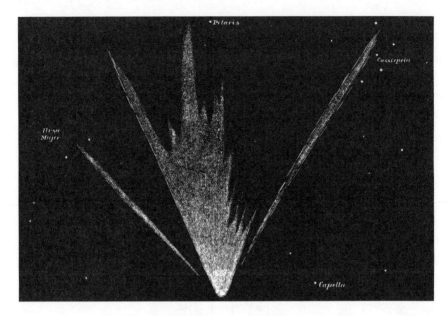

The immense fan of tail of the great comet of 1861.

saw shadows cast by the combined light of the head and the brightest part of the tail, then the total brightness would have to have been at least -1 or -2, and probably greater.

Did Earth pass through the comet's tail? On June 28, the comet had pierced up through the plane of Earth's orbit about 0.14 a.u. inside the point on Earth's orbit that the planet would reach just 2 days later. But the great dust tail—probably 0.34 a.u. in length—was lagging well behind, so there seems little doubt that we passed through it on June 30. John Russell Hind noted that he and other people saw a peculiar phosphorescence in the sky on the night of the 30th, which may or may not have been auroral (Northern Lights). British "meteorologist" E. J. Lowe recorded in his log that the sky had such a glow even before sunset that day, a yellowish tinge, and that although the Sun seemed feeble and the general illumination dimmed, the comet was first seen an hour earlier than on other evenings—actually before sunset.

Strange and magnificent views of the tail's structure were visible in those next few nights as the comet was pulling away. Father Angelo Secchi observed as many as twelve jets at one time from the central condensation, and Bredikhin saw five jets that he thought he could associate with five giant streamers he observed in the comet's vast fan of tail.

Those streamers were diagrammed by G. Williams in a drawing that shows a fan spreading across an arc of no less than 80 degrees of position angles and containing tail jets or streamers 40 to 50 degrees in length! At the time, Earth may have been within the southern part of the tail and Williams may have been viewing the head right through the tail's northern and central parts projected like a ledge toward the North Star.

The comet doubled its distance from Earth in the first week of July, and as it first pulled away, the fan closed but lengthened for geometrical reasons (our viewing angle), and observers saw it increasingly close to edge-on. Secchi reported a maximum tail length of 118 degrees on July 1 and 2, when the head was perhaps first magnitude. On one of these nights, Secchi saw the comet extended from next to Theta Aurigae (one of the shoulders of Auriga) and straight on for over 50 degrees to pass right over Polaris and then for almost 70 degrees more—through Draco and right over nearly all of Lyra except Vega and onward! Touching the north celestial pole a little less than midway along its length, this immense luminous bar must have appeared to rotate around the pole all night. Viewers at the latitude of England must have been able to see it all, even when the comet's head was directly below Polaris near the northern horizon and the end of the tail well on past the zenith in the southern sky! Curved and straight tails became visible but then shrank rapidly. Schimdt saw 57 degrees of dust tail by July 8 but only 21 degrees on the 12th, when the gas tail was still over 30 degrees. The magnitude dropped to third by July 17, and to fourth by July 24. It was last seen with the naked eye in mid-August but was viewed in telescopes until May 1, 1862, when it ought to have been approaching the orbit of Jupiter.

1862 II. SCHMIDT   Passed 0.10 a.u. from Earth.

Sketch of the coma of the great comet of 1861 by Brodie on July 2, 1861.

1862 III   Apparition of periodic comet Swift-Tuttle.

1864 II. TEMPEL   The first comet to have its spectrum observed (by Donati); passed only 0.10 a.u. from Earth, with coma half a degree wide. According to George Chambers, its central condensation looked telescopically very similar to "the well-known planetary nebula in Virgo." He must have been referring to one of what we now know are galaxies in Virgo.

1865 I. GREAT SOUTHERN COMET   A 26-degree tail, noticeably curved; passed just 0.03 a.u. from the Sun at perihelion, but not a Kreutz group member; visible almost solely in the Southern Hemisphere.

1874 III. COGGIA   This *great* comet had a maximum magnitude of about −0.5 and a tail 65.8 degrees long (as measured by Schmidt). The comet had simultaneously several marvelous asymmetrical halos or envelopes and possessed a prominent false nucleus and "shadow of the nucleus."

1880 I. GREAT SOUTHERN COMET   This was a Kreutz sungrazer (came to within 0.005 a.u. of the Sun) that faded quickly. But not before displaying a tail 50 degrees long.

1881 III   Total magnitude 1, false nucleus magnitude 2; tail 20 degrees long.

The Great September Comet of 1882 and the Moon.

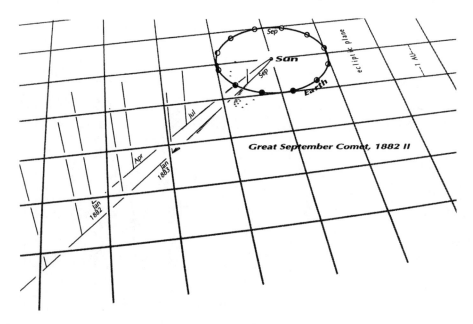

Orbit of the great comet of 1882.

1881 IV. SCHAEBERLE   About third magnitude, with a 10-degree tail; this and 1881 III were both nearly at their best when circling in the north circumpolar regions together.

1882 I. WELLS   Seen by Schmidt in daylight when 2.8 degrees from the Sun, at estimated magnitude of about −5. After perihelion (0.061 a.u.), this comet emerged to be seen as first magnitude with a 45-degree tail in twilight. This comet was not a Kreutz sungrazer, though it came in a time rich with them.

1882. OBJECT TEWFICK   See Chapter 4.

1882 II. GREAT COMET OF 1882 or GREAT SEPTEMBER COMET   First seen by Italian sailors on a ship in the Southern Hemisphere on September 1, the Great September Comet would become probably the brightest in at least 700 years when it was near the Sun. Some of its other attributes remain unmatched in the historical record.

Cruls, a professional astronomer in Rio de Janeiro, was the first to inform European astronomers of this great comet after he confirmed it on September 12. As early as September 5, it may have been as bright as Venus. It certainly must have been by September 14, when it was seen within 12 degrees of the Sun in broad daylight. On September 17, the comet passed only 464,000 kilometers from the Sun's surface—not so close as the Great March Comet of 1843, but the 1882 object was to be-

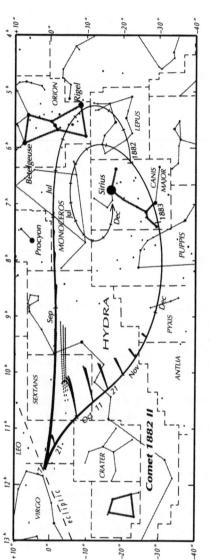

The great comet of 1882's path among the constellations.

Comet 1882 II

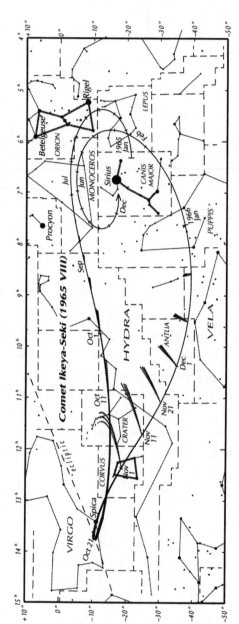

The great comet of 1965's (Ikeya-Seki's) path among the constellations.

Comet Ikeya-Seki (1965 VIII)

come much brighter! Like a few of the other great sungrazers, it could be seen right next to the Sun, with about half a degree of tail, by merely shading the eyes. But no other comet in history was followed all the way to the very disk of the Sun. Its surface brightness was roughly as great as the Sun's limb (edge), but its striking silver color contrasted strongly with the reddish-yellow of the solar limb. Elkin and Finlay compared the comet's passage in front of the Sun, as seen through a red filter, to the occultation of a fourth-magnitude star by the Moon: As the comet came to the Sun's edge, "it vanished as if annihilated." But the comet was transiting—passing in front of the Sun's face. Its nucleus was too small and far from Earth to see as a dark speck in front of the Sun.

How bright was the comet at perihelion? In light of the observations and comparison to 1965's sungrazer Ikeya-Seki, a magnitude of -15 or somewhat brighter is quite probable. As a matter of fact, brightness formula calculations suggest -17 as possible and Kronk mentions (but does not document) "laboratory studies of today" which suggest the maximum apparent brightness as between -15 and -20!

The next morning Sir David Gill at the Cape of Good Hope beheld a majestic scene, which he described in a now-famous passage:

> There was not a cloud in the sky, but looking due east one saw the tail of the Comet stretching upwards, nearly to the zenith, and spreading with a slight curve. Not a breath stirred, the sky was a dark blue almost to the horizon. The scene was impressive in its solemnity and grandeur. As the Comet rose, the widened extremity of its tail extended past the zenith and seemed to overhang the world. When dawn came, the dark blue of the sky near the point of sunrise began to change into a rich yellow, then gradually came a stronger light, and over the mountain and among the yellow, an ill-defined mass of golden glory rose, in surroundings of indescribable beauty. This was the nucleus of the Comet. A few minutes later, the sun appeared, but the Comet seemed in no way dimmed in brightness, and although in full sunlight the greater part of the tail disappeared, the Comet itself remained throughout the days easily visible to the naked eye; with a tail about as long as the moon is broad.

It is interesting that Gill's description suggests a tail length on the order of 90 degrees. Neither Vsekhsvyatskii nor Kronk mentions any estimates of total tail length for this comet in late September. Both, however, mention one of the most stunning facts of all about the comet: On the morning of September 27, a length of about 10 degrees of the tail was still visible when first-magnitude stars had already disappeared!

On October 5, Engelhardt judged the head considerably brighter than the first-magnitude false nucleus, which was still an intense yellow (presumably from sodium emission). Estimates of tail length during the month ranged from about 20 degrees on October 5 (with a dark lane down the middle) to as much as 30 degrees on October 23, by which time the comet was displaying Type I, Type II, and Type III tails. Beginning around October 16, an anti-tail was seen with the naked eye for a few days. Even as late as November 13, the brightness of the tail near the head was similar to that of out-of-focus stars of the third magnitude, and even in mid-January 1883, the tail was easily visible to the naked eye and 15 degrees long (well over 3 months with a tail in excess of 15 degrees long!). Extremely odd features of the tail included a luminous tube or envelope extending outward toward the Sun around October 9, and indeed a tube-like structure suggested by a glow around the tail was visible for quite a long time. Also unusual was the tail's shape, which was compared to the Greek letter gamma (γ): One of the ends of the tail's bifid structure was longer than the other and more curved. This shape is clear in Sir David Gill's photographs, the first really successful ones of a comet.

Also dramatic and odd to careful observers were the split nuclei. The major splitting has been calculated as having occurred on the day of perihelion, but it was not until October that many observers started seeing separate condensations. On October 13, C. L. Prince said the apparent nucleus was no longer oval but "had become a flickering column of light in the direction of the tail." The column turned out to be four or five nuclear condensations connected by a line, so that G. J. Symonds reported, "The nucleus was like a string of beads." Four of these objects were visible until the end of February 1883 and have had independent orbits calculated for them. These nuclei were not so astonishing as another feature that is thus far unique in the history of comets: the existence of fast-moving, separately tailed comets, apparently spawned—more violently than ordinary companion nuclei—when the 1882 comet passed perihelion. (There may be other explanations for these mini-comets, too, however.) Various observers who did not yet know about the multiple nuclei in the coma reported these little comets, which were (as we saw in Chapter 6) the almost unbelievable fulfillment of Barnard's dream. On October 9 to 11, Schmidt saw, southwest of the comet, "cloud-like formations" that were receding from it at about 1 degree (roughly 3 million kilometers!) per day. On October 9, Hartung saw a "cloud satellite" southwest of the comet. On October 14, Barnard saw six to eight objects that looked like miniature comets about 6 degrees southwest of the main comet, and some other observers glimpsed them too. Finally, some days after the appearance of the anti-tail, on October 21, Brooks saw a nebulous mass on the opposite side of the comet from where the others had been seen (though presumably this opposite-side position came about from a change of perspective only).

The Great September Comet remained visible to the naked eye until mid-February, but the comet was viewed in telescopes until June 1, 1883. The last view with the unaided eye came when the comet was over 3 a.u. from the Sun and $2^1/_2$ a.u. from Earth. On June 1, the figures were 4.42 a.u. from the Sun (almost out to the orbit of Jupiter) and 5.16 a.u. from Earth. Thus the Great Comet of 1882 was not only the brightest of modern times but also the most enduringly bright of the sungrazers.

1882. OBJECT SEEN NEAR SUN IN DECEMBER See Chapter 4.

1884 I Periodic comet Pons-Brooks appeared third magnitude with a 20-degree tail observed by Brooks.

1886 I. FABRY Nearly magnitude 0; 0.20 a.u. from Earth; perihelion distance, 0.64 a.u.; 10-degree tail.

1886 II. BARNARD Magnitude 3.5; 0.24 a.u. from Earth; perihelion distance, 0.48 a.u.; orbit very similar to that of 1886 I—apparently following 1 month behind that comet.

1886 IX. BARNARD-HARTWIG Coma 10 arc-minutes wide, though 1 a.u. from Sun and 1.4 a.u. from Earth; split tail, longer part about 7 degrees long; magnitude 3.3; small, bluish false nucleus.

1887 I. GREAT SOUTHERN COMET This comet, a possible Kreutz sungrazer, appeared *headless*! Tail 52 degrees long and very thin; about magnitude 1.5 (at sunward end of tail?) when first seen.

1888 I. SAWERTHAL Discovered with naked eye; double jet formation; later, split nucleus; magnitude up to about 3.5.

1889 II. RORDAME-QUENISSET 22 degrees long.

1895 IV. PERRINE First magnitude.

1901 I. VISCARA A *great* comet. Yellow coma; was glimpsed after sunrise—a daylight comet—and later had a central condensation estimated to be as bright as Sirius (-1.5). Two tails, 10 degrees and 30 degrees (the second of these shot out from nothing to full length in 3 days!). Possible split nucleus. This comet was discovered first by Viscara in

Uruguay on April 12 and then independently on April 24, the day of perihelion, by Halls in Tasmania and Tattersall in Australia. Later calculations show that it should have been discoverable several months earlier, so it must have undergone a great pre-perihelion surge in brightness not long before Viscara saw it. The perihelion distance was 0.24 a.u. It was still about first magnitude by the time it moved out of twilight.

1903 IV. BORRELLY   17 degrees long photographically; "ejected tail"—presumably a disconnection event.

1907 IV. DANIEL   Also 17 degrees long photographically, according to Kronk; second magnitude; possible disconnection events on two dates.

1908 III. MOREHOUSE   Yet another of the photographically excellent comets of these first days of really good comet photography (the earliest photographic images of comets, remember, were taken decades before). Photos of Comet Morehouse are still published to illustrate a structured plasma tail (there was apparently little dust in this comet to obscure that structure in the gas). Morehouse was most notable for remarkable disconnection events. Maximum magnitude was only 5.5.

1910 I. GREAT JANUARY COMET or GREAT DAYLIGHT COMET OF 1910 This comet truly was *great*. On the morning of January 13, 1910, workers arriving at the Transvaal Premier Diamond Mine in South Africa saw an object like a star with a tail just before sunrise. The comet was then in Sagittarius only 12 degrees southwest of the Sun. The rumor spread that Halley's Comet had been sighted, but R. T. A. Innes of the Transvaal Observatory knew that Halley was located in the opposite direction, and still much too dim to be seen under such circumstances. Weather finally cleared enough for Innes and W. M. Worssell to sight the new comet for themselves on January 17th. By then, it was just $4^1/_2$ degrees from the Sun—but it was visible throughout the day, a "snowy-white" object with a 1-degree long tail. That day it reached perihelion, 0.18 a.u. from the Sun.

A telegram from Johannesburg telling the Northern Hemisphere to get ready for a "Great Comet" was misheard by the telegraph operator, who thought it was "Drake's Comet." As the comet began to pull away from the Sun in the dusk sky it was seen around the world by not just astronomers but the public—thousands of peasants in Portugal gathered on the shore to watch it. Around January 23, various observers reported that the false nucleus was red with a yellow surrounding envelope, and that the tail was curving dramatically and was becoming bifurcated. Glorious striae like those in the dust tail of 1976's Comet West appeared in the Great January Comet (I saw rare lovely photos of this taken at Lowell Observatory when I did research at the Library of Congress some years ago), and its tail extended out to 30 degrees, with some estimates as high as 50 degrees, on January 30. The comet was still first or second magnitude, and just escaping from twilight, that day. But then a spectacular fading brought the magnitude down to 7.6 on February 4th. (Of all the great comet apparitions on Yeomans' list, only one—Halley in 1378—was recorded as visible to the naked eye for a shorter time.) The fading was more gradual after that but attention now turned to Halley. Max Wolf got one last look at the Great January Comet, shining at magnitude 16.5, on July 15, when it was out to 3.4 a.u. from the Sun. This comet is not expected to return for 4 million years.

1911 IV. BELJAWSKY   Maximum magnitude 1.0 or 1.5, tail 15 degrees long.

1911 V. BROOKS   Near second magnitude; tail 20 degrees long.

1914 IV. CAMPBELL   Fourth magnitude, with split nucleus.

1915 II. MELLISH   Fourth magnitude, 6 degrees long, with split nucleus. Not to be confused with periodic comet Mellish.

1917 I   Bright showing of periodic comet Mellish (see Chapter 7).

1927 III   Periodic comet Pons-Winnecke got bright because it passed only 0.04 a.u. from Earth.

1927 IX. SKJELLERUP-MARISTANY   A *great* comet, the last for almost 40 years (apparently the splendid 1947 XII doesn't quite qualify?). Observed in broad daylight 5 degrees from the Sun, magnitude about −6 at perihelion (0.18 a.u.); moving out into twilight, it briefly showed a tail 40 degrees long before dwindling and becoming a solely Southern Hemisphere object.

1930 VI   Periodic comet Schwassmann-Wachmann 3 passed only 0.06 a.u. from Earth.

1941 I. CUNNINGHAM   Expected to be brilliant, but turned out no brighter than magnitude 3.5—almost three magnitudes fainter than predicted. Became more and more diffuse as it moved out of view.

1945 VII. DU TOIT   The first Kreutz sungrazer in over half a century, but it must have been small because it didn't get very bright and was never observed emerging from the solar glare.

1947 XII. SOUTHERN COMET or GREAT SOUTHERN COMET   This object's perihelion distance was 0.11 a.u. Its maximum magnitude was 0, with a first-magnitude false nucleus. Its head was described as orange, its tail as 20 to 30 degrees long, its nucleus as split. When it moved into view of the Northern Hemisphere, photographs of its tail showed five components with lengths of slightly more than 4 degrees. It had faded to third magnitude by the time northerners saw it, and it continued to fade rapidly. This comet fascinated people in the Southern Hemisphere, including the child Rodney Austin in New Zealand, who grew up to discover three comets in the 1980s. Except for the next year's eclipse comet, this was the last anonymous comet to date with the word "great" in its name. It emerged suddenly at full brilliance just after sunset on December 7, so large numbers of people discovered it at the same time.

1948 XI. ECLIPSE COMET OF 1948   This comet passed perihelion— just 0.14 a.u. from the Sun—on October 27. It was first seen during the total eclipse of the Sun on November 1, when it was very bright (estimated magnitude −3) and just 1 degree 45 arc-minutes from the Sun's center with a long tail stretched toward the horizon—surely one of the most remarkable sights ever witnessed! The first person to see the comet after the eclipse was an airline pilot flying over Jamaica, when the magnitude was still probably greater than first and the tail 20 degrees long. The Southern Hemisphere was favored for viewing.

1957 III. AREND-ROLAND   Discovered by Silvio Arend and G. Roland in Belgium on photographic plates. This beautiful comet, which reached nearly first magnitude, had the most spectacular anti-tail of our century. At their longest, the regular tail stretched to 25–30 degrees long and the equally bright anti-tail to 15 degrees long!

1957 V. MRKOS   Discovered with the naked eye by Mrkos, who saw first the tail and then, just before sunrise, the head. Actually, Mrkos was only the first to get his discovery reported: Both a Japanese astronomer and an American airline pilot flying between Denver and Omaha saw it before him. Its magnitude at time of discovery was about 1. The comet maintained this magnitude for a while, but what was most remarkable about it was its dust tail. The gas tail grew to $9^1/_2$ degrees. The dust tail, up to 13 degrees long, was one of the most beautifully striated ever seen and photographed.

1961 V. WILSON-HUBBARD   The tail achieved maximum photographic length of 21 to 25 degrees, with an anti-tail of 2 to 3 degrees. This comet was independently discovered by numerous people around the world, but the first to report it were A. Stewart Wilson, who was piloting an airliner from Honolulu to Portland, Oregon, and William B. Hubbard, a professional astronomer at McDonald Observatory in Texas. It was later learned that a South African Airways stewardess named Anna Ras had seen the comet even before Wilson, while flying over Libya. At the time of discovery, the comet's magnitude was about 3 and the tail length 15 degrees. Ras and Wilson saw the comet on July 23 (Universal Time), 6 days after it had passed perihelion only 0.04 a.u. from the Sun.

1961 VIII. Seki   Passed 0.10 a.u. from Earth, appearing about fourth magnitude but very diffuse.

1962 III. Seki-Lines   Perihelion at 0.03 a.u. Another discovery by guitar instructor Tsutomu Seki, it was also discovered independently by Richard and Helen Lines from near Phoenix. The comet was not seen at its predicted magnitude of −7.5 when 2 degrees from the Sun, but it was seen just after sunset 2 days later at about −2.5. Ten days after this, the magnitude was already down to 3 and the tail length reached a maximum of 20 degrees.

1962 VIII. Humason   This comet showed coma and violent changes when over 5 a.u. from the Sun; later, it went through five disconnection events. Its apparent magnitude rose to about 5.5–6.0, indicating very high absolute magnitude from outbursts. But when it reached perihelion, well outside Mars's orbit, it was on the opposite side of the Sun from us and in any case had begun to fade.

1963 III. Alcock   Brightened from magnitude 7 to 4 in 2 days in a flare; multiple nucleus—maybe.

1963 V. Pereyra   A Kreutz sungrazer; may have been as bright as second magnitude when first seen; tail 10.5 degrees long; possible split nucleus.

1964 VI. Tomita-Gerber-Honda   Never more than 2 degrees long visually, but as much as 30 degrees photographically—with a kink in the tail at 24 degrees out. Maximum magnitude was 4.4.

1965 VIII. Ikeya-Seki   A *great* comet. This was by far the greatest Kreutz sungrazer since the Great Comet of 1882. It was discovered by Ikeya and Seki within 5 minutes of each other on September 8. By the end of September, after calculations had confirmed that it would make a spectacular passage by the Sun, professional astronomers began preparations. NASA even planned for the Gemini 6 astronauts to observe it, but the October 25 mission was delayed for several months.

Ikeya-Seki came to perihelion on October 21 at 0.0008 a.u. from the Sun. De Vaucouleurs saw it at noon that day with the naked eye when it was 2 degrees from the Sun. It then had "an intense nucleus" and shone at an estimated magnitude of −10 with a silvery tail 1 to 2 degrees long (de Vaucouleurs saw it straight, but Elizabeth Roemer reported a 2-degree tail strongly curved some hours earlier). The nucleus split into three parts, perhaps about 30 minutes before perihelion, but two of them quickly disappeared.

How bright did Ikeya-Seki get? Some tail was still visible when the head was 10 arc-minutes from the edge of the Sun's disk. But the incredible observation of seeing the head right up to the edge of the Sun could not be achieved as it had been with the Great Comet of 1882. Ikeya-Seki's head could no longer be seen when it was a quarter of an arc-minute from the Sun's edge. Nevertheless, some estimates have placed the maximum brightness of Ikeya-Seki at −15 (many times brighter than the full Moon—compressed into a tiny apparent area of coma).

As the comet hurtled away from the Sun, its head faded tremendously, but its ribbon of tail shone with great brilliance and shot out an enomous distance. The tail extended 20 degrees into a dark, pre-dawn sky by October 24. But this was just the beginning. By October 28, the tail was up to 45 degrees long—still not the longest it would get. On the morning of October 29, the comet and its majestic tail were not so far south yet as to be unviewable from Canada—for that is where, after many cloudy nights, young David Levy got a view of it, his first view ever of a comet.

But the comet was becoming a Southern Hemisphere marvel. According to Yeomans, the head was brightest outside of twilight back before perihelion on October 15, when it was second magnitude with a 10 degree tail, so the head must have been less bright as it moved out into a dark, pre-dawn sky after perihelion. But Australia's David Seargent has the head at something like second magnitude on October 31, when the tail reached a maximum angular length of up to 60 degrees, the first 30 degrees or so as bright as the head. The maximum physical length the tail achieved has been estimated at 1.3 a.u. Seargent says the tail had a bright central core in which there was a dark lane running for most of the tail's length. A faint anti-tail and some striae were also reported; a few weeks after perihelion, the primary nucleus was again joined by several companions. By 2 weeks after perihelion, most observers estimated the comet's head as already faded to fifth magnitude. But Seargent and other observers with excellent skies in the Southern Hemisphere kept the head in sight for more than a month after perihelion, apparently because they could detect more of the now diffusing outer coma against the sky background.

1969 IX. Tago-Sato-Kosaka My first comet. Third magnitude for a month (at first for Southern Hemisphere observers only), with a maximum tail length 15 degrees, probably a split nucleus. It was the first comet shown (by OAO-2 satellite) to have a huge hydrogen cloud visible in ultraviolet light.

1970 I. DAIDO-FUJIKAWA   Its perihelion distance was 0.066 a.u., but it was not a Kreutz sungrazer. Kronk reports that its orbit "bears a striking resemblance" to that of the Great Comet of 1577—but that historic comet did not come nearly so close to the Sun. Observations of Daido-Fujikawa were greatly hampered by its angular closeness to the Sun. It was never observed at great brightness, though estimates suggested it should have rivaled the brightness of Venus when it reached perihelion on February 15. If it had done so 3 weeks later, it might have been seen during a total eclipse of the Sun.

1970 II. BENNETT   This *great* comet was discovered on December 28, 1969, by John Caister Bennett in South Africa, who was using a 12-cm Moonwatch apogee telescope to search for possible Kreutz sungrazers when he came upon this small eighth magnitude object in the far south in constellation Tucana. The comet had brightened to fifth magnitude by early February and then third magnitude by early March. The tail had grown to 10 degrees by mid-March and then, as the northbound comet passed through twilight, it brightened to near magnitude 0 and reached perihelion, 0.54 a.u. from the Sun, on March 20th.

In the following week, Comet Bennett emerged from twilight and gained altitude rapidly for Northern Hemisphere viewers. It maintained its brightness and displayed spectacular orange pinwheel formations of jets to telescopic observers. In the first week of April, the first-magnitude comet displayed hoods which reminded John Bortle of those observers had sketched in Donati's Comet in 1858, and a "shadow of the nucleus" was prominent. Comet Bennett was now passing between the Great Square of Pegasus and Cygnus and although the length of its tails—about 10 degrees up to 19 degrees—was not as much as for some great comets, this was easily made up for by their brightness and spectacular structure. Beside the luxurious dust tail, the richly streamered gas tail from night to night appeared, disappeared, bent, twisted in the solar wind coming from a Sun near the maximum in the roughly 11-year cycle of solar activity. When the Moon waned in late April and early May, Comet Bennett was fourth magnitude but its tail was still long and the comet soared high among the main stars of Cassiopeia in the northeast for much of the night. During May, Bennett faded below naked-eye visibility. Comet Bennett was last photographed on February 27, 1971, when it was 4.9 a.u. from the Sun and 5.3 a.u. from Earth. Comet Bennett has been calculated to have an orbital period of about 1,680 years, which suggests that it may have been seen several times in history.

1970 VI. WHITE-ORTIZ-BOLLELLI   Co-discoverer Emilo Ortiz was an airplane pilot. This comet was a Kreutz member. It reached perihelion just 0.009 a.u. from the Sun on May 14. It was discovered 4 evenings later, only 12 degrees from the Sun, as a first-magnitude object with a 1-degree tail. Soon after, a 10-degree tail was seen, but the comet faded very quickly, and when it had become a telescopic object, only a tail (no distinct head) could be seen.

1973 XIII. KOHOUTEK   The most notorious disappointment in comet history. Even 22 years later, many people at a public observing session I attended immediately remembered this public relations fiasco. The comet had originally been projected as rivaling Venus in brightness, but even brighter claims were voiced. From Earth, a brightness as great as −0.5 was observed, but only when the comet was very near the Sun—and the comet had faded to third magnitude only 5 days later! Skylab astronauts were the only ones to get a view anything like what had been promised. They were able to observe it right through perihelion passage. With a coronagraph, they photographed Kohoutek at magnitude −3 just half a degree from the Sun's disk, 12 hours before it reached perihelion (0.14 a.u. on December 28, 1973). From Earth, tail lengths of up to 25 degrees were reported, but 10 degrees or less was more typical for even experienced observers with good skies. The gas tail was very interesting photographically, and a short anti-tail was reported from Earth and space. Professional astronomers' tremendous preparation for this comet paid off in information gained. But the public would never forget the disappointment of Comet Kohoutek.

1975 IX. KOBAYASHI-BERGER-MILON   13 degrees photographic length, over 8 degrees visual; interesting gas tail activity; magnitude up to about 4.5; passed very near Mizar and Alcor in the Big Dipper's handle.

1975 X. SUZUKI-SAIGUSA-MORI   See Chapter 6 for the unusual circumstances of the discovery. This comet was also notable because there was speculation that we might pass through its tail. When this intrinsically dim comet was nearest (0.104 a.u., on October 31), it was also passing almost directly between Earth and the Sun and could not be observed for several days around that time.

1976 VI. WEST   The discovery of the *great* Comet West was announced on November 5, 1975. The comet had been found by professional astronomer Richard M. West when he was looking over photos that had been taken back in August at the European Southern Observatory in Chile. The comet brightened but was badly situated relative to the Sun for Northern Hemisphere observers in January and February 1976. Nevertheless, as Comet West approached and passed perihelion, 0.20 a.u. from the Sun on February 25th, its brilliance became so outstanding that a number of observers were able to detect it in daylight with telescopes and binoculars. On perihelion day, several observers saw a short tail on the comet in daylight, and John Bortle was even able to spot Comet West with his naked eye seven minutes before sunset. The magnitude peaked at about −3.

Comet West got noticeably higher in the east before sunrise each day in early March for North Hemisphere observers. The comet came no closer than 0.8 a.u. from Earth, on March 4th, but by that date it was escaping from twilight, still shining at magnitude 0, and beginning to display a magnificent tail. The relatively broadside presentation of the tail made its gas and dust components visible separately, with the dust tail up to 30 degrees long during the next 10 days. Most remarkable was the apparent width of the fan of dust tail and its separation into synchronic bands and striae.

Comet West's dust tail structure and its unexpected brightness were apparently both derived at least in part from a break-up of its nucleus into four pieces, fragments which became visible to skilled telescopic observers in March. As the comet got higher and higher in the predawn sky, its brightness dropped to about magnitude 2 by March 10th, about $3^1/_2$ by March 24th, and $4^1/_2$ by month's end. In early April, it cast still a few degrees of dust and gas tail up into the little diamond of Delphinus the Dolphin. The comet faded below naked-eye visibility during the month. John Bortle made the last visual observation of Comet West, judging it to be magnitude 11.0, on August 25. The final photographs were obtained in the autumn. The orbital period of Comet West is about 558,000 years.

1976 XI   Periodic comet d'Arrest made its closest and brightest approach on record.

1978 XXI. MEIER   Early observations made this comet seem to be an intrinsically bright giant, though its large perihelion and perigee distances would prevent it from reaching great apparent brightness. The comet did not live up to expectations but still reached about fifth magnitude, which computes to a rather impressive absolute magnitude.

1979 XI. SOLWIND 1   The first known sunstriker and the first comet discovered by a spacecraft. Originally called Comet Howard-Koomen-Michels. See Chapter 4.

1983 V. Sugano-Saigusa-Fujikawa   Passed 0.0628 a.u. from Earth in June, little more than a month after IRAS-Araki-Alcock's even closer approach, but this comet had very low surface brightness and was difficult to observe. On its way toward Earth, it passed right in front of the Andromeda Galaxy, M31, but this sight was difficult.

1983 VII. IRAS-Araki-Alcock   See Chapter 1 for information about this, the closest comet in over 200 years.

1985   Periodic comet Giacobini-Zinner became the first comet to be visited by a spacecraft (see Chapter 7).

1987 XXX Levy and **1988 III Shoemaker-Holt**   The second was discovered accidentally while looking for the first. They turn out to be pursuing essentially the same orbit, obviously widely separated pieces of one original comet.

1989r. Okazaki-Levy-Rudenko   This fine comet was beautifully visible extremely close to the Moon.

C/1989 XI. Aarseth-Brewington   This comet got bright as it fell into morning twilight. It was bright but perhaps unobservable deep in twilight when it was occulted (hidden) by the Moon as seen from a remote part of the South Pacific.

C/1989 Y1. Austin   Originally expected to be a zero magnitude great comet, this comet fell far short of the promise of the early brightness it had when discovered out near the orbit of Jupiter. It was apparently "new in the Oort sense," like Kohoutek, but its activity levels stalled far more drastically and irregularly. Despite the disappointment it caused, Rodney Austin's third comet was a beautiful fourth magnitude object with an interesting tail, and cruised through some scenic regions.

C/1990 M1. Levy   This comet became as bright as third magnitude, produced a very interesting tail, and was well-placed for a long time (especially in summer).

D/1993 F2. Shoemaker-Levy 9   The periodic comet which crashed into Jupiter—see Chapter 7 for a full description.

1995   Three comets of interest, before the 1996 and 1997 great displays of Hyakutake and (we hope) Hale-Bopp: The recovery of long-gone periodic comet de Vico; the great outburst of SW-3 (see Chapter 7)—and Bill Bradfield's 17th comet.

THE LATEST
GREATS

Chapter 12

THE GRAND APPROACH OF COMET HALE-BOPP

Almost 20 years had passed since 1976's great Comet West. In that time, only two comets had even touched second-magnitude brightness outside of twilight. One was the intrinsically dim and weak IRAS-Araki-Alcock, whose ultra-close passage of Earth boosted its diffuse brightness to second magnitude for only about 2 days. The other comet was Halley. But it never appeared this bright for much of the world's population, because it had to be viewed low in the sky, where it was dimmed by the haze and thicker atmosphere.

Twenty years since the last great comet. Veteran observers were beginning to wonder how old they would be when they saw again a sight comparable to Comet West or Bennett. The younger generation must have wondered emotionally, if not intellectually, whether the accounts of past comets had been exaggerated, if perhaps there hardly was such a thing as a great comet.

Then, on July 23, 1995, something happened. Two amateur astronomers—one in New Mexico and one in Arizona—were looking through telescopes at the same attractive star cluster at almost the same time when they noticed a hazy blur of light off to one side. The cluster was a fairly popular one among amateurs in a very popular part of the sky at an excellent time of year to be looking. Thus both men, Alan Hale and Thomas Bopp, were fortunate to be observing through rather large amateur telescopes in good weather on the first night that this object nudged into the cluster's vicinity. It was almost as though the mysterious blur, so far missed, was finally drifting into a field of view where it could simply not be avoided, waiting for the exclamations to ring out when the world realized what it was.

What was it? It turned out to be probably (always, with comets, "probably") one of the largest and most intrinsically active comets destined to visit the vicinity of Earth's orbit in the past few thousand years. Was it really as big and/or active as it seemed? Would that activity continue to develop as it passed in toward Earth's orbit?

There could be informed opinions, strong likelihoods—but as usual with comets, no absolute guarantee. But when Hale and Bopp noticed that blur on that summer night in 1995, they were beginning a story which would grip the world's astronomers, both professional and amateur, and even percolate out to the public's attention before long. They were beginning a story of increasing anticipation and majestic 20-month-long approach to what could be a grand cometary climax in Earth's evening sky in March and April 1997.

They were beginning the story of Comet Hale-Bopp.

A Very Unexpected Discovery

Few comets these days are discovered visually by people who aren't looking for them. Comet-hunters are vigilant, and the competition among them is heated. Yet Comet Hale-Bopp's remarkable finding was incidental, a sidelight (literally) of the casual recreational observing of a well-known "deep-sky object" (object beyond our solar system).

Of course, we could argue that a comet so atypical was likely to be found in an atypical way. Comet hunters know they're most likely to find comets that are brightened by being fairly near the Sun and that therefore appear not too far from the Sun in the sky (either in twilight or just out of it). Hale-Bopp was found in the midnight sky, where comets are out beyond Earth's orbit and normally too faint and small to be detected by any means but professional photographic searches. So distant a comet must be unusually bright to be easily noticed in even a large amateur telescope.

There is another reason why few comet hunters would have been searching this part of the sky: It is on the edge of a rich region of the Milky Way in the constellation Sagittarius, a region with not only multitudes of stars but also lots of star clusters, many of which could easily be mistaken for a comet at low magnification. In fact, if the discovery region had not been on the *edge* of this richly starred, heavily clustered part of the sky, Hale and Bopp might never have noticed the comet or at least might never have investgated it. (This suggests a tip for would-be comet discoverers that I don't think I've read or heard elsewhere: Try looking just at or just past the edge of a celestial area so rich in clusters or galaxies that most comet hunters avoid it. You might find a comet that has just rapidly brightened there or just moved out of the rich area.)

Was it therefore no surprise that Hale-Bopp was found in circumstances other than comet hunting? No. Having said all this to help explain why Hale-Bopp was not found by comet hunting, I still must return to the assessment that it was a very unexpected discovery. Hale-Bopp had probably *not* just brightened greatly. And it was *not* just moving out of the bewilderingly rich Sagittarius Milky Way; it was moving into that region.

In addition, there is a further unexpected element of Hale-Bopp's discovery: its first discoverer was one of the world's most prolific comet observers, who for once was not trying to observe a comet!

The Discoverers

On one night in early 1996, Alan Hale visually observed eleven different comets—a feat known to have been equaled by only two other people (Reinder Bouma in 1987 and Charles Morris in 1996). Hale began observing comets in 1970, and by the summer of 1995 had viewed over 200—which must be either the record or close to it. He is thus one of the greatest comet observers in the world, a man whose tireless efforts have provided invaluable data for comet scientists. But like John Bortle and Charles Morris, the other greatest veteran observers who spend so much time estimating the brightness, size, degree of condensation, and other features of comets, Alan Hale had never discovered a comet of his own.

Hale considers himself both a professional and an amateur astronomer, and strangely enough, it is as an amateur that he is a world-class comet observer. (This demonstrates yet again that astronomy is one of the few remaining sciences in which amateurs can make tremendous contributions.) As a professional astronomer, Hale specializes in the study of Sunlike stars and the search for other planetary systems. His "side interests" are comets, near-Earth asteroids, and the development of space flight.

Hale is a native New Mexican who has intermittently spent time elsewhere, including 2$^{1}/_{2}$ years at that mecca of space science, the Jet Propulsion Laboratory in Pasadena, California. He was a Deep Space Network engineer at JPL and served as a visual observations recorder for the International Halley Watch Amateur Observation Network. Hale wrote the "Comets" section of Guy Ottewell's annual *Astronomical Calendar* from 1987 to 1992, providing rich perspectives on the observed comets of the previous year and on the predicted comets of the current year. During that time, he was a graduate student in astronomy at New Mexico State University, studying the formation and occurrence of planetary systems and eventually becoming Dr. Alan Hale. In 1991 his work in observing and publicizing comets was rewarded when asteroid 4151 Alanhale was named after him. More recently, he founded an independent research and education organization, the Southwest Institute for Space Research, at his home in Cloudcroft, New Mexico.

Tom Bopp has been a devoted observer of deep-sky objects for more than 25 years, yet at the time of discovery he not only didn't own a telescope—*incredibly, he had never before seen a comet!* Originally from Ohio, Bopp now lives in Glendale, Arizona, and is a shift supervisor in the parts department of a construction materials company in Phoenix. The fact that he was 90 miles from home for an all-night observing session with friends when he discovered the comet testifies to the depth of his interest in observing the universe.

Hale's Night of Discovery

The night of July 22–23, 1995, was a fine one in the Southwest—the first clear night for Alan Hale in a week and a half at his dark home site of Cloudcroft in the Sacramento Mountains of southern New Mexico. Hale says his usual routine is to go out about once a week to observe comets and estimate their brightness and other attributes. This particular night he had two comets on his agenda. The first was periodic Comet Clark, which he finished studying a little before midnight Daylight Saving Time. He figured he had about an hour and a half to wait before the second comet, periodic Comet d'Arrest, would get high enough for a good look. Consequently, Hale decided to observe some of the many fine deep-sky objects in Sagittarius with his rather large and powerful telescope, a Meade DS-16 (16-inch-diameter primary mirror). Far from the most impressive of the sights in Sagittarius is M70, a fairly small, eighth-magnitude globular cluster located just a few degrees from the similar cluster M69. (The M denotes an object originally listed by Charles Messier in his catalogue of deep-sky objects that comet hunters shouldn't mistake for comets.) M70 is located almost right on the line that forms the bottom of the Teapot pattern of Sagittarius and appears in a telescope as a round, condensed blob of pinprick stars. Hale had observed M70 as recently as 2 weeks earlier, and when he now turned his telescope to the cluster and looked through his eyepiece, he immediately noticed a fuzzy object that he was certain had not been there the last time.

Hale's first step was to check that he really was looking at M70; there are numerous globular clusters in Sagittarius, and many of them look much alike. Next Hale checked catalogues of deep-sky objects. Then he ran the comet-identification program at the IAU Central Bureau's computer in Cambridge, Massachusetts and sent an e-mail to Brian Marsden and Dan Green at the Central Bureau letting them know about a possible comet.

What did Hale need to turn his "possible comet" into a certain one? Movement. The comet should move in relation to the background stars. Sometimes, when an observer finds a comet just as it is about to be lost in bright morning twilight, set behind a mountain, or get covered up by clouds, there is not quite enough time to detect its motion. Movement is needed not only to demonstrate that the object is a comet but also to determine where the comet will be later—which way it is moving and at what rate of angular speed—so that the would-be discoverer can find it the next night. Even more important, recording and communicating the movement of a comet make it possible for other observers around the world to spot and confirm it within the next few hours. It's possible to lose a comet even after you've found it. But this was not going to happen to Hale's new comet. When he had verified that the fuzzy object was moving against the background of stars, he sent another e-mail to Marsden and Green. And then he was able to continue following the comet for about 3 hours before it set behind trees in the southwest. Finally, he sent yet another e-mail listing the object's initial and final positions.

What did Comet Hale-Bopp look like to Alan Hale on that night of discovery? I got several looks at the comet with smaller telescopes in the next few months, but not until October did I get a chance to observe it with precisely the same model of telescope that Hale used. It was no doubt a little brighter when I saw it and may have been in a different stage in its ongoing series of outbursts. But I probably got a view not greatly different from Hale's. Whereas with much smaller telescopes the fuzziness around the brighter center was difficult to detect, in the hefty DS-16 this coma was plainly visible. You wouldn't have had to be an expert observer to suspect you had something when you saw that view.

Hale himself appreciates what he calls the "irony" of the situation: he, a man who had spent thousands of hours looking for comets and at them, and about 400 hours specifically looking to discover a new comet, at last discovered a new comet—when he wasn't even looking for one.

Bopp's Tale of Discovery

That same evening of July 22, Tom Bopp had driven out with friends to a place called Vekol Ranch. This desert dark-sky site is about 90 miles south of Phoenix, just west of Stanfield, Arizona, and a few miles south of Interstate 8. It is very nearly the same latitude as Hale's Cloudcroft observatory. The group observing with Bopp included Jim Stevens, Kevin Gill, Bernie Sanden, and several other people. Bopp's friend Stevens had brought along his home-built $17^1/_2$-inch Dobsonian telescope. With this and other telescopes, the observers began looking at such famous deep-sky wonders as the Veil Nebula and North America Nebula in Cygnus. Then Jim Stevens said, "Let's look at some of the globulars in Sagittarius." The first ones they viewed were M22 (the most glorious globular in this wonderful region) and M28.

At about 11:00 P.M. local time, when M70 was in the field of the big $17^1/_2$-inch, the moment of discovery came. This was almost exactly the same time as Hale was first seeing the comet, because Hale's midnight Mountain Daylight Time was equivalent to 11 P.M. Mountain Standard Time (New Mexico uses Daylight Saving time, but Arizona does not).

Jim Stevens had gone to the star charts to figure out what the next deep-sky object to look at might be, and Bopp was at the eyepiece. Bopp watched M70 slowly float across the field (because of Earth's rotation). When M70 had drifted about three-quarters of the way across the field, a much dimmer patch of glow appeared at the eastern edge. Bopp moved the telescope to center the mystery object and found that he could not resolve it into individual stars. Was it a distant globular cluster or something else? Bopp called to Stevens and asked him whether he knew what this object was. The fuzzy spot of glow was only about 15 arc-minutes (about half a Moon diameter, well within a medium-power field of view) to the east-northeast of M70 that night. Stevens looked and announced that he wasn't familiar with this object. He then proceeded to check two star atlases to see whether there was supposed to be any cluster, galaxy, or other faint deep-sky object in this position. There wasn't. Bopp writes,

> The moment Jim said "We might have something" excitement began to grow among our group and I breathed a silent prayer thanking God for his wondrous creation.

Bopp's friend Kevin Gill next used his digital setting circles to determine a position for the mystery object.

But Bopp cautioned that they needed to check the object for motion. It was now 11:15 p.m. MDT, and they decided to watch for motion for an hour. That was all they needed. At 12:25 a.m. Bopp decided he should drive the 90 miles northeast to his home in Glendale to try to report to the IAU Central Bureau what they all knew must be a comet. Unfortunately, once home, his first attempts to send a telegram failed because he was using an incomplete address. But a search of his library eventually turned up the full address and Bopp got his message off.

At 8:25 the next morning—July 23, 1995—Dan Green telephoned and said, "Congratulations, Tom, I believe you discovered a new comet." And, writes Bopp, "that was one of the most exciting moments of my life."

Giant Comet or Temporary Outburst?

Anybody would be thrilled to learn that he or she had really discovered a comet. But the thrills were just beginning, for the new Comet Hale-Bopp was more than just an ordinary comet.

Hale, the comet expert, must have noted that first night that the comet's motion was unusually slow; Bopp and his friends probably noticed this, too. And they must have known that this probably meant the comet was a very distant object. Yet it was between about tenth and eleventh magnitude in brightness, already within reach of fairly small amateur telescopes. How distant could it be?

Later on July 23, IAU Circular 6187 announced the discovery of C/1995 O1, Comet Hale-Bopp. Observations were already coming in from Japan and Australia, and Dan Green was able to calculate a preliminary orbit. It turned out that Comet Hale-Bopp had been discovered at the incredible distance of over 7.2 a.u. (about 670 million miles) from the Sun and 6.2 a.u. (about 570 million miles) from Earth. That 7.2 a.u. is the farthest out a comet has ever been discovered by amateur astronomers—almost halfway between the orbits of Jupiter and Saturn. Yet Comet Hale-Bopp was easily bright enough to be detected visually at that distance! Such remarkably bright-early comets as Comet Kohoutek in 1973 and Comet Austin in 1989 paled by comparison. But the names of Comets Kohoutek and Austin immediately evoked the embarrassing fizzles those comets made after initial great promise. Could Hale-Bopp, despite its tremendously promising beginning, also end up a flop?

We are getting ahead of ourselves here. The first question about Hale-Bopp was how close it would get to the Sun. Might it be like the intrinsically brilliant Comet Sarabat of 1739 and never even go very far within the orbit of Jupiter?

The good news was that its perihelion distance would be about 0.91 a.u.—slightly within Earth's orbit. This is somewhat farther out than we'd ideally like if we want Hale-Bopp to get most vigorously stirred and lit by the Sun. (One reason Kohoutek was expected to get so bright was that it was to travel far within the orbit of Mercury.) But another way Hale-Bopp could get extremely bright would be to pass close to Earth. Will it? No. The early calculations, now verified, showed that the closest we would come to Hale-Bopp was 1.31 a.u.—pretty disappointing.

But the calculations of Hale-Bopp's probable apparent brightness were not at all disappointing. So great was the intrinsic brightness at discovery—due to size and/or activity—that the standard brightness formula predicted the comet could in late March 1997 peak at either magnitude −2 (as bright as Jupiter!) or magnitude 0 (rivaling all but the brightest star). The formula predicted nearly as great a brightness for Hale-Bopp for several straight months. And the orbital calculations showed that all this time the comet would be far enough from the Sun in our sky to be visible, with the best part of the show probably coming when the comet was conveniently located in the evening sky for several hours after sunset each night.

A number of us comet enthusiasts looked at these figures—the distance from Earth and Sun contrasted with the expected apparent brightness, certain aspects of the orbit—and immediately thought of only one other comet: the Great Comet of 1811. Yet at least one hurdle had to be cleared before we could even begin to speculate that Hale-Bopp's showing would ever resemble the Great Comet's: We had to make sure that Hale-Bopp was not undergoing at the time of discovery a major outburst in brightness from which it would quickly fade and stay many magnitudes fainter.

Not Just an Outburst

Had any comet ever exhibited occasional giant outbursts and flares of brightness at a distance similar to that at which Hale-Bopp was discovered? Yes indeed. Comet Schwassmann-Wachmann 1, which you may recall from several chapters back. SW-1 is believed to be a very large comet, but as it pursues its orbit, more circular than almost any other comet's, between the orbits of Jupiter and Saturn, it occasionally undergoes outbursts which increase its brightness tremendously—for a short while. As a matter of fact, the maximum brightness—about magnitude 10—that SW-1 can hit is approximately that which Hale-Bopp showed at discovery.

The longer Hale-Bopp maintained its brightness, the more likely it wasn't undergoing an outburst of the kind that SW-1 exhibits. It might take many weeks, we all knew, before we could relax about this concern. But then, only about 10 days after discovery, another event increased our confidence that Hale-Bopp was not just flaring temporarily.

Brian Marsden had put out a call for prediscovery images of Hale-Bopp. There was such an image on a May 28 photograph by astronomy writer and editor Terence Dickinson. But on August 2, 1995, IAU Circular 6198 announced that well-known comet discoverer and observer Robert H. McNaught had found what seemed to be an image of Hale-Bopp on a photograph taken back on April 27—in 1993! The observatory photographic plate was a 50-minute exposure showing an elogated blur of just the length and direction that Comet Hale-Bopp would have made, and it was bright enough to suggest that Hale-Bopp was not undergoing an outburst in July 1995.

Even back in 1993, the comet showed coma—about $0'.4$ arc-minutes (a little less than the apparent size of Jupiter) in diameter. The comet's total magnitude back then was about 18, with the star-like nucleus or central region about magnitude 19. All this from an object that was 13.1 a.u. from the Sun, well beyond the orbit of Saturn. McNaught could not locate the comet on a plate taken on September 1, 1991, when it should have been only about a magnitude dimmer than in 1993 and conceivably detectable on that plate.

Was the 1993 image McNaught found really Hale-Bopp? It was slightly off from where the preliminary orbit of Hale-Bopp suggested it should be. This created some controversy for months, until, in January 1996, Marsden released a statement reiterating that the image was consistent enough with an orbit derived purely from 1995 observations—there was no reason to think that this was not Hale-Bopp.

The 1993 position, when used by Marsden in an early determination of the orbit, also offered further reassurance that Hale-Bopp was likely to become a bright object in 1997. Kohoutek and Austin had fizzled because they were "new in the Oort sense": first-time visitors to the inner solar system that appeared deceptively bright at first because of the vaporization of a virgin surface supply of volatile compunds. Marsden was able to show that Hale-Bopp was not new in the Oort sense. It has a much less elongated elliptical orbit that would have previously brought it into the inner solar system "about 3000 years ago."

Comparison to the Great Comet of 1811

In early August 1995, Marsden himself went on record comparing Hale-Bopp to the Great Comet of 1811. "Taking the situation at face value, " he wrote, "[This] comet is in many respects similar to the great comet C/1811 F1 (early absolute brightness, perihelion distance, orbital inclination, 3000-year revolution period, placement beyond the sun at perihelion) and *may* perform as spectacularly." Marsden even went so far as to quote from Tolstoy's grandiloquent descriptions of the Great Comet of 1811 in *War and Peace!*

If you look again at Donald Yeomans's list of great comets (see Appendix 2), you'll see that the Great Comet of 1811 is unique in the historical record. No other comet was so far from both Sun and Earth and yet managed to satisfy—easily—the criteria of visual greatness. No other great comet has been visible to the naked eye for more than about *half* the time this one was. I quote from Yeomans's paper on great cometary apparitions:

The great comet of 1811, whose influence was said to have created a memorable vintage that year, did not make a particularly close pass to the Sun or to the Earth. However, it was an intrinsically active comet whose perihelion distance was just outside the Earth's orbit and whose highly inclined orbit kept it well above the Earth's orbital plane when it was its brightest. Thus, it was easily sighted in the more populous northern hemisphere, remaining in view to the naked eye over a period spanning 260 days.

The orbital period of Hale-Bopp, as later refined by Marsden, turns out to be not so close to that of the 1811 comet but close enough. According to Marsden, Comet Hale-Bopp last visited the inner solar system approximately 4200 years ago. Presumably because it passed about 0.75 a.u. from Jupiter in February 1996, its orbit is changed slightly so that its next visit after the present will occur about 3400 years from now. Unlike nearly all other comets, the Great Comet of 1811—and Hale-Bopp, if it is as bright as we think—must be seen with the naked eye whenever they return (assuming they do not have long inactive spells at some returns). Thus we can say with some hope of correctness that the 1811 comet was noticed at its previous return in about 1000 B.C. and Hale-Bopp in about 2200 B.C.

This is a bracing thought. Might Moses or the combatants in the Trojan War have watched the Great Comet of 1811 in their own time? More to the point, when we see Hale-Bopp we ought to be looking at a comet last seen when the Old Kingdom of Egypt was collapsing, the Bronze Age culture in Europe beginning, the Indus civilization at Mohenjo-Daro near its end, and the city-states of southern Mesopotamia only recently united under the rule of Sargon of Agade. Most exciting of all is to think of humankind's oldest monuments in relation to Hale-Bopp. When Hale-Bopp last flamed in the skies of Earth, the main phase of Stonehenge's building may have been in progress and the Great Pyramid at Khufu still fresh—only about 350 years old!

But all of these speculations become legitimate only if Hale-Bopp maintains an intrinsic brightness and activity comparable to those of the 1811 comet. (And even the dependability of the 1811 comet's intrinsic brightness has recently been questioned.)

The Boisterous Hale-Bopp

Why was Hale-Bopp so bright at such huge distances from the Sun? It was obviously bright due to the presence of its surprisingly big and substantial coma. Observations in the first weeks after discovery suggested minimal gas production but almost unprecedented levels of dust production. Of course, water ice would not yet be sublimating so far out from the Sun. But some gas must have been carrying the dust out from the nucleus. By late August 1995, it was established that, as expected, molecules of both CO and CN were present. Nevertheless, most of the light must have been coming from sunlight reflected off dust grains (possibly ice-covered ones). The dust coma extended to an elongated 2.5 million kilometers, and a separate structure that might be considered a dust tail was observed.

But what was most amazing from the opening weeks of Hale-Bopp observation onward through the autumn was the way the dust coma was produced and maintained: through semi-regular outbursts. Rather than flaring in brightness from a single or infrequent outbursts, Hale-Bopp seemed to be puffing them out one after another like a locomotive.

Professional imaging established five major outburst episodes starting at approximately the following dates: August 16, September 9, September 24, October 14, and October 31. Perhaps one of these outbursts was in progress at the time of discovery (or was ending, with the next outburst a "dud" at the end of August), and presumably more occurred after the comet began, in November, getting too close to the solar glare to be visible for a while. The comet's total visual magnitude fluctuated between about 11 and 10 during these episodes, the brightness of a false nucleus or innermost coma increasing by two or three magnitudes to about $11^{1}/_{2}$ before diffusing and fading. The first time this happened, John Bortle quite understandably thought the comet's appearance might be changed forever. But the same process happened again and again. In each episode an intense center appeared, followed by a spectacular jet that curved around to assume a spiral form—obviously as a result of the rotation of the comet's nucleus.

The jet emerging from the outburst which began around September 24, 1995, was especially impressive on Hubble Space Telescope images. The dust from this outburst, as from the other ones that autumn, is estimated to have shot out at somewhere between 100 and 175 mph (probably closer to the low end of this range). The distance out to the sharp bend in the spiral jet was about 2.7 arc-seconds (12, 800 kilometers) late on September 28 (almost 5 days after birth early on September 24) and was 4.9 arc-seconds (23,300 kilometers) late on October 2 (almost 9 days after birth). On October 5, the Wide Field and Planetary Camera of the Hubble Space Telescope obtained spectacular images of the counter-clockwise spiral jet sporting a section larger and brighter than the nuclear region itself (see this book's color insert for the image). There was speculation that this section might be a piece of the nucleus's icy crust broken off and flung out at about 77 mph by sublimation and centrifugal force. (If this speculation was verified there would be some concern that Hale-Bopp might use up its resources before it reached the inner solar system, but I don't think it has been verified). By October 14 and 15, a new jet was seen emerging from the nuclear region.

The Nucleus of Comet Hale-Bopp

If the dates of outbursts given above are correct, what is Hale-Bopp's rotation period? In the autumn of 1995, this remained uncertain.

For instance, the intervals between the proposed outburst dates were 24, 15, 20, and 17 days. This gave a mean interval of about 19 days plus or minus 4–5 days. Could the rotation period in any case be a submultiple of this ($1/2$ of 19 = $9^1/_2$ days; $1/3$ of 19 = $6^1/_3$ days, and so on)? Zdenek Sekanina suggested this is highly unlikely because it would require too many "duds" (times when the active area producing the outbursts came under the Sun but failed to fire off a jet large enough to be noticed or to brighten the comet). But might there be more than one strongly active region?

After analysis of the August outburst, Sekanina thought there was evidence of another active area, but later outbursts convinced him that such a jet was not necessary. Instead, by late October, he interpreted the data to mean that there was just one active region, located near the nucleus's equator, and that shortly after an outburst's onset around local noon (that is, when the Sun is highest as seen from the active region on the nucleus), the dust emission from this region peaked sharply and then gradually subsided, terminating at local sunset.

More revisions were necessary. On the basis of the August outburst, Sekanina calculated that one end of the spin axis of Hale-Bopp's nucleus was aimed at a point a bit southeast of the head of Cetus the Whale. But by late October, it became clear that at each outburst the position angle of the straight (initial) section of the jet was migrating, which indicated that the spin axis was precessing. (The alternative interpretation that the active area producing the jets was migrating along the surface in an orderly fashion, was considered very unlikely). Thus, like many comet nuclei, that of Hale-Bopp is apparently wobbling as it spins.

How big is the nucleus of Hale-Bopp? In the midst of the densest central cloud of dust, the true nucleus must always since discovery (and perhaps for years before) have been hidden. Our estimates are based largely on the observed production rates of dust and gas. Early estimates suggested that Hale-Bopp's nucleus could be as huge as 125 kilometers across—10 times the size of Halley's. Later, Hubble Space Telescope imagery suggested that its maximum size could be 70 kilometers but that 40 kilometers is more likely. Other measurements have indicated that it need not be more than about 16 kilometers wide (comparable to Halley's) but could be several times larger.

The Hype for Hale-Bopp

So Hale-Bopp's nucleus is not the hugest we know. The comet still may have a very large nucleus, and its combination of size and activity may turn out to be tremendous. But even if that proves not to be true, the groundswell of anticipation for this comet may not die easily.

To put it succinctly, the hype about Comet Hale-Bopp has been great. Consider that all of the following occurred 12 to 18 months before the comet's peak.

- Soon after its discovery, several print tabloids and at least one syndicated tabloid TV show did stories on Hale-Bopp. One of them mentioned a belief that every time you look at a comet, you should say a prayer or you will be cursed. A few of these tabloids alleged that Hale-Bopp might hit the Earth (it will miss us by over 120 million miles). In April 1996, someone calling in to a radio show I was on asked me about "the giant comet that is coming in 15 years and is going to hit the Earth." I said I knew of no such comet and didn't think there was one. But a minute later, when I mentioned that Hale-Bopp was coming in 1997, he interrupted—"Yeah, that's the one."
- The cover of the November 1995 issue of the venerable, dignified *Sky & Telescope* featured an early photograph of Hale-Bopp with the huge headline "Could This Become the Next Great Comet?" and in smaller type: "Believed to be 10 times larger than Halley's Comet, newly discovered Comet Hale-Bopp may grow 100 times brighter." All of this was perfectly true at the time it was written (the second half is still true as I write this page).
- The first ad for a cruise to see Hale-Bopp appeared just a few months after the comet's discovery. What would be the need for a cruise when

most of the world's population would be at a suitable latitude for a very good view of the comet? Well, there would be the benefits of escaping light pollution, getting to a clearer climate, and gathering interesting speakers. And cruises can be fun. But the first ads for the cruise concentrated on taking passengers to the optimum latitude—a northerly one—for viewing the comet. Thus they offered viewers a chance to observe the comet in March 1997 from "the pitch-dark skies" of—the raw and windy North Atlantic. By spring 1996, the destination had been changed to the Caribbean. Another company's Caribbean comet cruise, also announced in spring 1996, featured "Guests: Hale & Bopp."

- I first saw an ad for a Hale-Bopp T-shirt in early 1996. By spring, the telescope companies' ads were beginning to tie the comet to their products in a big way. One company's selection of 123 different models of binoculars offered "The Sure Cure for Comet Fever!" By spring of 1996, the first book devoted entirely to the comet had appeared. At least three others might see print.

- Richard West, writing for the European Southern Observatory on August 25, 1995, quoted a headline (source unidentified) calling Hale-Bopp "the comet of the century, if not of the milennium."

- In a fine article on Comet Hyakutake in the generally accurate and laudable magazine *Science News* (June 1, 1996 issue), Ron Cowen concluded by writing, without reserve or qualification that "Astronomers . . . predict [Hale-Bopp] . . . will be the comet of the century."

How Bright Will Hale-Bopp Get?

The question asked again and again by everyone from layperson to amateur astronomer to professional comet scientist has of course been this: How bright will Comet Hale-Bopp become?

It is a vital question. But there is no simple or secure answer. The discussion a few pages back about the comet's nucleus, size, rotation, and active areas gives you a glimspe of the complexity and mystery involved. The difficulty is greater because Hale-Bopp is such an unusual comet. Even the comparison to the Great Comet of 1811 doesn't help much, because we don't know how similar the two objects really are. The 1811 comet was not noticed until it was bright enough to be seen with the naked eye and was already in to the distance of the main asteroid belt.

And magnitude estimates were so imprecise in 1811 that it is possible that the supposedly dependable and steady comet behaved very irregularly, perhaps undergoing a major outburst at the key time for visibility.

In general, any discussion about the possible brightness of Hale-Bopp should begin with the disclaimer that we can never be absolutely sure about a comet's brightness. "Never bet on a comet" is a slogan of "Mr. Comet" himself, Fred Whipple. And David Levy made the following specific statement (which was published on Charles Morris's Comet Observation Home Page):

> Regarding people who insist that it (Hale-Bopp) will be magnitude −1.7 in April 1997: Comets are like cats. They have tails, and they do precisely what they want.

Levy was not necessarily implying that −1.7 is a particularly inaccurate figure, only that no one is justified in predicting any comet's brightness with a high degree of confidence, especially that of an unusual comet so far in advance.

Bearing in mind this disclaimer, let us consider how bright Hale-Bopp *might* get.

Although they are not the only possibilities, two different most-likely scenarios for Hale-Bopp's brightness emerged early and have persisted through May 1996. The more favorable is based on the idea that the comet will brighten with the "n" factor in its brightness equation being equal to about 4, the average for comets; the less favorable has Hale-Bopp with $n = 3$, more typical of comets that appear unusually bright at large distances from the Sun. The brighter scenario has the comet attaining a brightness of about −2, similar to the brightness of what is normally the second brightest object in the night sky, Jupiter. The fainter scenario has Hale-Bopp achieving a maximum brightness of magnitude 0—about as bright as Vega or Arcturus. Either way, Comet Hale-Bopp would easily qualify as a "great" comet by Yeomans's criteria. If it behaved according to the brighter scenario, it would be the brightest comet outside of twilight (though not far outside) since the Great Comet of 1882 and would be one of only three or four so bright outside of twilight since the invention of the telescope. To the spring of 1995, the comet tended to be as bright as the bright-scenario prediction would have it. On the other hand, a comet's rate of increase in brightness often slows as it approaches the Sun.

To keep its rate of brightening from slowing drastically, Hale-Bopp would have to (1) keep producing a lot of dust and (2) start, once it comes close enough to the Sun, to produce lots of water vapor from frozen water. When the initial estimates of huge nucleus size were reduced, some people who had hoped for a very bright Hale-Bopp in 1997 became worried. They worried more when Sekanina suggested that there seemed to be only one major active area on the precessing nucleus: Suppose that as the comet neared perihelion, the active area got precessed to a latitude that never saw the Sun (like the 6-month nights of Earth's polar regions)? Sekanina's studies of the comet prompted him to respond to the first of these concerns:

> One can infer that C/1995 O1 [Hale-Bopp] might be (relatively) CO-rich and, at the same time, dust-rich. If such CO supplies would last until perihelion, then the comet could indeed become very bright. Yet, it does not have to have an excessively large nucleus.

The answer to the second concern was that even if the initial active area became quiet (for whatever reason), there was a good chance that other regions would be sufficiently heated by the Sun to produce vigorous jets. (By summer 1996, numerous jets had in fact become visible.)

Despite Sekanina's comment, the lesser size of the nucleus and perhaps a few other factors had induced some commentators by the end of 1995 to conclude that the second scenario, a maximum brightness of magnitude 0 or 1, was more likely. But dimmer—or brighter—scenarios are not impossible. If Hale-Bopp behaves as Comet Austin did in 1990, it might be of only modest naked-eye brightness. No one really understands fizzles as extreme and erratic as Austin's. But it proved that comets can, in rare cases, nearly shut down their gas and dust production even when they are near the Sun. On the other hand, imagine the boisterous and ongoing outbursts of Hale-Bopp in 1995 continuing or even intensifying as the comet nears climax. Or imagine the fireworks (brightening, dust tail enhancement, and so on) if the nucleus were to split. If a comet already at magnitude 0 had a major flare, the display would be staggering.

Another way to assess Hale-Bopp's probable brightness was (and is) to see what happens at a series of critical points in the comet's approach. The most important checkpoint of all was earlier thought to be when Hale-Bopp would reach a distance of about 3.5 a.u. from the Sun. At that point, the surface of the nucleus should be warm enough for sublimation of water ice to begin. If the production of water vapor started feeble and remained that way, the comet would surely not follow the brighter scenario to climax. Hale-Bopp reached this key point on the inbound leg of its orbit in early August 1996. But it wasn't such a critical juncture after all: the comet's ice (perhaps ice on dust grains, which can sublime more easily) had begun to sublime in spring and early summer. This was encouraging. Charles Morris issued a new brightness prediction in late June 1996 suggesting that it was likely Hale-Bopp would reach a peak brightness between -2 and -0.5. John Bortle also sounded optimistic in the early summer.

Back and Bright

There was an earlier checkpoint at which those interested in Hale-Bopp held their breath: the return to visibility after solar conjunction. In late November 1995, the last visual observations of Hale-Bopp were made before the comet started setting too soon after the Sun to be visible in the solar glare. The comet remained hidden in that glare throughout December 1995 and January 1996, and some commentators predicted it wouldn't be recovered visually in the dawn sky until late February.

They were wrong. Gordon Garradd obtained an image of Hale-Bopp with a CCD camera and telescope on February 1, and the next morning his keen-eyed fellow Australian amateur astronomer Terry Lovejoy spotted Hale-Bopp visually. Lovejoy tends to be at the bright end of estimates among his fellow expert comet observers. But even taking this factor into account, Hale-Bopp was reappearing after solar conjunction not having faded, and not merely as bright as expected but brighter.

"Hale-Bopp is Back and It's Bright!" proclaimed the headline on Charles Morris's Comet Observation Home Page. But even as Hale-Bopp was making news with this glad return, a newly discovered comet was having its orbit calculated. As we'll see in the next chapter, it was about to steal the show from even Hale-Bopp for a few months . . . !

Chapter 13
HYAKUTAKE LIGHTNING

First it was lightning to everyone's ears.

Even as we heard that Hale-Bopp was back out of the solar glare and brighter than expected, the news came: A new comet was predicted to become a bright naked-eye object and pass 0.10 a.u. from Earth when it was visible all night and overhead for most of the world's population. While making its close pass, it would race right by the brilliant star Arcturus (as Donati's Comet and the great comet of 1618 both did), right by the Big Dipper (might it be as long as the Big Dipper then?), and right by the North Star (around which its tail would rotate like the hour hand of a 24-hour clock). And, oh yes: After the Earth-pass, its head, bigger than the Moon's apparent size, would be visible during the April 3 total eclipse of the Moon and would plunge down, with nearly optimum visibility and perhaps eventually Jupiter-rivaling brilliance, to pass 0.23 a.u. from the Sun.

If it brightened as expected, this major comet—not coming anywhere near so close as 1983's IRAS-Araki-Alcock but not a pale, diffuse, weak thing either—would be the brightest comet to pass this near Earth in 440 years.

That was the news, the news of how close and bright and midnight-passing and north-o'erleaping the comet should get to Earth and Sun in just the next few months. So first the new Comet Hyakutake—however you pronounced it—was lightning to everyone's ears.

But it would not be long before it would also be lightning to everyone's eyes. *If you could call something lasting for weeks, millions of miles long in space, seen by untold millions of people, and heart-piercingly icy blue . . . "lightning."*

The Early Approach

JANUARY 30, 1996—DISCOVERY   While I was preparing for an important meeting on light pollution, Japanese amateur astronomer Yuji Hyakutake was discovering the second comet in his life—and his second comet in 5 weeks.

He had found the first Comet Hyakutake on Christmas Day, 1995, using his enormous and powerful mounted Fujinon binoculars (each of the main lenses is 6 inches in diameter). The first comet had been a respectable one, never approaching naked-eye visibility but bright enough to be seen well in small telescopes. But the more magnificent of Yuji Hyakutake's two comets was to become an overnight wonder to the world.

Hyakutake is a photoengraver who lives in Hayato, Japan. He says his name is pronounced "hyah-koo-tah-kay"—four syllables with no greater

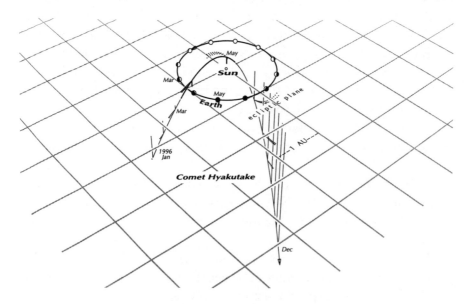

Orbit of Comet Hyakutake.

stress on any of them. Many people in Western countries had difficulty figuring out how to pronounce the name and sidestepped the issue by referring to "*the* comet" (which it certainly did become). But the modest Yuji eventually said that he didn't care how people pronounced the comet's name, only that they got out and enjoyed it.

That we would do.

FEBRUARY 2   It took a little while to confirm Hyakutake's discovery and calculate a preliminary orbit.

I first heard about the new comet in a barely audible answering machine message from a friend and then quickly got the details on *Sky & Telescope*'s "Skyline" weekly recorded phone message. I had briefly broken my habit of frequently looking at Charles Morris's superb Comet Observation Home Page, perhaps because I had to prepare extensively for another important meeting of the official New Jersey Light Pollution Study Commission. After an encouraging but exhausting day in the state capitol and a long ride home—altogether a 6 A.M. to 6 P.M. experience—I came home, hit the answering machine, and heard (just barely) how close to Earth and Sun a supposedly major comet was coming.

There are no guarantees with comets. And I was so tired that the thing felt unreal. But I sank into a chair in a state of joyful bewilderment. Hale-Bopp was coming, too, so now—after 20 years of no visually "great" comets—we had a potential two on tap within a year. Hyakutake could be (as electronic "HB" magazine publisher–editor Russell Sipe later wrote) "the herald of Hale-Bopp." It could be the stingingly good aperitif to the full-course feast of Hale-Bopp. And, I warily reminded myself, it doubled our chances of not getting shot down by another comet disappointment.

FEBRUARY 25—A FIRST LOOK   Not all the implications of Hyakutake's path (or, for that matter, exactly what its path was—for instance, how close it would come to the north celestial pole) were understood right away. After a few weeks, however, most of them were becoming evident. Now the key question was what was coming on that path: whether it would be the truly big and/or active comet everyone hoped for. The way to start answering the question was to start looking. After several weeks of bad weather, bright Moon, and other excuses, I went out this morning to get my own first look at Comet Hyakutake.

I was to meet a friend down at the dark site at East Point. But as I drove down the road at about 3:30 A.M., I couldn't resist pulling off to the shoulder to try to get a first look through a finderscope similar to 10 × 80

Table 7.   **Major events in apparition of Comet Hyakutake**

Dec. 25, 1995. Yuji Hyakutake discovers his first comet, which will become an 8th magnitude object in February.

Jan 1, 1996. Pre-discovery images of 1996 C/B2.

Jan 31. Discovery of 1996 C/B2, the great Comet Hyakutake.

Feb 24. 1 a.u. from Earth, and approaching.

Feb 27. First naked-eye sighting, by Terry Lovejoy in Australia.

Mar 12. Passes through ascending node, 1.3 a.u. from Earth.

Mar 20. Passes north over celestial equator.

Mar 21–22. American night that a jet emitted from back of central condensation is first spectacular—"a comet within a comet" for a number of days.

Mar 22–23. American night that comet is fairly near Arcturus in the evening and passing almost right over the star Izar (Eta Bootis).

Mar 23. Luminous knots of material (possible small pieces of nucleus?) first observed moving back from nucleus.

Mar 23–24. American night that the comet passes directly *overhead* as seen from near 40° N latitude.

Mar 24–25. American night that the comet is nearest the handle of the Big Dipper.

Mar 24–25. Radar signals bounced off Hyakutake's nucleus indicate it is amazingly small, 1 to 3 km in diameter.

Mar 24–26. Major ion tail disconnection event.

Mar 25–26. American night that the comet is nearest the bowl of the Little Dipper.

Mar 25. Comet passes *nearest to Earth*—0.1019 a.u., 15 million km, 9.5 million miles.

Mar 25–26. Brightest (about magnitude 0).

Mar 26–27. American night that comet passed *farthest north* (+86° 30') and *closest to the North Star.*

Mar 26–28. *Longest angular tail length* (80° to > 100°!).

Apr 3. American evening of *total lunar eclipse* and Venus-Pleiades conjunction with comet simultaneously visible.

Apr 8–16. *Great tail lengths* seen again, this time by a very few people who detected the tail out to about 60° and possibly even 80° or more.

Apr 16. Within about 1 day plus or minus of this date, a change in the behavior of the nucleus, resulting in less increase of activity and of brightness (the change is noted in both the light curve and spectroscopic observations).

Apr 24. 1 a.u. from Earth and receding.

Apr 27. Last pre-perihelion visual observation, by John Bortle of U.S.

May 1. Perihelion, 0.23 a.u. from Sun.

May 1. *Superior conjunction.*

May 5. Passes through *descending node,* 0.28 a.u. from Sun.

May 7. Farthest from Earth (temporarily), 1.24 a.u.

(continued)

Table 7. (**continued**)

May 9. First post-perihelion visual observation, by Gordon Garradd of Australia.
May 15. Passes south over celestial equator.
June 5. 1 a.u. from Sun and receding.
June 8. Secondary minimum of closeness to Earth, 1.06 a.u.
June (Late). *Last naked-eye sighting*, by Terry Lovejoy in Australia.
July 31. Maximum declination south (−70° 11′).
Circa 16,000 A.D. Comet Hyakutake (1996 C/B2) makes its next return to the
   inner solar system.

binoculars. If the comet was living up to predictions, it should be pretty easy to find. I hopped out of the car. This was a first moment of truth— the first test . . . which the comet quickly passed. There it was in a pretty wasteland of Libra, about magnitude 7.2, a surprisingly large puff of light, already showing a trace of tail. I leaped back in the car and headed on-ward with buoyant spirit to East Point.

There, where we had enjoyed the outburst of Comet Schwassmann-Wachmann 3 more than 4 months before, my friend and I now exulted about the cometary treat that was destined to come so close. (The comet had been discovered about 2 a.u. from the Sun and about $1^3/_4$ a.u. from Earth and had already closed to just under 1 a.u. from us, but it would come about 10 times closer still and, supposedly, would shine hundreds of times brighter). We looked at two other comets that night, too: the first Comet Hyakutake and Comet Szczepanski. We did not have the forti-tude to stay there for the remainder of the February night and try catch-ing Hale-Bopp just before dawn. But it was enough to see for myself that the second Comet Hyakutake was—so far—for real.

FEBRUARY 29   The comet seemed less condensed and not much brighter. Perhaps the night was poorer, but this observation concerned me a bit. Hyakutake should be getting noticeably brighter every few days as it headed for us. Fortunately, a look online at what other observers were reporting bolstered my confidence that Hyakutake was coming along nicely—always at least a little brighter than even the original opti-mistic forecast.

MARCH 4   A difficult observation in very bright Moon, but it was there.

MARCH 10–11   A much brightened Hyakutake in my telescope un-derlined all that I was hearing from expert observers and from profes-sional studies. Some of the astonishing conclusions I had been drawing about what the comet would look like near close approach, which were so remarkable I could hardly believe them, were now communicated on-line to the world, in more detail and with immensely more experience and authority, by no less than John Bortle. Hyakutake was producing roughly as much water vapor as Halley does at a comparable distance from the Sun on its way in. The comet didn't even have to increase its intrinsic brightness after this point (which it reasonably could be expected to do as it neared the Sun). Even if Hyakutake just maintained its current level of activity, its swiftly decreasing distance should make it a fine object by late March. Hyakutake looked well concentrated and continuingly active. John Bortle was beginning to say that a fine close-approach showing looked definite.

If Hyakutake was comparable to Halley and was going to come one-third closer than Halley's legendary 1910 visit—then what would be *your* conclusion? Of course, Hyakutake's would be a pre-perihelion close approach, so the dust tail would probably not be much developed. And the prominence of Hyakutake's gas tail around close approach remained a question mark. The eye is much less sensitive to the wavelengths of light emitted by gas tails than to those of the light reflected by dust tails. Most authorities were cautiously saying that a fairly long gas tail—20 degrees, with a possibility of more—might be glimpsed, but only by the most experienced observers at the darkest sites.

Even so, I think it was around now that Bortle stated, "I believe a really amazing, historic display is in the offing."

MARCH 12　My first naked-eye glimpse of Hyakutake, low just before moonrise!

MARCH 13　I had been intending to go out just before moonrise, but I fell asleep on the couch. When I woke up it was 4 A.M. The fatter than half-lit Moon was high in the sky, but would it greatly disturb the viewing of the comet? I stepped out sleepily to the deserted country road in front of my house. With all that moonlight and streetlight, what did I see down the road? The big puff of the comet was easily visible to the naked eye a few degrees from the lovely, wide double star Alpha Librae. I was no longer sleepy—although there was a dream-like majesty to the sight of the comet, with stronger tail in view. The comet had become a definite presence in the naked-eye scene of the heavens. I thought it was almost as bright as magnitude 4.0.

MARCH 14　Magnitude 4.0 with a 2 degree tail.

MARCH 15　Through and between clouds; promising.

March 16 Getting noticeably larger, magnitude 3.1. This is what a comet looks like as it heads almost straight toward Earth. Hyakutake was not much varying its position in the sky, but clearly was brightening and enlarging, suggesting that it was heading almost straight for us. Only within the final 10 days or so before reaching Earth—starting about now—would the comet suddenly seem to lift up and head north at dramatic speed to pass above us.

March 18–19 Magnitude 2.4 and a 10-degree very dim tail. A few skilled observers at very dark sites were now seeing tails tens of degrees long!

The Threshold Night

March 20–21 For me, this was the defining night that the comet proved itself and made the transition from good to great. It was the night I had, at last, a Comet West– or Comet Bennett–level experience, although that is not to say I think this comet was quite their all-around equal. To compare great comets is far more difficult than—and perhaps as ultimately inadvisable as—comparing "apples to oranges." To both beginning and veteran comet watchers there is probably nothing like a great dust tail. But Hyakutake would offer rare sights—some might not be seen again for centuries.

What did I see this night? In the evening, just a glimpse through clouds, which cost an hour of waiting at a site a mile from home. Then, after midnight, I stepped out yet again from home and, like so many people these nights, turned around to find—a giant comet flying high over my house. The words *monster comet*—entirely laudatory—came into my mind. I called my wife to step out. She said a peculiar and strangely fitting thing—the comet looked like a horseshoe crab in the sky. This region we live in is one of the best areas in the world for those odd, ancient creatures. What my wife was getting at was that the head of the comet was big and sort of elliptical to the naked eye, with a much thinner and, in the bright light streaming out from the house, short tail.

The comet was surprisingly much brighter than before. I estimated it was magnitude 0.8—brighter than Spica. There seemed to be a new giant asterism (star pattern) in the sky—not the well-known Summer Triangle, but a "Spring Triangle" however short-lived, consisting of Arcturus, Comet Hyakutake, and Spica. The rarity of a comet at this level of brightness high in a dark midnight sky gave me new appreciation for how wondrously bright those two familiar stars really are.

Now things got crazy. The clouds had broken, but they were still coming and going. For the rest of the night, the cloud coverage seemed to vary from about 30% to 90%, though some of these clouds were thin. At first I feared the entire sky would soon disappear behind clouds. While the southeast was still clear I decided to drive a mile down the road to a nice site with no nearby street lights. The view with naked eye and small optical instruments was tremendous down there, but soon the clouds were streaking into this area of the heavens, and I thought it was time for me to race home and get out one of my large telescopes. I don't usually do "absent-minded professor" things, (or at least not outstanding ones), but now I did. I drove home at more than 50 mph—over a mile—and then got out and looked in the back seat for my 7 × 50 binoculars. Where were they? They were sitting, upside-down in a precarious position, on the trunk lid of my car. How they failed to fly off I do not know. I had to drive back to the site to find their case, which *had* blown off. But all this just added to the sense of wild abandon and joy of the night!

For the 4 hours between about 12:30 A.M. and 4:30 A.M., I plied my craft with four telescopes, the rescued set of binoculars, and my naked eyes to see an incredible variety of sights. Just a few days earlier, someone had asked me what the best way to look at a great comet is—with binoculars or telescope? I told him it is best to use *all* the optical instruments, including the unaided eye; the views complement each other. But nothing could have prepared me for how true that was with Comet Hyakutake. With the naked eye, I could see the giant, already slightly blue head and a tail 18 degrees long. With low-power wide-field instruments, the comet's head was a richly blue ball of remarkable structure. For some reason, "ion rays" in the tail really stood out now, fine lines radiating back from the head. (John Bortle told me he too thought the rays were most prominent around this date. Perhaps it was simply because not much of the comet's dust had yet gotten back into the tail.)

Then I climbed up a ladder to look through my 10-inch f/7 telescope (about 8 feet tall, altogether). What would I see when I stared into it? I couldn't begin to guess. Answer: a view such as no one currently alive had previously experienced before this night. The coma washed across more than my entire field of view, so spread out that it lacked much blue. But it was bright, a luminous fog; and in the midst of it, with no apparent fanfare, there was a starlike nucleus. It seemed tremendously lonely. It had little tufts sticking out from it—giant jets. The apparent proportions of this comet's features were unlike anything this generation had ever experienced, so it was often hard to interpret what we were seeing. The longest tuft, pointing back into the tail, was not overly long or bright. But the star-like central region was golden, except during moments of excellent "seeing" or steadiness. At such times, this "false nucleus" had a distinct orange hue!

I ended the night with some naked-eye looks at the tail and the entire scene, including the head and tail and the landscape. The amazing aspect of that scene, as dawn approached, was that the tail was pointed not up or even sideways but somewhat *down* to the right—the head had just passed the central north–south meridian of the sky.

I climbed into bed weary and in one of the truest and rarest states of enchantment possible. Twenty years of waiting were over. Once again, a great comet was in the sky. And I thought how, for even the most experienced observer, a great comet can truly rewrite the book (it has literally made me rewrite this one!). It shows even the most veteran observers sights they could never have imagined were part of the visual universe.

The Nights of Close Approach

The close approach came with some of us estimating the comet as brighter than Arcturus (I judged it to be about −0.4 on two nights). One night I stood in a field with our fairly small local astronomy club and, in our sparsely populated rural area, found that we were surrounded by a crowd of about 460 people. We showed free sights through numerous telescopes large and small to more than 1000 people over that 3-day "close-approach" weekend. I did several radio interviews. I could hardly have imagined, a few weeks earlier, that I would be standing in a remote field with hundreds of people looking at an enormous, vivid blue comet while being interviewed via cellular phone by a Philadephia radio station.

Before this weekend, perhaps only thousands of living people in all the world had gotten a fairly good look at a great comet (West and Bennett were bright in the lonely hours before dawn, remember). After this weekend millions of people, perhaps a thousandfold increase, had gotten such a view. Those people had seen the Sun and Moon, had seen stars and (even if unknowingly) planets. But now for the first time, they stood there and saw, naked-eye, right in their local sky, an example of a whole other major class of solar system object—a *comet*.

For a few days, the head of the comet was bright enough for its blue to be easily visible to anyone in the country. Many people independently described the hue as "aquamarine." What in blue blazes is that? the uninformed may have asked. That blue simply burned right into your soul with an icy burn. I found this to be especially true—and the tail especially longer—late at night when the comet was most nearly overhead. Perhaps the dimming effect of the atmosphere made that significant a difference even very high, as opposed to moderately high, up the sky. Though maybe in part it was because late I left the crowds (or they left me) and I could concentrate more. It helped not to be in too wide-open a field, where too much light from the sky as a whole and even from distant city skyglows down low in the sky, could slightly hamper night vision. In any case, I will never forget how, around the time of close approach (around 2 to 3 A.M. EST on March 25), I stood in my yard after my friends had left, staring up at that icy blue head. Days later I heard or read comments by other veteran observers (such as John Bortle) that suggested a similar experience: Never before had they been so transfixed by a comet that they simply had to stand for considerable lengths of time and stare. "The most unbelievable sight I've ever seen in the sky!" said seasoned comet observer Charles Morris.

Telescopes showed various jets spouting from the comet's center, but there was one that dominated the entire view. I still don't know how to categorize it. It was always pointing back into the tail, so I suppose it was formed by the combined action of much activity on the nucleus not just one localized emission area. How amazing the change was between my quiet orange, star-like nucleus with a few tufts on the morning of March 21 and, after a cloudy night when I just glimpsed the comet through clouds briefly with low power, the evening of March 22, when the tailward jet was a giant golden streamer. It looked, many people said, like "a comet within a comet." This was also the night that sometime between 9 and 9:30 P.M. Eastern Standard Time, we saw the comet's already screamingly fast central condensation approaching a rather bright star—perhaps sixth magnitude, bright in a telescope. I was talking to some people for almost too long. "Fred, it's going to cover the star!" a friend shouted. I immediately looked in my prime telescope and saw the star go either behind or grazingly along the edge of the central condensation and then come back out.

The comet's position, high in a midnight sky, made for hours of passage through starfields of great beauty.

Other marvelous sights I saw long after midnight on the close approach nights: the comet's coma, 1 to $1^1/_2$ degrees wide, almost enveloping the lovely double star Izar as it visibly moved past it . . . the comet standing on its tail high in the west beside the constellation Bootes, and proving even taller than Bootes . . . the nights of the comet's tail searching the Big Dipper like a flashlight or Luke Skywalker's blue "light-saber," a celestial artifact even longer than the great Dipper (for a while one part of the tail seemed stuck on one part of the Big Dipper and appeared to pivot around that point, until the giant thick finger of comet began moving down the Dipper's length again. Another night the comet stretched overhead—as long as 45 degrees now—and seemingly extended right down the country road back toward my home. The changing orientation of the tail was unique in our experience. Instead of the typical sticking up from out of twilight, it seemed we saw Hyakutake's tail going every way imaginable: trekking sideways, pointing straight up, tilted slightly down at end, and so on.

And in association with these changing directions, there was the comet's victorious and decisive march across the heavens—from the southeast after midnight; to high in the east near Arcturus in late evening; to the northeast, east, and northwest through the whole night with the Dippers; and finally, in April, into the northwest just after nightfall.

The first few degrees of the comet's tail were bright enough to be seen from the largest cities. On one of the closest nights, clouds came in and made the limiting magnitude around the comet's tail about $3^1/_2$, but a few degrees of the tail were indeed still detectable to my naked eye. On the night of closest approach, the Moon was bright in the evening, yet even a nearly half Moon did not prevent me from seeing about 20 to 24 degrees of tail. In my front yard, bathed with streetlight and half-moonlight illumination, I beheld the comet's head to one side of a tall familar pine and the tail extending strongly right behind the tree-top and then out the other side—and onward. To see more than that 24 degrees of tail, one needed rather dark skies, concentration, and experience at using averted vision (see Chapter 6).

On the night the comet was nearest Polaris, I drove about 30 minutes with two friends to what we thought was the darkest field in our state. It was bumpy going, with severe ruts and ditches. But practical matters like damaging a car muffler have little hold on one's mind on such a venture.

After the Moon set, the sky proved almost as clear as it ever gets. I saw the faint counterglow, or gegenschein; a section of the "zodiacal light bridge"; and, for a while, my maximum of 70 to 75 degrees of Hyakutake's tail. Other observers saw an even longer tail. Charles Morris, in the mountains of California, saw over 90 degrees, and apparently Steve O'Meara and David Levy believe they glimpsed the tail extending for over 100° (Brian Marsden pointed out that a few of these lengths were geometrically impossible on the nights reported—but Morris has countered that the gas tail may have deviated from straightness, or the farthest section have been of a disconnected tail.) The greatest angular lengths worked out to something like 20 million miles or more of gas tail (but a physically even longer gas tail would be seen later!). The night I saw this tail, it followed a line not far from one of the sky's hour circles of right ascension—pointing almost due south from the head near Polaris. It was rather high in the west, but in seeing it, I was for a while looking west at a luminous band that ran almost horizontally, like the daytime halo phenomenon called the parhelic circle. It seemed to curve, but most of this was the result of the illusion that the sky is a dome.

Scientific Wonders and Departure

A visually observable kink raced down the plasma tail at incredible angular speed—maybe the apparent width of the moon each hour—on closest-approach night. The "kink" turned out to be an awesome disconnection event, one gas tail replacing another. But photographs from several great imaging telescopes showed clumps of material moving brightly down the tail-jet on several nights. (Eagle-eyed Steve O'Meara apparently glimpsed a beadedness to the tail-jet, so perhaps this is what he was seeing.) The two biggest shockers were the size of the nucleus and the X-rays. Nobody expected to detect a significant amount of X-rays coming from the coma—but they were detected, and at the time of this writing no theory to explain them has proved adequate. The Hubble Space Telescope might have been able to photograph the nucleus of Hyakutake if it had been about 8 kilometers across or larger. But radar images, formed by bouncing radar beams off the nucleus, revealed that it was only about 1–3 kilometers wide! Since Halley's nucleus has an average diameter of about 10 kilometers, we must assume that far more of Hyakutake's nucleus is reflective and active (or was, before its April slowdown).

What could be more astounding than knowing that all these wonders of Hyakutake came from a strange chunk of ice perhaps only a mile or two across?

What might be more astounding are some of the unusual compounds and proportions of compounds detected in Hyakutake. Ethane and acetylene in the spectrum of the comet have provoked some dramatic speculation about where and how Hyakutake formed. It has been suggested that Hyakutake may belong to a whole new class of comets, and may rewrite the book of what we believe comets are.

A few days after closest approach, the comet's brightness began decreasing noticeably, then dramatically more. On April 3, instead of being bigger than the eclipsed Moon and almost rivaling it, the comet's head was only about magnitude $2-2^1/_2$ and was surprisingly small—well, smaller. I was rather frustrated that night: not only lots of questions to answer at a public observing session but also three different sights to admire in so little time. The third (in addition to the lunar eclipse and the comet) was Venus at the very edge of the main gathering of the Pleiades star cluster. This event, which happens once every 8 years, was that night in rich-field telescopes ravishingly beautiful: a gold doubloon beside a pile of sapphires.

The comet had been expected to dim first as it moved away from Earth, but then, as it approached the Sun, it was supposed to start brightening again. John Bortle had hoped that by the last week of April and perhaps well into May—on either side of its May 1 perihelion—Hyakutake might be visible in daylight.

But it was not to be. The comet dropped past the lovely clusters of Perseus, very near one fairly bright Perseus star, then fairly near the great variable star of Perseus, Algol. A few degrees of dust tail was strengthening, and beyond it extended the gas tail—5 or 10 degrees for me under mediocre sky conditions, but for many other observers tremendously longer. I got a few fairly good nights and finally was able to trace as much as 25 degrees of very faint tail. But people like Charles Morris and especially an observer at a superb site in Hawaii were detecting, with naked eye and binoculars, 50 degrees, 60 degrees, perhaps even 80 degrees. Unlike such lengths at the close approach, these were from a comet tens of millions of miles away. Thus the physical length they suggest would be on the order of 50 million miles or more.

However much of the comet's tail you saw on April nights, the comet did form a wonderful assortment of lovely tableaus with Venus and Mercury, the passing crescent Moon, and lots of bright stars.

One final beautiful phenomenon of some of those April nights I mustn't fail to mention: I finally saw red in a comet. Initially, the color in the first few degrees of tail seemed more like brown and was more obvious in an 8-inch reflector than in small rich-field telescopes or binoculars. But on the best nights, it had a considerable ruddiness to it, a hue presumably caused by its particular combination of dust and gas. In medium-size telescopes it reminded me—with a chill of wonder—of Tycho's description of naked-eye reddishness in the comet of 1577 as resembling flame seen through smoke. That is an excellent description of what I saw.

There was still a chance that Hyakutake would start to brighten nicely. But then came April 16. That day shows up as a turning point in light-curve diagrams of the comet and also marks when certain changes in Hyakutake's spectrum occurred. The comet, despite getting closer and closer to the Sun, remained stuck at about magnitude 2 to $2^1/_2$. Predictions had been for it to reach 0 or -1 or even -2 near perihelion. A few people started to worry that it would never appear after perihelion.

But John Bortle made a sighting of the comet at magnitude 2.5 or 3 when it was only 12 degrees from the Sun. Then there was a period of unviewability around perihelion, May 1. But then recovery—at a similar brightness—from Australia, for now the comet had become (and would stay) a purely Southern Hemisphere object. Terry Lovejoy could still glimpse it with his naked eye in late June.

As I write these words, I can't imagine that I'll be in the Southern Hemisphere in the next few months. With great good-weather luck and a lot of work, I managed to see Hyakutake on 10 nights in a row and, I think, 18 of 20 nights in its best period. I saw it on a total of 32 nights.

On the evening of April 24, I and two friends at East Point, with binoculars and rich-field telescope, through some thin cirrus and therefore intermittently, beheld a ghostly pale degree to degree and a half of tail and head in the twilight glow.

And that, after two of the most astronomically electrifying months of our century, was the last I ever saw of the second and magnificent of the comets found by Yuji Hyakutake.

Chapter 14

A NARRATIVE CALENDAR FOR OBSERVING HALE-BOPP

To get a look at the much ballyhooed Comet Hale-Bopp, you may need to do nothing more than step outside on any clear night in late March or April of 1997 and gaze approximately northwest as evening twilight fades.

If Comet Hyakutake was seen by millions of people and was visible even from the middle of big cities, there's a pretty good chance that Hale-Bopp will be, too. Indeed, although Hale-Bopp when bright will not be anywhere near overhead in a fully dark sky, it will probably outshine greatly—perhaps tremendously—the display that Hyakutake put on in the northwest after sunset in April 1996.

But just stepping outside will get you only *minimal* sight and satisfaction. To capture a thousand more of the great comet's beauties and see it flame forth in as many as possible of its guises, you will need three things: good information, good observing conditions, and—if possible—some good optical instruments.

The preceding chapter about my observations of Hyakutake should give the novice some idea of the difference it can make to see a comet in a clear, moonless country sky and with a variety of optical instruments ranging from telescopes to binoculars to the most wonderful of all (especially for great comets): the naked eye. Chapter 6 gives some information on how to observe comets in general (with a few notes about features of bright comets in particular). It also recommends that every really avid comet observer become familiar with the basics of astronomy and learn how to choose, use, and care for binoculars and telescopes. Is it necessary to become a full-fledged amateur astronomer just to enjoy a great comet like Hale-Bopp? Not really, but there is an entire universe of magnificent sights that will be there long after Hale-Bopp has faded from view.

Much of this chapter is devoted to providing specific information about where and when to look for Comet Hale-Bopp, and what you should try to see. Table 8 lists some of the most important orbital and observational events. But the essence of what you need to know is in the maps and text that follow and in Appendix 8, which lists HB's[1] position, possible brightness, possible tail length, tail direction, altitude at the best time to see it, time of comet-rise or comet-set, and more, for 40° N altitude during the prime months of its visibility.

Do I really think Comet Hale-Bopp is likely to be worth all of this preparation, information, and general wordwide hubbub? I sure do, and you can read some of the reasons why in the pages just before the section of this chapter covering January 1997—the month when Hale-Bopp is not certain, but is likely, to start the 4-month (or longer) period that will establish it as a historic and splendid comet.

From Discovery Through April 1996

First of all, we will survey the period between the discovery of the comet and the initial writing of this book. Whenever you are reading this, or starting to observe Comet Hale-Bopp, you need to gain the perspective afforded by the full story of the comet's path through the sky and what was seen of it.

JULY–NOVEMBER 1995   In this period, Hale-Bopp wandered up through the Teapot pattern of Sagittarius. The comet fluctuated between about tenth and eleventh magnitude with an alternately brighter-concentrated and dimmer-diffuse center, producing jets visible in large amateur telescopes. With smaller telescopes a good finder chart was needed, for the coma often was scarcely visible and the otherwise star-like comet was lost in tremendously rich starfields. In September, the comet was highest in the south (not very high) at sunset, but by late November, it started set-

---

[1]For convenience, I've used the abbreviation HB for Hale-Bopp sporadically throughout this chapter.

Table 8.   **Hale-Bopp timetable of major events**

Circa 2,200 B.C. Previous Return of Comet Hale-Bopp.

September 1, 1991. Date of photograph which does not show HB, indicating its magnitude must have been dimmer than 19.

April 27, 1993. Date of *first pre-discovery photograph* of Hale-Bopp, at 18th magnitude (and 13 a.u. from the Sun), identified by Robert McNaught.

May 28, 1995. Date of pre-discovery photograph taken by Terence Dickinson.

July 23, 1995. H-B *discovered* by Alan Hale and Thomas Bopp, 7 a.u. out from the Sun but magnitude 10 1/2, with coma.

January 3, 1996. Conjunction (in right ascension) with the Sun for first of four important times (only 2° south of Sun this time). [See January 3, 1997 and March 3, 1997 and July 4, 1997.]

February 24. Passes through ascending node at 5.23 a.u. from Sun, 0.75 a.u. in front of Jupiter. [See May 6, 1997.]

April 14. West quadrature (90° west of Sun, in south at dawn).

May 8. Occulted by Moon as seen from much of western U.S., Mexico, Central America.

May 10–31. International Hale-Bopp Days: Period 1.

May 18. First naked-eye sighting, at estimated magnitude 7.2, by Steve O'Meara.

June 11. 6th-magnitude Comet Hale-Bopp passes 3° from 8th-magnitude Comet Kopff.

July 4. Opposition at 168.5° from the Sun.

August 3. Nearest to Earth (first time), 2.74 a.u.

September 15. East quadrature (90° west of the Sun, in south at dusk).

September 29–October 19. International Hale-Bopp Days: Period 2.

October 5. Occults 9th magnitude star as seen from western U.S.

October 28. Farthest from Earth (temporary), 3.06 a.u.

October 28–31. Brushes south of globular cluster M14.

December 5. Passes north over celestial equator. [See June 26, 1997.]

January 3, 1997. Conjunction (in right ascension) with Sun for second of four important times (28° north of Sun this time). [See January 3, 1996 and March 3, 1997 and July 4, 1997.,]

February 6. Passes south of globular cluster M71.

February 22–24. Skims southern edge of the Veil Nebula.

March 3. Conjunction (in right ascension) with Sun for third of four important times (46° north of the Sun this time). [See January 3, 1996 and January 3, 1997 and July 4, 1997.]

March 3–April 12. International Hale-Bopp Days: Period 3.

March 4. Maximum latitude north of the ecliptic, 46°.

March 9. 1 a.u. *from Sun* and heading in. [See April 23, 1997.]

March 9. Visible during total eclipse of the Sun seen from parts of Asia.

March 21. Equally high at dawn and dusk (from 40° N).

March 22. Nearest to Earth, 1.31 a.u.

(continued)

Table 8.   **(continued)**

March 23. American evening: near-total eclipse of the Moon, in some places with comet simultaneously visible.

March 23–27. Passes 5° north of M31, the Great Galaxy in Andromeda.

March 25. *Maximum declination north,* +45° 48′.

March 27. *Brightest (predicted)*, –2.3.

March 31. *Perihelion (nearest to Sun)*, .914 a.u.

April 4. Skims north of the edge-on galaxy NGC 891.

April 6–7. Passes near open cluster M34.

April 14. Skirts the southern edge of open cluster NGC 1342.

April 19. Passes between the California Nebula (NGC 1499) and the Pleiades, but in bright moonlight.

April 23. *1 a.u. from Sun* and heading out. [See first entry for March 9, 1997.]

May 6. Passes through *descending node* at 1.11 a.u. from Sun. [See February 24, 1996.]

May 8. Only 3° to 5° from crescent Moon at dusk in America.

June 26. Passes south over celestial equator. [See December 5, 1996.]

July 4. *Conjunction (in right ascension) with Sun* for fourth of four important times (26° south of the Sun this time).

August 26–September 16. International Hale-Bopp Days: Period 4.

November 29. West quadrature.

Circa 5400 A.D. *Next return of Comet Hale-Bopp.*

ting too soon after the Sun to be seen. The last visual observations before the comet became unviewable may have been Terry Lovejoy's on November 23 and Alan Hale's on November 23 and 24.

DECEMBER 1995–JANUARY 1996   HB was lost in the solar glare on the other side of the Sun.

FEBRUARY–MARCH 1996   Terry Lovejoy recovered HB low in the dawn sky on February 2, finding it easily visible in a 10-inch reflector at an estimated magnitude of 8.8—significantly brighter than expected.

HB wandered northeast under the Teaspoon asterism in northeastern Sagitarius. It passed on February 24 through ascending node. That means it went up, northward, through the extended plane of Earth's orbit—at a point about 5.2 a.u. from the Sun. (It will not drop back through descending node until May 1997, just outside Earth's orbit—so in all the interim, HB is north of the plane of Earth's orbit, a benefit to viewers in the Northern Hemisphere). HB's passage through ascending node happened to be almost right at Jupiter's orbit and, to make things far more interesting, only about 0.75 a.u. ahead of Jupiter. That is not so large a distance for a comet to pass from the massive planet, and indeed Jupiter's gravity did perturb HB's orbit slightly. (How long does Jupiter take to cover this distance, arrive where HB was, and perhaps encounter some of the comet's meteors? A little more than 3 months, I think, so any Hale-Bop-pid meteor shower in the upper atmosphere of Jupiter might have oc-curred in early June 1996.)

Coincidentally, HB passed through ascending node at a point in the sky almost exactly where Comet Kopff would go through descending node (dropping south of the plane of Earth's orbit in June); see the June 1996 entry for the conjunction of the two comets.

During February–March 1996, Comet Hale-Bopp brightened from about magnitude 9.0 to 8.0 and became clearly identifiable in 10 × 50 binoculars under good conditions. Alan Hale with his 16-inch reflector observed a faint, broad, featureless tail extending about 10 arc-minutes to the west. In much of March, comet observers were so busy watching Comet Hyakutake's close approach to Earth that they neglected HB.

APRIL 1996   HB was highest in the south at sunrise by mid-April. It began a retrograde curve first to the north only a few degrees to the east of the Teaspoon asterism (star pattern) in Sagittarius and came its closest in the sky to Neptune and Uranus.

On April 1, 1996, Alan Hale rated the comet as being magnitude 8.2 in 10 × 50 binoculars and announced "T minus one year and counting!" (to the day that Hale-Bopp reaches perihelion). By mid-April several ob-servers were seeing the tail Hale had noted as early as March. The comet took on a dramatic appearance in the second half of April. On April 18, John Bortle reported a strongly fan-shaped coma spreading out to the north with an almost star-like false nucleus at its apex and with a tail emerging slightly west of north before bending to the west. In the days that followed, observers with binoculars in very good sky conditions saw the westward tail about half a degree long (as long as the apparent width of the Moon), and the head and tail formed a surprisingly prominent V. Had one of the semiregular outbursts seen in 1995 just occurred? Presum-ably, but through spring 1996 the same experienced observers had not seen the previous year's brightness fluctuations, and it became clear that more consistent activity from a number of regions on the nucleus was now occurring.

In the latter half of April, most observers were rating HB's brightness as between 7.5 and 8.0, but in 10 × 50 binoculars Terry Lovejoy found it closer to 7.0. On April 28, Lovejoy rated the 15 arc-minute-wide coma at 6.9 in his 10 × 50 binoculars and made a remarkable statement:

"With naked eye the comet combined with a magnitude 8 star can be just glimsped." Unfortunately, in the nights after this, bright moonlight started interfering. But there seemed little doubt that the record-shatteringly long span of HB's naked-eye visibility was about to begin.

May–October 1996

The next period is that between the initial completion of this book's manuscript and its publication. I will describe it in the past tense, although as I write these words some of it is in the future, and I cannot give details on what was actually seen. The magnitude values I suggest will be rather optimistic, with occasional reminders about the more pessimistic possibilities that exist.

MAY 1996   Now that Hyakutake had fled south through the solar glare, Northern Hemisphere observers were able to turn more of their attention to Hale-Bopp. On May 8, a remarkably rare event occurred: The Moon passed in front of a seventh-magnitude Comet Hale-Bopp.

Period 1 of International Hale-Bopp Days was held from May 10 to 31. The NASA steering committee selected these periods because they found it "scientifically urgent to suggest some key periods during which systematic observations would best be carried out with multiple techniques, from the ultraviolet regime to the radio regime." The effort would be made to coordinate intensive observations of many kinds from around the world in these periods, which were selected "partly for their scientific interest (*i.e.*, for the phenomena that should occur at the particular heliocentric distances), but . . . also . . . for observability" (including the pointing constraints of various scientific satellites such as the Hubble Space Telescope).

The comet crept northwest, brightening to the naked-eye limit for skilled observers. It was first seen with the naked eye by Steve O'Meara on May 18 from Hawaii, at a magnitude of 7.2. Thus began what would almost surely be a record-long period of naked-eye visibility.

JUNE 1996   An exciting observational event—and photographic opportunity—of this month was the conjunction of Comet Hale-Bopp with Periodic Comet Kopff (an interesting comet which was originally planned to have a spacecraft following it at close range this summer!) For a few nights around June 11, the two comets were only 3 degrees apart but Kopff was about a magnitude dimmer than its predicted magnitude of about 7.3, with Hale-Bopp about 6.5 or maybe a bit brighter. Kopff was heading south and HB northwest, the former pretty far out (just beyond the orbit of Mars) but the latter several times farther. HB shot across a re-maining section of Sagittarius the Archer, leaving the constellation for the first time since discovery and entering the constellation Scutum. Near month's end, Charles Morris issued a rather reassuring prediction: Hale-Bopp seemed to be performing reliably and might be expected to reach a peak brightness of somewhere between −2 and −0.5 in March 1997.

JULY 1996   On July 4, Comet Hale-Bopp reached opposition with the Sun—meaning that it was opposite the Sun in the sky, rising around sunset, highest in the south around midnight, and setting around sunrise. The comet became brighter than magnitude 6.0 during the month, and there were numerous naked-eye sightings. However, any further tail or any new gas tail that had been developing was mostly foreshortened, pointed more or less away from Earth, behind the comet's head. During the few weeks after opposition, HB crossed through the constellation Scutum and the bright Scutum star cloud of the Milky Way, passing only about 5 degrees from the beautiful, rich, sixth-magnitude open star clus-ter M11.

AUGUST 1996   The comet was highest in the south by late evening as it continued to head northwest, now leaving Serpens Cauda and entering Ophiuchus. The comet brightened slightly, to about 5$^1/_2$ this month. The coma might have looked big because HB's distance from Earth shrank to 2.74 a.u. on August 3 as a result of the combination of the comet's head-ing inward and Earth's being on the near side of its orbit to the comet. After this, Earth started speeding away toward the far side of its orbit from the comet and would not get as close again until December.

SEPTEMBER 1996   HB was at its highest in the early evening, setting by the middle of the night. The brightness may have been about 5. In September, the comet's apparent motion in the sky slowed (this effect was due to perspective as Earth headed more nearly directly away from it toward the far side of our orbit). By month's end, HB had gone as far west in the heavens as it will ever go, to a point about 5 degrees south of the celestial equator in Ophiuchus. Observations of it were possible from much of the Americas during the total eclipse of the Harvest Moon (with Saturn at brightest near the Moon) on the evening of September 26. On September 29, the second period of International Hale-Bopp Days began, timed to coincide with some important physical events on the comet.

OCTOBER 1996   HB was highest at nightfall and was setting by the middle of the night, brightening from about 5 to 4.5, becoming a puff of light easily visible to the naked eye in really dark skies, and perhaps showing something new in binoculars and telescopes: possibly a straight gas tail along with its broader (probably still short) dust tail. International Hale-Bopp Days continued until October 19, during which time HB moved from 2.9 to 2.6 a.u. from the Sun, the distance at which Halley's Comet seemed to get activated when inbound and at which the initiation of plasma phenomena can typically start. On October 5, the comet was due to pass directly in front of a 9th-magnitude star as seen from the western U.S. HB slowly picked up apparent speed in the sky as it headed northeast in Ophiuchus. At month's end, HB brushed near the attractive but much dimmer globular cluster M14 and was at a temporary maximum distance from Earth, 3.06 a.u.

November 1996: Brightening at the Edge of Dusk

We now move to the future tense.

During November 1996, Comet Hale-Bopp will get lower in the south-southwest at each nightfall. But even near month's end, it will be about 20 degrees (about two fist-widths at arm's length) high as evening twilight ends around 6 P.M. (assuming that you live in the middle latitudes of the Northern Hemisphere—the United States or southern Canada, Europe, Russia, China, or Japan). How impressive it will look depends on three things: how bright it is (of course), whether the dust tail of this apparently very dusty comet has grown a bit longer, and whether the gas production has been robust enough to shoot out an already lengthy gas tail.

An optimistic projection would have HB brightening from about 4.5 to 3.8 in November. That would make its head in late November considerably dimmer than Hyakutake's was when the latter was about this high at dusk's end around April 11, 1996. But Hale-Bopp will still be well over 2 a.u. from the Sun, as far out as the main asteroid belt, and 2.9 a.u. from Earth.

There's little doubt that HB will be a fascinating object in binoculars and in the smallest of telescopes in November. Conspicuous changes in the head and tail should be visible each week, maybe even each night. Will a parabolic shape to the coma already be evident? Will various

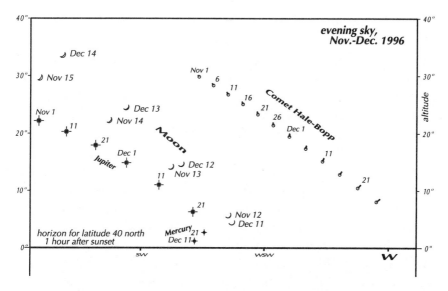

November–December 1996 sky locations of Hale-Bopp and bright solar system objects an hour after sunset at 40 degrees N (exact positions are plotted for 75 degrees W longitude, center of Eastern Time Zone). Tail lengths are fairly speculative.

bright parts of the head and tail look golden or blue in medium-size telescopes? On the pessimistic side, if HB does not have at least a few degrees of prominent dust tail or a large coma, the novice will have trouble finding it without careful reference to the accompanying diagram. The comet spends November about 8 degrees south of the magnitude-2.8 star Beta Ophiuchi (also called Cebalrai).

December 1996: First Evening Exit

The brightness of HB this month is absolutely vital to its visibility. On the American evening of December 6, the comet lies 13 degrees (more than a fist-width) high in the sky just south of due west as evening twilight ends (here taken to mean that the Sun is about 16 degrees below the horizon). But it's only 3 degrees high at that time (roughly 6 P.M. local time) on the evening of December 27. Three degrees high is about the lower limit for viewing even a zero-magnitude planet with the naked eye under any but extraordinarily clear atmospheric conditions—and even a very optimistic prediction would have Hale-Bopp no brighter than about 2.0 at this point. The best time to look by late in December is actually a little earlier, before twilight is quite finished but while the comet is a little higher. HB might then be almost as bright and high as Hyakutake was around April 23, 1996—perhaps ghostly pale with a degree or so of tail in binoculars. Again, there's always a chance it will be decidedly brighter (comets can do almost anything!), but the chances seem better that it will be somewhat dimmer (about magnitude 3.0?), in which case all but the most veteran observers will have lost it by near month's end.

Early in the month HB will still be a fascinating object (especially with some optical aid), and its disappearance from visibility will be short-lived. Although we are moving around to the far side of our orbit from Hale-Bopp, as we did last winter, this time the comet is much farther above the plane of our orbit around the Sun (and appears at a higher angle above that plane because it is now much closer). In other words, HB has lifted up high enough above our line of sight to the Sun to pass well north of the Sun in the heavens. HB is heading to the dawn sky, where it will pull away from the Sun and back into visibility—brightening visibility.

Hale-Bopp passes north of the celestial equator on December 5 and then spends the month cruising northeast along the border of Ophiuchus and Serpens Cauda—with few stars brighter than itself nearby. By month's end, the comet has pulled to within 2.6 a.u. of Earth (that's mighty far) and to within 1.8 a.u. of the Sun. Any gas tail is likely to be short and will point roughly northeast in early December and north in late December.

Why Comet Hale-Bopp Is Likely to Be Grand

Let's pause at the turn of the year to consider what will lie ahead—for we are about to enter into Hale-Bopp's long period of prime display.

Can I begin by guaranteeing that Hale-Bopp will put on a good show from January or February to May? No, there is no guarantee, written or otherwise, that comes with these things called comets. What I think I can do is to argue that *if* the comet doesn't have a lengthy version of one of those rare slowdowns of overall activity that do sometimes occur, then it will put on a very good show. "A very good show" means at least weeks of prominent naked-eye visibility with at least a short length of bright tail visible to anyone with fairly favorable sky conditions (no worse than moderate light pollution, moonlight, or haze). For viewers at mid-northern latitudes, the comet's positioning will be good enough for long enough to ensure that even with the more pessimistic of the two major brightness forecasts, we will be more than all right: We will be treated to prominent and beautiful sights for several memorable months.

The only thing that could spoil this forecast would be Hale-Bopp's shutting down the way Hyakutake seems to have done in mid-April. If Hale-Bopp shows a similar reduced activity level during its entire prime period, then it will never be more than a modest naked-eye object. That's possible. But such a shutdown would have to last not just for a few weeks but for the entire period HB is, let's say, less than 1.5 a.u. from the Sun— and that is from late January to mid-June! (Actually, the comet gets too close to our line of sight to the Sun to see properly by the end of May, but we are still talking about 3 to 4 months.)

How impressive will HB appear? For now, we'll use only the mildly pessimistic brightness predictions (in which HB never gets any brighter than about magnitude 0.4). Ignore, for the moment, the magnitude −1.7 or −2.3 that some optimistic projections have suggested HB will reach (even though, as I write these words, these greater brightnesses can by no means be ruled out). Now ponder the following, calculated for 40° N latitude but roughly similar for all mid-northern latitudes:

- Hyakutake was brighter than magnitude 1.0 for roughly 7 days, but Hale-Bopp should be that bright (according to the more pessimistic formula) for about 7 *weeks*. (Admittedly, the earlier comet's display was high in the sky, but even if we dim Hale-Bopp's magnitude by about 0.5 to 0.7 to account for "atmospheric extinction" down at HB's altitude, HB still manages to appear nearly as bright as 1.0 for about a month.)
- Hyakutake was brighter than magnitude 2.0 for about 12 days. Hale-Bopp should be that bright for about 70 days or, factoring in atmospheric extinction, about 60 days.
- Forget Hyakutake (maybe you never saw it, didn't get a good look at it, or never even heard of it until you read the previous chapter!) and consider Hale-Bopp's visibility from January to June in absolute terms. HB rises more than 1 hour before the start of morning twilight or sets more than 1 hour after the end of evening twilight (twilight being defined here as when the Sun is between 0 and 16 degrees below the horizon) all the way from mid-January to early May; more than 2 hours, from early February to late April; more than 3 hours at best.

Remember that the foregoing comparison to Hyakutake was based on a mildly pessimistic brightness forecast. If HB follows the optimistic forecast, it's really not possible for anyone alive today to imagine the wonders that will be visible. (I hasten to add that if HB falls quite a bit short of the mildly pessimistic forecast, it will seem like a mighty poor—albeit long-visiting—cousin of the dusty giants of the 1970s, West and Bennett.)

West and Bennett had tails: my, they had tails. To veteran comet enthusiasts as well as to the public, the tail could make or break Hale-Bopp as an object of great wonder. Whether HB will have a very prominent tail or tails is hard to predict. At the time of this writing, the answer is that yes, the dust tail is *likely* to be very prominent, bright, and dense, though it is far less likely to be long. But as I think back to 1970's great Comet Bennett with its 10- to 20-degree tail, I realize that if Hale-Bopp has a tail anywhere near that bright and dense and of no greater length, absolutely no one will be unhappy. (Hale-Bopp *might* produce a dust tail longer than 20 degrees, but only if that tail gets an early start and grows very long in space before it gets foreshortened by our viewing angle later on.)

To summarize, if you demand negative-magnitude brightness, an extremely long (as opposed to a quite bright) tail, or certain other marvels, HB may (or may not) disappoint you. But would you be satisfied with three or more solid months of a bright comet prominent (in fairly good conditions) for a few hours before dawn and/or after dusk every clear day, with an ever-changing, sometimes "jewel"-bearing, possibly colorful, large head and a rich, densely luminous, structured, two-component tail? If so, you're probably going to love Comet Hale-Bopp!

January 1997: Dawn Opening to the Great Main Act

On January 3, 1997, Comet Hale-Bopp is at conjunction with the Sun. But the comet has lofted tens of millions of miles above the plane of the ecliptic and is over $2^{1}/_{2}$ a.u. away from us and $1^{3}/_{4}$ a.u. from the Sun.

We stare over the Sun to see the comet appear 28 degrees due north of the Sun on January 3. Or *do* we see it? I think expert observers with telescopes may keep it in sight as it shifts from dusk to dawn visibility in early January. And the good news is that as January progresses, Hale-Bopp will get much higher and should brighten by something like a magnitude and a half as its months-long tenure of grandeur opens.

HB brightens from either magnitude 4 to 2¹/₂ or 3 to 1¹/₂, depending on which formula you use. Either way, what we're seeing is the effect of the comet nearing both Sun and Earth: HB's solar distance decreases from about 1.75 to 1.37 a.u. (it comes closer to the Sun than the average distance of Mars's orbit), and its distance from Earth declines from 2.6 to 2.0 a.u. This is the month when Hale-Bopp's coma may be largest in actual physical dimensions, because when a comet gets any closer to the Sun, the solar wind usually compresses the coma. But it is the gas coma that is so affected. What might the dust coma of HB now look like? The distance between Earth and comet is still great, so the coma is not likely to appear gigantic—though if Hale-Bopp's visible coma is as large as that of the Great Comet of 1811, it could look half the apparent size of the Moon in late January.

HB will be coming up almost due east in Aquila the Eagle for most of January, and if the fainter forecast is right, beginners may have trouble finding the comet with the naked eye until late in the month. The best time to look will probably be just as morning twilight is beginning, which will be a little before 6 A.M. for most readers of this book, but our diagrams and the statistics (including rise-times) in Appendix 8 will help you fine-tune your schedule and search. Most observers will have to brave the cold to see Hale-Bopp this month—but even though brighter, grander views of it are on the way, the really avid comet observer will not want to miss this month's unique sights. Interesting toward month's end will be HB's fairly distant passage by the very bright star Altair. If the comet is anywhere near as bright as Altair by the end of January, we'll know that a very bright climax is on the way. And even if the comet is greatly fainter than that, a dust tail a few degrees long may already be plainly visible. Our diagram shows that a very low Venus and Mercury will be far to the lower right of HB in January.

February 1997: A Great Comet Before Dawn

The comet moves in from 1.37 a.u. to 1.07 a.u. from the Sun in February, and from 2.01 to 1.50 a.u. from Earth.

Unless Hale-Bopp is considerably dimmer than expected, it should this month become a quite obvious object, with a bright dust tail growing to perhaps 10 degrees and pointing northwest (or will the dust tail bend far away from the gas tail pointing in this direction?) The slightly pes-

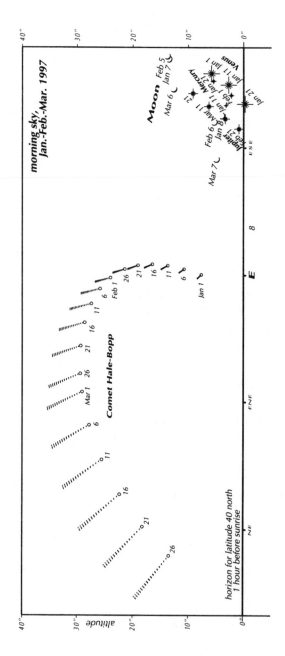

January–March 1997 sky locations of Hale-Bopp and bright solar system objects an hour before sunrise at 40 degrees N (exact positions are plotted for 75 degrees W longitude, center of Eastern Time Zone). Tail lengths are fairly speculative.

simistic formula predicts HB will brighten from magnitude $2^1/_2$ to nearly 1 (optimistically, it could brighten from 1.5 to 0!), qualifying it as a great comet by Yeomans's criteria, for it will be visible outside of twilight. In fact, HB's rising time (as seen from 40° N) goes back from around 4 A.M. to 3 A.M., and despite twilight getting rapidly earlier, the comet's altitude at the first trace of dawn (between 5 A.M. and 5:30 A.M.) increases from about 20 to 25 degrees.

The vista of constellations and Milky Way toward which Comet Hale-Bopp stretches its tail on these cold pre-dawns could hardly be more beautiful. It is unfortunate that the bright Moon will diminish the effect on mornings after February 20 or 21, as the comet's head glides along the edge of the Veil Nebula and through the outstretched eastward wing of Cygnus the Swan. But before this, many magnificent photographs should be possible, and if the comet follows the optimistic forecast for it—reaching magnitude 0 by mid-month!—who knows what sights and images of its head and tail might be possible even in bright moonlight? Before full Moon, will the comet not quite rival Deneb's brightness, or will it rival Vega's? It is likely to make a diamond out of the famous Summer Triangle of Vega, Altair, and Deneb. Hale-Bopp will not need a tail much longer than the 10 degrees modestly predicted in order to penetrate deep into some of the brightest star clouds of the Milky Way. It travels much closer to Cygnus and the Cygnus Milky Way in late February and early March than Comet Bennett did in April 1970. If moonlight weren't going to interfere, there would even be a fair chance of seeing Hale-Bopp's tail reach right to Deneb late in the month (but see next month).

March 1997: Closest, Brightest at Both Dawn and Dusk!

How many non-astronomers will have gotten up to see Comet Hale-Bopp at 5:30 A.M. or earlier last month? All too few, I fear. But this month Hale-Bopp passes from the dawn to the dusk sky, for a few weeks being easily visible in both—and even visible the whole night if you live far enough north. This transition to the evening sky, where millions of people can conveniently see it, occurs late in the month when Hale-Bopp is closest to Earth, closest to the Sun, and brightest—as it finishes passing just 5 degrees north of the famous Great Andromeda Galaxy!

Comet Hale-Bopp starts March 1997 about 1.5 a.u. from Earth and 1.1 a.u. from the Sun. Technically, it is due north of the Sun—at "superior conjunction"—on March 3, but it is so far north of the Sun that it is visible during many nighttime hours. Depending on your brightness formula, it could shine at about magnitude 1 or −1, the difference between marvelous and almost staggering if there is a well-developed dust tail. By the end of the first week of March, when the Moon will have dwindled to a slender crescent, the 15-degree long (or much longer?) dust tail of Hale-Bopp may curve gloriously toward (over?) Deneb and the Milky Way. Then, on March 9, avid amateur astronomers and thrill seekers who have traveled to Mongolia will see there, weather permitting, a total eclipse of the Sun low in their sky—and high in their sky, easily visible if the bright forecasts hold, Comet Hale-Bopp and some of its tail! The comet is exactly 1 a.u. from the Sun mere hours after the eclipse.

Do you regret that you won't be one of the lucky travelers to see HB during a total eclipse of the Sun? Well, if you live in the Americas, you'll have the consolation of a nearly total eclipse of the Moon on the night of March 23–24—and many viewers will be able to see HB during that lunar eclipse and its darkened sky! (This is just about when HB becomes higher at dusk than dawn, a slower version of the entrance of Comet Halley into the evening sky to meet a total lunar eclipse on May 23 in 1910—but HB may be brighter at this juncture!)

For the 2 weeks between the solar and lunar eclipse, the Moon brightens in the evening sky, making the hour before morning twilight the prime dark time to observe the comet in the northeast (noticeably, but not drastically, lower by mid-month), far north of the rising Great Square of Pegasus, to the right of the main pattern of Cassiopeia, and far to the lower left of the uplifted Deneb in Cygnus and the entire Summer Triangle. The comet's dust tail, growing to 20 degrees or perhaps more, fountains up across a rich and strange region of the Milky Way in Lacerta and Cepheus. How separated is the gas tail from the dust? They should be considerably separated, and the dust tail presented fairly broadside—how will it compare with the tail of Comet West? The comet will surely have a very distinct parabolic shape to its head by now, but what kind of naked-eye "false nucleus" or telescopically observable jets or expanding halos may be evolving is, at this writing, anybody's guess. Whatever amazing details of the head and tails there may be, the prime opportunities to see them will move to the after-dusk sky starting about the American evening of March 21—the evening when the comet first is higher at the end of evening twilight than at the start of morning twilight the next day.

The timing of this clever shift could not be better. The full Moon's light drenches the whole night of March 23–24 (or would if there weren't a lunar eclipse), but then the next few nights there is a window just after nightfall and before moonrise in which to see the comet (in a fully dark sky) (this Full Moon is the one opposite from Harvest Moon and thus rises much later each night—just what Hale-Bopp watchers will want). Hale-Bopp will be seen in the northwest about 20 degrees high a little before 8 P.M., the end of evening twilight. The comet will not set until about 3 hours later as seen from 40° N latitude. Above about 45° N, in southern Canada and northern Europe, Hale-Bopp will not set at all for a number of nights, though it will skirt very low to the north horizon in the middle of the night. Of course, if Hale-Bopp actually is as bright at this time as optimistic forecasts predict, its head should be visible very close to the horizon indeed. Furthermore, the comet's tail now points almost due north—thus up from the north horizon toward Polaris and, if long and bright enough, perhaps glimpsible from farther south, in the United States. And this final week of March, when the comet has zoomed up to its highest in the after-dusk sky, when it is visible all night long from far-northern locations (however marginally), is the very week it is closest to Earth and Sun and is probably brightest.

How bright will Hale-Bopp be? Both optimistic and pessimistic formulas agree that with its distance from Earth and Sun remaining nearly the same for the second half of March, the comet should stay at about the same brightness during this period. This, of course, is independent of any brightness fluctuations, or perhaps major flares or dimmings, that may result from idiosyncrasies of the comet's activity. Theoretically, the peak brightness might occur about March 27. The original conservative formula predicts that this brightness will be 0.4, the liberal that it will be −1.7 (or even −2.3—bright as Jupiter!—according to one variation). If HB is as bright as −2, the comet might even be observable with optical aid in broad daylight, about 43 degrees from the Sun.

At halfway between the two brightness values just mentioned, the comet would still outshine every point of light in the evening sky except the brightest star, Sirius, in the southwest, and the planet Mars in the east. "Only" 61 million miles from Earth in late March 1997—almost exactly half the distance of Comet Hale-Bopp—Mars will be the brightest it's been for several years. The comet may still outshine it. In the morning sky, Hale-Bopp's competition for brightness is Jupiter (difficult to beat)—and by the way, Hale-Bopp will still be visible, a few degrees high, at the start of morning twilight on the last day of March as seen from 40° N (it will still be 10 degrees high at that time on that day as seen from 50° N!). The brightest of all planets, Venus, is near superior conjunction, on the far side of the Sun from Earth, and unviewable this week. (Venus is then on the *same* side of the Sun as Hale-Bopp, so the comet would be a magnificent sight in the midnight sky if seen from Venus—or rather from above the thick Venusian clouds). Where are the remaining naked-eye planets at this time of Hale-Bopp's maximum brightness? Saturn is at superior conjunction on March 30, just 3 days before Venus is. But elusive Mercury is peeking up low in the west-northwest in the final week of March, best visible in the first 10 days of April.

If Hale-Bopp shines at magnitude 0.4 or somewhat less, there will be a few equally bright stars high in the west and southwest sky.

But a closer look at the vicinity of Hale-Bopp in the last week of March finds it going only about 5 degrees north of M31, the Great Galaxy in Andromeda, on March 25, when the comet is farthest north in the heavens. Farther to its south (left) at dusk are 2nd-magnitude stars of Andromeda, and above it is the prominent pattern of Cassiopeia—with the comet's tail perhaps extending right up into it. Far to the comet's upper left is Perseus; even farther up, Capella.

Finally, let's consider the most important orbital landmarks that occur in this final week of March. Important to us denizens of Earth is that Hale-Bopp is closest to our planet on March 22—1.314 a.u., a full 123 million miles. But of all positions that a comet reaches, the most important in an absolute sense is perihelion—its closest venture to the Sun that activates it. Almost 20 months after discovery, almost 4 years after the McNaught-found photo of it from April 1993, Comet Hale-Bopp comes to perihelion on the American evening of March 31 at a distance of 0.914 a.u., or about 85 million miles, from the Sun.

The third of four periods of International Hale-Bopp Days will be held from March 3 to April 12. HB will reach the part of its orbit farthest above the plane of Earth's orbit in early March, and one of the things professional astronomers will use the comet to study in the weeks that follow is the latitude dependence of the solar wind. For the comet will be taking a rapid plunge south, sampling the solar wind and showing us, as it goes, the solar wind's effects with plasma phenomena of head and gas tail. The inclination of HB's orbit to that of Earth is 89°—almost exactly perpendicular. The place at which Hale-Bopp pierced up through the extended plane of Earth's orbit was at the orbit of Jupiter, back in late February of

1996. The place where it will pierce down through the extended plane of Earth's orbit is a point just beyond Earth's orbit, which the comet will reach on May 6.

April 1997: Final Heyday and Start of Fall

Although Hale-Bopp does not beat a hasty retreat from its closest approach to the Sun, it does start moving away from Earth fairly rapidly. Nevertheless, the conditions of its visibility linger at near their late-March best—and now entirely in the evening sky—for most of April. The comet loses only about half a magnitude by April 20 and is virtually as high (about 20 degrees) up—in the west-northwest, not the northwest—as evening twilight ends. But HB does set about 2 hours after twilight's end by April 20, compared to 3 hours at the start of the month. Beginning in mid-April, the brightening Moon also reduces the splendor of a now northeastward-pointing tail, which may start shrinking to less than 15 degrees (even though it is probably growing greatly in physical length in space) because our angle of view rapidly increases the foreshortening.

The final week of April finds the bright moonlight gone from the early evening sky but the comet fading by almost another half-magnitude, down to 1.4 ... or 0! If HB is really still as bright as this latter figure, it will have been brighter than this, says the formula, for about 2 months—just about unprecedented in the historical record and productive of so many stunning sights as to stagger the imagination.

If Hale-Bopp is at only magnitude 1.4 at the end of April 1997, it still will have a brighter head—probably a much denser, brighter, and longer (about 10 degrees?) dust tail—than Hyakutake had in mid-April of 1996. Amazingly, on April 11, 1997, Hale-Bopp will be almost exactly where Comet Hyakutake was in the heavens on April 11, 1996, and possibly two (but maybe as many as four!) magnitudes brighter. Earlier in April, HB will have passed quite near to the second-magnitude Gamma Andromedae (Almak), the loose, bright, open cluster M34 in Perseus, and the famous variable star Algol. On April 23, just after full Moon, Hale-Bopp will enter Taurus and will have nudged back out to 1 a.u. from the Sun (but still some distance above the plane of Earth's orbit). At month's end, the view around 8:30 to 8:45 P.M. (daylight saving time) will encompass, low in the sky from southwest to west-northwest, the following:

Sirius, Betelgeuse, and Hale-Bopp—with Hale-Bopp possibly still brighter than Betelgeuse.

Is it possible that Hale-Bopp will be brighter through much of April than it was in late March? Many comets are somewhat brighter after perihelion. In this case, however, the increasing distance from Earth (1.36 to 1.77 a.u.) will be a problem. The head and tail could continue to offer remarkable phenomena; what they will look like at this point is wide open to conjecture. The Great Comet of 1811's tail was longest in true dimensions—about 1.2 a.u.—a month or so after perihelion, and it will be awesome to observe Hale-Bopp's when we know it to be anything like that long in space, but even the longest tail imaginable would be severely foreshortened at this point.

May 1997: Fade, Still Bright, Into the Sunset

Hale-Bopp stands about 13 degrees high at the end of twilight in the west-northwest sky on May 1. But by May 10, it is only half that high then, and by mid-May the comet sets at twilight's end. On May 20, Hale-Bopp sets little more than an hour after the Sun. If it is at about magni-

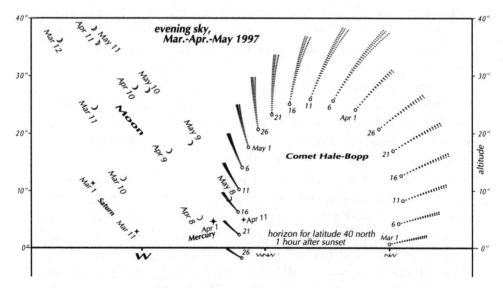

March–May 1997 sky locations of Hale-Bopp and bright solar system objects an hour after sunset at 40 degrees N (exact positions are plotted for 75 degrees W longitude, center of Eastern Time Zone). Tail lengths are fairly speculative.

tude 2, we may need binoculars or a very clear sky for anyone to glimpse it (much like Hyakutake around April 26, 1996). It will be lost from view even in telescopes a few days later. Of course, if the comet is brighter than first magnitude on May 20, it will probably be plainly visible to the naked eye under good conditions. But even if it is brighter than 1.0 at month's end (which the optimistic formula predicts), nobody is likely to see it, because it will be setting only 20 minutes after the Sun.

All these time figures are for an observer at 40° N. But as the comet's elongation from the Sun decreases to 25 degrees on May 20 and to 23 degrees on May 31, there's nowhere in the world that offers an easy view.

Hale-Bopp cuts across Taurus in May. On May 6, the comet reaches the descending node of its orbit, passing only about 10 million miles outside Earth's orbit—at a point we won't reach for months. By mid-May, a way to find Betelgeuse, and to its right the comet, is to look straight down from the far, far higher Gemini stars Pollux and Castor.

During May, HB recedes from 1.05 a.u. to 1.38 a.u. from the Sun and from 1.77 a.u. to 2.26 a.u. from Earth. But vigorous activity might still make the head very structured. The few degrees of tail we see left as May progresses are pointed first east and then, by later in the month, a little south of east.

And May will bring a potentially stunning sight: all across the U.S. and Canada on May 8, a very slender crescent Moon will shine only about 3 to 5 degrees to the lower left of the comet!

June–December 1997: Southward Departure and Last Naked-Eye Sighting

The end of May 1997 is not the end for observation of Hale-Bopp—certainly not for professional astronomers, probably not for Southern Hemisphere amateurs, and maybe not for Northern Hemisphere amateurs. After a break from visual observations while the comet is hidden behind the solar glare, HB emerges far south of the Sun in the dawn sky. Then, the question will be: Even if the comet has proved as bright as we could have hoped through perihelion and the 2 months thereafter, will it fade to a pale diffuse ghost of a ghost early on its way out (as many truly great comets have)? Or will it maintain some vigorous activity and keep to its brightness curve?

In the opening days of June 1997, HB slips just a few degrees past the bright, golden-orange star Betelgeuse, though it is by no means certain that anyone will be able to get a visual observation—even in the unlikely event that Hale-Bopp is as bright as Betelgeuse at this point. It's more likely to be a second- or third-magnitude patch of fuzz hopelessly washed out by bright sky. In late June, the comet passes south of the celestial equator in Monoceros. On July 4, 1997—one year after it was in opposition to the Sun—HB is at conjunction with the Sun, 26 degrees south of it.

Sometime in August, observations of Hale-Bopp should resume in the Southern Hemisphere, as the comet comes up a while after Sirius. It is conceivable that it could still be a naked-eye object then! Even if not, its naked-eye visibility will have lasted from May 1996 until May 1997 (with perhaps a small break around New Year's Day)—giving Hale-Bopp the longest naked-eye visibility in history. Of course, our early awareness of the comet and the detailed knowledge of its position and movement will give it an advantage over comets of the past—such as the record-holding Great Comet of 1811, seen with the naked eye for about $8^{1}/_{2}$ months (including a sizable break when the comet was in the solar glare).

If Hale-Bopp has exhausted itself by the climax of its performance (exhausted itself for this time—it probably will be back and bright at the next visit), it might be a very difficult binocular or telescopic object for southerners in late summer 1997 and even more so for northerners in September. September might be the last month people at mid-northern latitudes can hope to see it, considering that by early October it is passing near the star Zeta Puppis (Naos), very low in the south.

But professional observations from the Southern Hemisphere and outer space may yield some of their most valuable information about the comet in the fourth and final period of International Hale-Bopp Days—from August 26 to September 16. During this period, Hale-Bopp is about $2^{1}/_{2}$ a.u. outbound (about as far out as the main asteroid belt, though already getting far south of the plane of Earth and most of the planets' and asteroids' orbits). This is thought to be the distance from the Sun at which the comet's plasma phenomena might turn off and sublimation of water from the nucleus ought to decline rapidly. The elongation of Hale-Bopp in this period is again more than 45 degrees from the Sun, making it observable by the HST (Hubble Space Telescope) and IUE (International Ultraviolet Explorer) for the first time since early April. The HST will have no trouble getting interesting images of Hale-Bopp even though the comet will by this last period be moving out to almost 3 a.u. from Earth.

Theoretically, Hale-Bopp could be visible with the naked eye from places like Australia even in November and December 1997 as it cruises, below the due south horizon for mid-northern latitudes, through the constellation Carina in the hour before dawn. At the very least, telescopes in southern lands will be trained on it.

1998 and Beyond

On New Year's Day of 1998, Comet Hale-Bopp will lie near the Pictor-Dorado border (declination −64 degrees), more than a fist-width (held out at arm's length) south of the second brightest star in the heavens, Canopus. One version of the optimistic brightness formula for it would have its magnitude then at 6.3—just brighter than the naked-eye limit! On that day, the comet will lie 3.772 a.u. from Earth and 3.941 a.u. from the Sun.

Will the last telescopic observation of Hale-Bopp be late in 1998? When will it at last be time to say not "Hale and Bopp" but "Hail, and farewell"?

I don't know. But if you consider how early this comet was detectable on photographic plates, you come up with yet another remarkable estimate: It may last be detected (as opposed to seen) by even modest Earth-based professional efforts in 2001 or 2002.

Flying onward, Comet Hale-Bopp will reach its cold aphelion, the far point of its orbit, in about A.D. 3400. It will be over 100 times farther out than Pluto. Then, slowly, imperceptibly at first, the mountain of dark, dusty ice will begin its fall back toward the Sun. Recorded human history will be more than half again as long—we hope—when Comet Hale-Bopp returns and fires up another grand show for Earth in about A.D. 5700.

Chapter 15
COMET OF THE CENTURY

And so, at the time of this book's writing, Comet Hale-Bopp is promising to become the latest great—the latest candidate to be our "comet of the century." But now, in light of all we have surveyed in the course of this book, what do we think about that title?

First of all, we can now understand in much greater depth, illuminated by many examples, why it is never possible to be completely certain that an approaching object will become a great comet, let alone the "comet of the century." A complex variety of interdependent factors work together to make a comet visually spectacular—or not. Many comets initially seemed sure bets for greatness but flopped. And others, which no one thought would burst into glorious prominence and beauty, proved awesome.

The next question to ask is obvious: is it even *possible* to single out an individual comet as being the "best"—most visually gratifying—of a century? Surprisingly, despite the tremendous variety of comets and the fact that not everyone would find the same features and attributes of comets most appealing, I think it might be possible to come up with some clear-cut winners if we held a vote among reasonably knowledgeable comet enthusiasts. Perhaps I'm wrong. But I would venture to guess that the 1577, 1680, 1744, and 1882 comets would win such a vote and be selected the comets of their respective centuries. They would win by a combination of brute brightness and widespread accessibility (mostly the former)—but also by their impressive performances in other categories: tail length, tail brightness, tail structure, duration of naked-eye visibility and naked-eye prominance. (After all, these other attributes are not unrelated to brute brightness.)

Suppose, however, that we chose these comets as being overpoweringly great and, by making that choice, tended to neglect the marvelous comets of other years. If so, we would miss the drama of closeness and tail passage and the amazing tail length and breadth of the Great Comet of 1861; the exquisite beauty of form and placement of Donati's Comet; the stupendous true length of the Great March Comet of 1843 and Messier's Comet of 1769; the intrinsic brightness and duration of the Great Comet of 1811; and much more.

This brings us to the question of the twentieth century's greatest comet. Unlike any other century since the invention of the telescope, the choice for this "comet of the century" is, I think, unclear. No comet has had "brute brightness"—we're talking better than borderline negative magnitude here!—for long enough when visible enough. Hale-Bopp has an excellent good chance of displaying the same virtues as the 1811 comet: great intrinsic brightness and a very long period of impressive (magnitude 0 or 1) if perhaps not mind-boggling (magnitude −1 or −2) brightness. But if the optimistic brightness forecasts for Hale-Bopp pan out, and it enters for quite a while into the realm of mind-boggling brightness . . . ? Then I think we *would* have (by the rules of this artificial game I'm playing!) our "comet of the century."

Under those conditions, Hale-Bopp would be the "winner." But let's not forget what the other great comets of the twentiethth century have had to offer.

Even excluding the rather brief and most southerly of the comets that are "great" by Yeomans's criteria (those of 1901 and 1927), we have had five other great comets.

The Great January Comet of 1910 offered brilliance (though rather brief), evening visibility, and a long, beautifully striated tail.

Halley in 1910 brought the drama of closeness, the through-tail passage, good visibility in both morning and evening skies, and the longest tail on record. (If forced to select the comet of the twentieth century at this very moment (before Hale-Bopp), by the way, I'd still give the nod to Halley-1910.)

The stunning Ikeya-Seki, in 1965, would be a "comet of the century" if it had not escaped Northern Hemisphere skies so quickly and faded even in Southern Hemisphere skies almost as quickly. (I'm not sure I'd be able to deny it the title if I had seen its almost dazzling tail as a long beam of light that for weeks stretched so far in the Australian sky—or perhaps even if I had just seen it for one morning in the New Jersey sky.)

Though beauty is to some extent subjective, there was overwhelming agreement that 1970's Comet Bennett, with its bright, enduring, scenically placed tail was about as beautiful as any sky sight could get—even if Bennett was visible only in the pre-dawn sky and was not so bright-headed as a few other great comets.

In 1976, Comet West showed us greater (even daylight) brightness, though briefly, and its tail, though perhaps less enduring and less concentrated as a whole than Bennett's, offered a *combination* of breadth, structure and brightness that maybe few dust tails over the centuries have matched.

In 1996, Comet Hyakutake lacked a great dust tail, but its gas tail stretched to enormous length as its head grew almost as big as, and perhaps more blue to the naked eye than, any bright comet in centuries—during a brief period of all-night, high-up visibility that was followed by a longer period of respectable evening display.

Perhaps Hale-Bopp has the potential to become the most spectacular of the twentieth-century comets. But if Hale-Bopp merely gives us its special variations on the master theme of beauty and does no more than join the glorious ranks of these five, we will be foolish indeed to lament its not becoming the "comet of the century."

PERIODIC COMET NUMBERS

The assignment of periodic comet numbers is the responsibility of the Minor Planet Center. The current list is as follows, first listed alphabetically and then numerically.

Alphabetical Order

Name	Number	Name	Number
Arend	50P	Metcalf-Brewington	97P
Arend-Rigaux	49P	Mrkos	124P
Ashbrook-Jackson	47P	Mueller 1	120P
Biela	3D	Neujmin 1	28P
Boethin	85P	Neujmin 2	25D
Borrelly	19P	Neujmin 3	42P
Brooks 2	16P	Olbers	13P
Brorsen	5D	Oterma	39P
Brorsen-Metcalf	23P	Parker-Hartley	119P
Bus	87P	Perrine-Mrkos	18P
Chernykh	101P	Peters-Hartley	80P
Chiron	95P	Pons-Brooks	12P
Churyumov-Gerasimenko	67P	Pons-Winnecke	7P
Ciffreo	108P	Reinmuth 1	30P
Clark	71P	Reinmuth 2	44P
Comas Sola	32P	Russell 1	83P

(continued)

Name	Number	Name	Number
Crommelin	27P	Russell 2	89P
d'Arrest	6P	Russell 3	91P
Daniel	33P	Russell 4	94P
de Vico	122P	Sanguin	92P
de Vico-Swift	54P	Schaumasse	24P
Denning-Fujikawa	72P	Schuster	106P
du Toit	66P	Schwassmann-Wachmann 1	29P
du Toit-Hartley	79P	Schwassmann-Wachmann 2	31P
du Toit-Neujmin-Delporte	57P	Schwassmann-Wachman 3	73P
Encke	2P	Shajn-Schaldach	61P
Faye	4P	Shoemaker 1	102P
Finlay	15P	Shoemaker-Holt 2	121P
Forbes	37P	Shoemaker-Levy 4	118P
Gale	34P	Singer-Brewster	105P
Gehrels 1	90P	Slaughter-Burnham	56P
Gehrels 2	78P	Smirnova-Chernykh	74P
Gehrels 3	82P	Spitaler	113P
Giacobini-Zinner	21P	Stephan-Oterma	38P
Giclas	84P	Swift-Gehrels	64P
Grigg-Skjellerup	26P	Swift-Tuttle	109P
Gunn	65P	Takamizawa	98P
Halley	1P	Taylor	69P
Harrington	51P	Tempel 1	9P
Harrington-Abell	52P	Tempel 2	10P
Hartley 1	100P	Tempel-Swift	11D
Hartley 2	103P	Tempel-Tuttle	55P
Hartley 3	110P	Tsuchinshan 1	62P
Helin-Roman-Alu 1	117P	Tsuchinshan 2	60P
Helin-Roman-Crockett	111P	Tuttle	8P
Herschel-Rigollet	35P	Tuttle-Giacobini-Kresak	41P
Holmes	17P	Urata-Niijima	112P
Honda-Mrkos-Pajduskova	45P	Vaisala 1	40P
Howell	88P	Van Biesbroeck	53P
Jackson-Neujmin	58P	West-Hartley	123P
Johnson	48P	West-Kohoutek-Ikemura	76P
Kearns-Kwee	59P	Westphal	20D
Klemola	68P	Whipple	36P
Kohoutek	75P	Wild 1	63P
Kojima	70P	Wild 2	81P
Kopff	22P	Wild 3	86P
Kowal 1	99P	Wild 4	116P
Kowal 2	104P	Wilson-Harrington	107P

(continued)

Name	Number	Name	Number
Longmore	77P	Wirtanen	46P
Lovas 1	93P	Wiseman-Skiff	114P
Machholz 1	96P	Wolf	14P
Maury	115P	Wolf-Harrington	43P

Numerical Order

Number	Name	Number	Name
1P	Halley	63P	Wild 1
2P	Encke	64P	Swift-Gehrels
3D	Biela	65P	Gunn
4P	Faye	66P	du Toit
5D	Brorsen	67P	Churyumov-Gerasimenko
6P	d'Arrest	68P	Klemola
7P	Pons-Winnecke	69P	Taylor
8P	Tuttle	70P	Kojima
9P	Tempel 1	71P	Clark
10P	Tempel 2	72P	Denning-Fujikawa
11D	Tempel-Swift	73P	Schwassmann-Wachmann 3
12P	Pons-Brooks	74P	Smirnova-Chernykh
13P	Olbers	75P	Kohoutek
14P	Wolf	76P	West-Kohoutek-Ikemura
15P	Finlay	77P	Longmore
16P	Brooks 2	78P	Gehrels 2
17P	Holmes	79P	du Toit-Hartley
18P	Perrine-Mrkos	80P	Peters-Hartley
19P	Borrelly	81P	Wild 2
20D	Westphal	82P	Gehrels 3
21P	Giacobini-Zinner	83P	Russell 1
22P	Kopff	84P	Giclas
23P	Brorsen-Metcalf	85P	Boethin
24P	Schaumasse	86P	Wild 3
25D	Neujmin 2	87P	Bus
26P	Grigg-Skjellerup	88P	Howell
27P	Crommelin	89P	Russell 2
28P	Neujmin 1	90P	Gehrels 1
29P	Schwassmann-Wachmann 1	91P	Russell 3
30P	Reinmuith 1	92P	Sanguin
31P	Schwassmann-Wachmann 2	93P	Lovas 1
32P	Comas Sola	94P	Russell 4

(continued)

Number	Name	Number	Name
33P	Daniel	95P	Chiron
34P	Gale	96P	Machholz 1
35P	Herschel-Rigollet	97P	Metcalf-Brewington
36P	Whipple	98P	Takamizawa
37P	Forbes	99P	Kowal 1
38P	Stephan-Oterma	100P	Hartley 1
39P	Oterma	101P	Chernykh
40P	Vaisala 1	102P	Shoemaker 1
41P	Tuttle-Giacobini-Kresak	103P	Hartley 2
42P	Neujmin 3	104P	Kowal 2
43P	Wolf-Harrington	105P	Singer-Brewster
44P	Reinmuth 2	106P	Schuster
45P	Honda-Mrkos-Pajdusakova	107P	Wilson-Harrington
46P	Wirtanen	108P	Ciffreo
47P	Ashbrook-Jackson	109P	Swift-Tuttle
48P	Johnson	110P	Hartley 3
49P	Arend-Rigaux	111P	Helin-Roman-Crockett
50P	Arend	112P	Urata-Niijima
51P	Harrington	113P	Spitaler
52P	Harrington-Abell	114P	Wiseman-Skiff
53P	Van Biesbroeck	115P	Maury
54P	de Vico-Swift	116P	Wild 4
55P	Tempel-Tuttle	117P	Helin-Roman-Alu 1
56P	Slaughter-Burnham	118P	Shoemaker-Levy 4
57P	du Toit-Neujmin-Delporte	119P	Parker-Hartley
58P	Jackson-Neujmin	120P	Mueller 1
59P	Kearns-Kwee	121P	Shoemaker-Holt 2
60P	Tsuchinshan 2	122P	de Vico
61P	Shajn-Schaldach	123P	West-Hartley
62P	Tsuchinshan 1	124P	Mrkos

Additions to this list will appear in the Minor Planet Circulars. This list was last updated on January 31, 1996.

Appendix B
GREAT COMETS IN HISTORY

The following table is from Donald Yeomans' paper "Cometary Apparitions: Great Comets in History." Additions by the author of this book appear in brackets.

1st Date Reported	Obs. Int.	Perihelion		Perigee		Brightest Outside Twilight		
		Date	Dist.	Date	Dist.	Date	Mag.	Notes
Julian calendar								
B.C. Dates								
373–372 Winter								
147 Aug. 6	32	6/28	0.43	8/4	0.15			
87 July	35	8/6	0.59	7/27	0.44	7/27	2	Halley
12 Aug. 26	56	10/10	0.59	9/10	0.16	9/10	1	Halley
A.D. Dates								
66 Jan. 31	69	1/26	0.59	3/20	0.25	3/20	1	Halley
141 Mar. 27	30	3/22	0.58	4/22	0.17	4/22	−1	Halley
178 Sep.	80							
191 Oct.								
218 May	40	5/17	0.58	5/30	0.42	5/30	0	Halley
240 Nov. 10	40	11/10	0.37	11/30	1.00	11/20	1–2	

(continued)

1st Date Reported	Obs. Int.	Perihelion		Perigee		Brightest Outside Twilight		
		Date	Dist.	Date	Dist.	Date	Mag.	Notes
295 May	30	4/20	0.58	5/12	0.32	5/12	0	Halley
374 Mar. 4	30	2/16	0.58	4/2	0.09	4/2	−1	Halley
390 Aug. 7	40	9/5	0.92	8/18	0.10	8/18	−1	
400 Mar. 19	30	2/25	0.21	3/31	0.08	3/19	0	
442 Nov. 10	100	12/15	1.53	12/7	0.58	12/7	1–2	
451 June 10	60	6/28	0.58	6/30	0.49	6/30	0	Halley
530 Aug. 29	30	9/27	0.58	9/3	0.28	9/3	1–2	Halley
565 July 22	100	7/15	0.82	9/13	0.54	9/13	0–1	
568 July 28	100	8/27	0.87	9/25	0.09	9/25	0	
607 Mar.–Apr.	30	3/15	0.58	4/19	0.09	4/19	−2	Halley
684 Sep. 6	33	10/2	0.58	9/7	0.26	9/7	1–2	Halley
760 May 17	50	5/20	0.58	6/3	0.41	6/3	0	Halley
770 May 26	60	6/5	0.58	7/10	0.30	7/10	1–2	
837 Mar. 22	46	2/28	0.58	4/11	0.03	4/11	−3	Halley
891 May 12	50							
905 May 18	25	4/26	0.20	5/25	0.21	5/23	0	
912 July		7/18	0.58	7/16	0.49	7/19	0	Halley
962 Jan. 28	64	12/28/61	0.63	2/24	0.35	2/21	1	
989 Aug. 12	30	9/5	0.58	8/20	0.39	8/20	1–2	Halley
1066 Apr. 3	60	3/20	0.58	4/24	0.10	4/24	−1	Halley
1106 Feb. 4	40							
1132 Oct. 5	20	8/30	0.74	10/7	0.04	10/7	−1	
1145 Apr. 15	81	4/18	0.58	5/12	0.27	5/12	0	Halley
1222 Sep. 3	35	9/28	0.58	9/6	0.31	9/24	1–2	Halley
1240 Jan. 27	64	1/21	0.67	2/2	0.36	2/2	0	
1264 July 17	80	7/20	0.82	7/29	0.18	7/29	0	
1301 Sep. 1	61	10/25	0.58	9/23	0.18	9/23	1–2	Halley
1378 Sep. 26	15	11/10	0.58	10/3	0.12	10/3	1	Halley
1402 Feb. 8	70	3/21	0.38	2/19	0.71	3/12	−3	
1456 May 27	42	6/9	0.58	6/19	0.45	6/19	0	Halley
1468 Sep. 18	81	10/7	0.85	10/2	0.67	10/2	1–2	
1471 Dec. 25	58	3/1/72	0.49	1/23	0.07	1/23	−3	
1531 Aug. 5	34	8/26	0.58	8/14	0.44	8/27	1	Halley
1532 Sep. 2	119	10/18	0.52	9/21	0.67	10/13	−1	
1533 June 27	81	6/15	0.25	8/2	0.42	6/27	0	
1556 Feb. 27	73	4/22	0.49	3/13	0.08	3/14	−2	
1577 Nov. 1	79	10/27	0.18	11/10	0.63	11/8	−3	Great Comet

(continued)

1st Date Reported	Obs. Int.	Perihelion Date	Perihelion Dist.	Perigee Date	Perigee Dist.	Brightest Outside Twilight Date	Brightest Outside Twilight Mag.	Notes
Gregorian calendar								
1607 Sep. 21	35	10/27	0.58	9/29	0.24	9/29	1–2	Halley
1618 Nov. 16	67	11/8	0.40	12/6	0.36	11/29	0–1	Great Comet
1664 Nov. 17	75	12/5	1.03	12/29	0.17	12/29	−1	
1665 Mar. 27	24	4/24	0.11	4/4	0.57	4/20	−1	
1668 Mar. 3	27	2/28	0.07	3/5	0.80	3/8	1–2	
1680 Nov. 23	80	12/18	0.01	11/30	0.42	12/29	1–2	Great Comet
1682 Aug. 15	40	9/15	0.58	8/31	0.42	8/31	0–1	Halley
1686 Aug. 12	30	9/16	0.34	8/16	0.32	8/27	1–2	
1743 Nov. 29	45 [or ~140–150]	3/1	0.22	2/27	0.83	2/20	−3	DeCheseaux
1769 Aug. 15	100	10/8	0.12	9/10	0.32	9/22	0	Messier
1807 Sep. 9	90	9/19	0.65	9/27	1.15	9/20	1–2	Great Comet
1811 Apr. 11	260	9/12	1.04	10/16	1.22	10/20	0	Great Comet
1843 Feb. 5	48	2/27	0.006	3/6	0.84	3/7	1	Great March Comet
1858 Aug. 20	80	9/30	0.58	10/11	0.54	10/7	0–1	Donati
1861 May 13	90	6/12	0.82	6/30	0.13	6/27	0 [or −2?]	Great Comet (Tebbutt)
1865 Jan. 17	36	1/14	0.03	1/16	0.94	1/24	1	
1874 June 10	50	7/9	0.68	7/23	0.29	7/13	0–1	Coggia
1882 Sep. 1	135	9/17	0.008	9/16	0.99	9/8	−2	Great September Comet
1901 Apr. 12	38	4/24	0.24	4/30	0.83	5/5	1	Viscara
1910 Jan. 13	17	1/17	0.13	1/18	0.86	1/30	1–2	Great January Comet
1910 Apr. 10	80	4/20	0.59	5/20	0.15	5/20	0–1	Halley
1927 Nov. 27	32	12/18	0.18	12/12	0.75	12/8	1	Skjellerup-Maristany
1965 Oct. 3	30	10/21	0.008	10/17	0.91	10/14	2	Ikeya-Seki
1970 Feb. 10	80	3/20	0.54	3/26	0.69	3/20	0–1	Bennett
1976 Feb. 5	55	2/25	0.20	2/29	0.79	3/1	0	West
[1996 Feb. 27	~100	5/1	0.23	3/25	0.10	3/25	0	Hyakutake]

The first tabular entry gives the approximate date when the comet was first reported as a naked-eye object. The following entry is the approximate observation interval (in days) during which the comet remained a naked-eye object. The next two entries give, respectively, the month and day of perihelion passage, and the distance between the comet and Sun at that time. Next follows the approximate date when the comet passed perigee and the minimum separation distance, the date when the comet appeared brightest in a darkened sky, and the apparent magnitude at that time.

(*continued*)

The perihelion and perigee distances are given in astronomical units. An object with an apparent magnitude of 6 is just visible to the naked eye in a clear, dark sky. Compared with a comet whose apparent magnitude is 6, a magnitude 5 comet is 2.5 times brighter and a magnitude 4 comet is 2.5 × 2.5 = 6.3 times brighter still, and so on. The bright star Vega has an apparent magnitude of 0 while the brightest star in the sky (Sirius) has an apparent magnitude of −1.5. The planet Jupiter appears at magnitude −2.7 when at its brightest.

CLOSEST COMETS IN HISTORY

The following tables are adapted from the paper "Encounters and Collisions with Comets," by Z. Sekanina and D. K. Yeomans. Additions by the author appear within brackets.

Comets known to have approached the Earth within 2500 Earth radii, listed chronologically by perihelion passage.

Comet[a]		Perihelion distance (AU)	Orbital Period (yr)	Orbit inclination[b]
374	P/Halley	0.577	78.8	163°5
390		0.92		36
400		0.21		32
568		0.87		4
607	P/Halley	0.581	77.5	163.5
837	P/Halley	0.582	76.9	163.4
868		0.42		65.1
1014		0.56		117
1066	P/Halley	0.574	79.3	163.1
1080		0.681		6.9
1132		0.736		106.3
1345		0.89		23

(*continued*)

Comet[a]		Perihelion distance (AU)	Orbital Period (yr)	Orbit inclination[b]
1351		1.01		7
1366	P/Tempel-Tuttle	0.976	33.7	162.2
1472		0.486		170.9
1491 II		0.85		9
1499		0.95		16
1556		0.491		32.4
1702		0.647		4.4
1718		1.025		148.8
1723		0.999		130.0
1743 I		0.838		2.3
1759 III	Great Comet	0.966		175.1
1763	Messier	0.498		72.5
1770 I	P/Lexell	0.674	5.60	1.6
1797	Bouvard-Herschel-Lee	0.525		129.4
1806 I	P/Biela	0.907	6.74	13.6
1853 II	Schweizer	0.909	781	122.2
1862 II	Schmidt	0.981		172.1
1864 II	Tempel	0.909	3930	178.1
1927 VII	P/Pons-Winnecke	1.039	6.01	18.9
1930 VI	P/Schwassmann-Wachmann 3	1.011	5.43	17.4
1961 VIII	Seki	0.681	759	155.7
1975 X	Suzuki-Saigusa-Mori	0.838	446	118.2
1983e	Sugano-Saigusa-Fujikawa	0.471	(hyp.)	96.6
1983d	IRAS-Araki-Alcock	0.991	1020	73.4
[C/1996 B2	Hyakutake	0.230	14,000	124.9]

[a]Prefix P/ signifies a short-period comet (orbital period of less than 200 years).
[b]Standard equinox and equator of 1950.0 (except for Hyakutake, which is epoch 2000.0).

Comets known to have approached the Earth within 2500 Earth radii, listed by increasing miss distance.

| Comet | Time of closest approach | | Minimum separation | | | Equatorial coordinates[c] | | Elongation[d] | Moon age (days) |
	Date[a] (UT)	ΔT^b (days)	astron. units	Earth radii	mill. km	R.A.	Decl.		
1491 II[e]	1491 Feb. 20.0	+28.3	0.0094	220	1.41	6^{h}7	+40°	112°	11.2
1770 I	1770 July 1.7	−43.3	0.0151	354	2.26	19.9	+68	88	8.3
1366	1366 Oct. 26.4	+7.9	0.0229	537	3.43	18.1	+51	79	21.3
1983d	1983 May 11.5	−9.7	0.0313	734	4.68	9.3	+41	80	28.2
837	837 Apr. 10.5	+41.3	0.0334	783	5.00	16.0	−52	139	1.6
1806 I	1805 Dec. 9.9	−23.5	0.0366	859	5.48	23.3	−37	78	18.7
1743 I	1743 Feb. 8.9	+28.6	0.0390	915	5.83	20.6	+80	95	14.1
1927 VII	1927 June 26.8	+5.8	0.0394	924	5.89	21.0	+8	130	26.9
1014	1014 Feb. 24.9	−40.1	0.0407	955	6.09	8.7	+2	136	22.4
1702	1702 Apr. 20.2	+37.1	0.0437	1025	6.54	20.7	+26	76	22.7
1132	1132 Oct. 7.2	+37.5	0.0447	1048	6.69	2.9	+33	156	25.8
1351	1351 Nov. 29.4	+10.4	0.0479	1123	7.17	3.4	+19	150	9.7
1345	1345 July 31.9	−22.1	0.0485	1138	7.26	10.2	+54	41	2.3
1499	1499 Aug. 17.1	−22.9	0.0588	1379	8.80	18.7	+30	111	10.0
1930 VI	1930 May 31.7	−13.5	0.0617	1447	9.23	21.4	+8	102	3.5
1983e	1983 June 12.8	+42.5	0.0628	1473	9.39	19.4	−2	143	1.6
1080	1080 Aug. 5.7	−36.2	0.0641	1503	9.59	15.1	−6	75	17.1
1759 III	1760 Jan. 8.2	+21.8	0.0681	1598	10.19	7.5	−16	142	19.6

(continued)

Comet	Time of closest approach		Minimum separation			Equatorial coordinates[c]		Elongation[d]	Moon age (days)
	Date[a] (UT)	ΔT^b (days)	astron. units	Earth radii	mill. km	R.A.	Decl.		
1472	1472 Jan. 22.9	−38.5	0.0690	1618	10.32	18.5	+77	96	12.4
400	400 Mar. 31.1	+35.1	0.0767	1799	11.47	12.6	+76	91	18.9
1556	1556 Mar. 13.0	−40.7	0.0835	1959	12.49	14.8	+58	113	1.5
1853 II	1853 Apr. 29.1	−11.2	0.0839	1968	12.55	1.7	−4	22	20.6
1797	1797 Aug. 16.5	+37.9	0.0879	2062	13.15	10.9	+83	71	23.5
374	374 Apr. 1.9	+44.6	0.0884	2073	13.22	15.8	−32	150	3.1
607	607 Apr. 19.2	+34.7	0.0898	2106	13.43	4.6	+38	28	16.8
568	568 Sept. 25.7	+29.0	0.0918	2153	13.73	20.0	+29	102	18.3
1763	1763 Sept. 23.7	−39.7	0.0934	2190	13.97	18.9	−51	96	16.0
1864 II	1864 Aug. 8.4	−7.7	0.0964	2261	14.42	9.4	+28	12	5.8
1862 II	1862 July 4.6	+12.1	0.0982	2303	14.69	18.0	+77	79	7.3
390	390 Aug. 18.9	−17.1	0.1002	2350	14.99	6.8	+6	66	21.4
[C/1996 B2f	1996 Mar. 25.3	−36.1	0.1019	2390	15.2	14.5	+60	113	6]
1961 VIII	1961 Nov. 15.2	+35.5	0.1019	2390	15.24	7.9	−60	84	6.8
1718	1718 Jan. 18.1	+2.7	0.1031	2418	15.42	13.6	+61	112	16.4
1723	1723 Oct. 14.5	+16.4	0.1036	2430	15.50	1.1	−60	111	15.1

(continued)

Comet	Time of closest approach		Minimum separation			Equatorial coordinates[c]		Elongation[d]	Moon age (days)
	Date[a] (UT)	ΔT^b (days)	astron. units	Earth radii	mill. km	R.A.	Decl.		
1975 X	1975 Oct. 31.6	+16.2	0.1040	2439	15.56	15.0	−19	11	26.5
1066	1066 Apr. 23.6	+33.6	0.1043	2446	15.60	4.7	+27	22	26.0
868	868 Jan. 25.2	−38.8	0.1050	2463	15.71	3.7	+44	98	26.3

[a]Julian calendar until 1582, Gregorian calendar from 1582 on.
[b]Measured from the time of perihelion passage: negative value means before perihelion; positive, after perihelion.
[c]Standard equinox and equator of 1950 (except for Comet Hyakutake—2000.0).
[d]Angle Sun-Earth-comet.
[e]Minimum separation distance likely to be significantly in error.
[f]Figures for Comet Hyakutake are approximate.

Appendix D

CELESTIAL COORDINATES

The diagram below, adapted from Donald Yeomans's *Comets: A Chronological History of Observation, Science, Myth & Folklore*, shows the "equatorial coordinate system" of celestial coordinates.

We imagine the Earth surrounded by a "celestial sphere" whose inner surface is the dome of the heavens above our horizon and the bowl of the

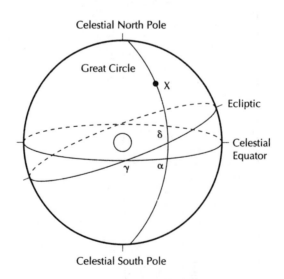

Equatorial coordinate system.

heavens underneath our horizon (underneath the Earth). The "celestial equator" is formed by extending the Earth's equator onto the celestial sphere. The "ecliptic" is the Sun's apparent path through the constellations. The point where the Sun appears to pass north over the celestial equator is the "vernal equinox," which the Sun reaches around March 21, making the start of spring in Earth's northern hemisphere.

For a celestial object located at point "X," its "right ascension," α, is measured eastward along the celestial equator from the vernal equinox point to the great circle passing through the object and the north and south celestial poles. The object's "declination," β, is measured along the object's great circle, north or south from the celestial equator to the object.

ORBITAL ELEMENTS

Six orbital elements of a comet (or other astronomical object) are needed to define the characteristics of its orbit uniquely, both in space and time. If you have these six elements for a comet, you can calculate where it is going to be on any date. Other orbital elements supply additional information which may be of interest.

The six essential orbital elements are:

Perihelion date, T. This is the date and time that the comet is closest to the Sun in space.

Perihelion distance, q. The distance between the Sun and the object at perihelion, measured in astronomical units.

Eccentricity, e. The extent to which an orbit is elongated or open in comparison to a circle ($e = 0$), parabola ($e = 1$), or hyperbola ($e \geq 1$). For a new comet, the eccentricity is initially set at 1 (parabola). Then the comet's motion is watched to see how it departs from a parabolic orbit. (Almost always, the e turns out to be less than 1 and turns out to be a more or less elongated ellipse.)

Inclination, i. The angle in degrees that an orbit is tilted from that of the Earth. The figure can range from 0° to 180°, with anything over 90° meaning that the orbit has toppled past perpendicular and the object is moving retrograde—that is, in the direction opposite to that of the Earth and most solar system bodies.

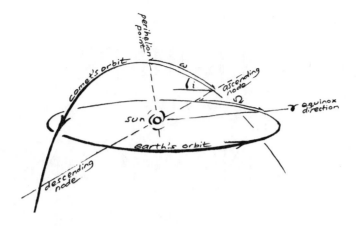

The essential orbital elements.

Longitude of the ascending node, Ω. The angle in degrees (measured forward in the direction of Earth's motion) from the vernal equinox point of Earth's orbit to the object's ascending node. (The ascending node is the place where the plane of the object's orbit passes northward through the plane of Earth's orbit.)

Argument of the perihelion, μ. The angle in degrees from the ascending node of the object's orbit to the perihelion point of its orbit.

Our frame of reference, Earth's orbit, moves and changes its shape slightly, so it is necessary to refer orbital elements to an epoch (epoch of the year 1950, for instance).

And there are several other orbital elements which, although they can be derived from the essential six, are useful to have conveniently supplied. Most interesting of these is the *P*—the orbital period in years. Also interesting is the *Q*—the aphelion (outermost point from the Sun) distance in astronomical units.

Appendix F

COMET BRIGHTNESS FORMULA AND ESTIMATING BRIGHTNESS

The standard formula for predicting a comet's brightness is usually written:

$$m_1 = m_0 + 5 \log \Delta + 2.5n \log \mathbf{r}$$

where m_1 is the apparent visual magnitude of the comet, m_0 is the absolute magnitude of the comet, Δ (delta) is the comet's distance from the Earth expressed in astronomical units (a.u.), n is what might be called "the brightening factor," and r is the comet's distance from the Sun expressed in astronomical units.

The absolute magnitude of a comet is the apparent magnitude it would have if placed both 1 a.u. from the Sun and 1 a.u. from the Earth. The brightening factor n is the figure that expresses how rapidly a comet is brightening as it approaches the Sun. The higher the value of n, the more rapidly a comet brightens. A poor n would be 2 (identical with that of a merely reflecting body); a very high n would be 8.

When a comet is discovered and we have the apparent magnitude at the sighting, the remaining unknowns are m_0 and n, and the practice is to assume $n = 4$ (approximately the average for comets) and solve the equation for m_0. Therefore, $2.5n$ in the equation becomes $2.5 \times 4 = 10$. The next step is to use the tentative m_0 to calculate m_1 in the days ahead and see roughly how bright the comet ought to become.

The only problem is that m_0 and n are not really constants. Neither the intrinsic brightness not the brightening factor of a comet really behaves in a completely regular fashion. For that matter, when a comet has only been observed for a relatively short time there can be a number of combinations of m_0 and n that can fit the observations equally well. Absolute magnitude and the brightening factor can only be regarded as useful, not perfect, guides for helping to predict and characterize a comet's brightness and brightness behavior during an apparition, or during part of one.

Predicting cometary brightness can be complicated and imperfect, but estimating observed brightness is not simple or foolproof either. The observer is comparing the brightness of a usually diffuse object (a comet) to point sources (stars). In the text we noted that the "Bobrovnikoff method" of brightness estimation was a common and relatively easy one. In addition to this method, however, several others have been devised: all are typically for use with binoculars or a telescope, and all involve placing at least one of the objects, star or comet, out of focus:

Sidgwick Method (In–Out). Memorize the appearance of the in-focus comet and compare it to stars placed out of focus until their blurred images attain a size similar to that of the in-focus comet.

Bobrovnikoff Method (Out–Out). Comet and stars are both placed out of focus until they are nearly the same size, and then their brightness is compared.

Morris Method. Memorize the slightly out-of-focus image of the comet and then compare it to out-of-focus images of stars when they become the same size as the memorized image of the comet.

Beyer Method. Both comet and star are put out of focus until they disappear, and the brightness comparison is then made according to the order and timing of their disappearances.

The Beyer method is the most difficult to use, most sensitive to sky brightness, and is neither used or recommended by today's leading observers. Of the other methods, the Bobrovnikoff is the only one that does not require memorization of images.

Finally, the matter of how bright a comet will be to an observer is complicated by what is called "aperture effect." This is the dependence of perceived cometary brightness on the aperture (i.e., diameter of the light-gathering mirror or lens) of an optical instrument. The larger the aperture, the dimmer a comet appears relative to the background stars. It is not really the larger aperture per se but the higher magnification that goes (almost) inevitably with larger aperture which causes this effect. The higher magnification spreads the already diffuse and soft-edged image over a larger apparent visual angle, reducing the apparent surface brightness and rendering the outer coma invisible. Technically, brightness predictions are for an aperture of 6.8 cm (because that was the average of the apertures used in a study of the aperture effect made by Bobrovnikoff in the 1940s).

The practical upshot of all of this about the aperture effect is that comet observers should always make magnitude estimates using the lowest-magnification optical instrument capable of detecting the comet. If the comet becomes bright enough, the ultimate instrument for making brightness estimates is the naked eye.

Appendix G
HOW TO REPORT A COMET DISCOVERY

So you want to know how to report a comet you've discovered? First you need to know when to report a comet you've discovered. Or, rather, when *not* to report.

When Not to Report a "New Comet"

Suppose you are observing a fuzzy patch of light among sharp images of stars in your telescope. You have moved your telescope a bit and the fuzzy patch doesn't suddenly change its position radically compared to the stars, so it is not a ghost image of a reflection from some brighter object in your field of view (or just outside your field of view). You have tried higher magnifications and the patch doesn't turn out to be a close-together pair or group of faint stars. You have consulted a star atlas (like *Sky Atlas 2000.0* or *Uranometria*)—preferably two (all atlases have errors and omissions)—but find that the atlas does not show any "deep-sky object" (any galaxy, nebula, double star, or star cluster) in your mystery object's location. Now comes the most important test of all: motion. After 15, 30, maybe 60 minutes, do you detect a change in the position of the fuzzy patch in relation to the surrounding stars? If so, you can be virtually certain you are seeing a comet.

But there is still a chance it is not a *new* comet. It could be a known comet of which you simply were unaware. There are a number of sources you can consult to see if any known comet is supposed to be in your mystery object's location that night: IAU Circulars, International Comet Quarterly's *Comet Handbook*, Charles Morris's JPL *Comet Observation Home Page,* and (at least for comets brighter than 11th magnitude) *Comet Rapid Announcement Service* (see Sources of Information for the addresses of some of these). Remember that it is always possible that a known comet predicted to be faint could have undergone an unexpected outburst and brightened enough to be detected by you (of course, a major outburst that you might have been the first to see is itself worth reporting). The most definitive way to determine if your mystery object is a known comet is to subscribe to the CBAT (Central Bureau for Astronomical Telegrams) computer service so that you can (as Alan Hale did when he saw Hale-Bopp) access the program to identify all known comets and asteroids in the area of your object.

How far down this checklist should you go before reporting your mystery object as a comet you've discovered?

Every person must make this decision for him- or herself. But whether you are seeking glory or to help science, you will find that not checking your object carefully is very likely to lead to the opposite of both: you'll get shame (or at least some tarnish on whatever reputation you hope to establish) and you will impede the work of scientists by wasting their time.

Remember that about 90 percent of the reports of new comets coming in to the CBAT prove to be false—ghost images, sightings of two different deep-sky objects believed to have been one object that moved, etc.

Are you satisfactorily warned now? You will be careful before claiming a comet. But let's say your mystery patch of light does pass all the checks. How do you report your discovery?

How to Report a New Comet

The place to contact is the CBAT, the Central Bureau for Astronomical Telegrams, at the Smithsonian Astrophysical Observatory of Harvard University in Cambridge, Massachusetts. You should not make a phone call. The prime method for contacting the bureau used to be by sending a telegram. But the bureau's TWX number was discontinued for lack of use on July 1, 1995. Now the appropriate way to send notice of what you think is a new comet is by e-mail. You send your message to both:

marsden@cfa.harvard.edu
and green@cfa (.span or .harvard.edu)

Suppose you have no access to a computer or e-mail? Then you will need to contact a local observatory or planetarium or college astronomy or physics department, or contact a fellow amateur astronomer who does have e-mail. Actually, to contact other astronomers about your alleged discovery is a very good idea even if you can send e-mail. Doing so permits other observers to check your object and confirm what you think you are seeing, before the CBAT is contacted.

What information should you give in your e-mail to Brian Marsden and Dan Green?

First, state which kind of object you suspect you have found (comet, nova, asteroid, etc.).

Second, the object's right ascension and declination, preferably in epoch 2000 coordinates (you should specify the epoch of the star atlas or other source you used for determining the right ascension and declination).

Third, the direction and rate of motion. The best way to give this is with a second position, preferably as many minutes as possible after the first position.

Fourth, your estimate of the comet's magnitude.

Fifth, whether the discovery was visual or photographic.

Sixth, a description of the object, including any comments about angular size, degree of condensation, or the tail.

Seventh, the time and date of the observation, preferably given in Universal Time.

Eighth, the optical instrument used: include the aperture (diameter of the primary lens or mirror), type (refractor, reflector, etc.), magnification used.

Ninth, your full name, address, and phone number.

Appendix H

EPHEMERIS FOR COMET HALE-BOPP

The following is a special ephemeris for Comet Hale-Bopp calculated by Steve Albers. Tail lengths and magnitudes are speculative.

Explanation:

DATE Universal time (UT) of date. Time given is either 0h UT or the time of the day that is most favorable for observing the comet from the central U.S.

DEC,RA Declination and right ascension: Epoch 2000.

DELTA,R Comet–Earth and Comet–Sun distances in astronomical units (au).

MAG Apparent magnitude.

ELONG Elongation of Comet from Sun in degrees.

TAIL Tail length in degrees assuming actual length according to a more recent version of Andreas Kammerer's formula (ICQ 16, pp. 144–148, Oct. 1994).

TPA Position angle toward which the tail should point (assuming it points directly away from the Sun).

TLAU Tail length in AU according to the calculations in TAIL.

ALDF Comet minus Sun altitude when comet is 3 degrees above horizon. Useful for bright comets embedded in twilight.

AZDF Comet minus Sun azimuth when comet is 3 degrees above horizon.

RISE Rise time (LST: Local Standard Time)

TRAN   Transit Time (LST): often the best viewing time.

SET   Set Time (LST)

NOTE: RISE, TRAN, and SET times are valid at the listed UT. These may have to be interpolated to the actual UT of the RISE, TRAN, and SET to obtain more accurate values.

TWI   Start or end time (LST) of twilight if there is a limiting factor in the comet's visibility. Solar depression angle of 16 degrees is used. "S" or "R" denotes evening or morning object, respectively.

MOON   Moonrise or moonset time (LST) if moon interferes in the comet's visibility. Moon is considered bright when EMC < 1.8 × (EMS 40) or EMC < 6.0 × (EMS 60). "R"/"S" denotes moon rise/set, "T" denotes that due to the moon's interference the best viewing occurs at the "opposite" twilight from what one would expect if the moon were not a factor. The **best** time for viewing a comet is TRANsit time. This is superseded by TWIlight time (if given). The first two are superseded by MOON time (if given). This is when the comet has the highest altitude (AL2) in a dark moonless sky. The best time is generally the same as the UT time.

AL1   Altitude of a comet at best time for viewing, taking twilight into account. Highest altitude comet can be seen in a twilight-free sky.

AL2   Altitude of a comet at best time for viewing, taking twilight and moonlight into account. Highest altitude comet can be seen in a sky free of both twilight and moonlight. If AL1 > 0 and AL2 = 0, then the comet's visibility is strongly hampered by moonlight.

AZ1   Azimuth of a comet at the best time for viewing, taking twilight into account.

AZ2   Azimuth of a comet at the best time for viewing, taking twilight and moonlight into account (given when comet's visibility is strongly hampered by moonlight).

EMS   Elongation of the moon from the sun (corresponds to moon's phase).

EMC   Elongation of the moon from the Comet.

(Ephemeris calculated by Steve Albers. Orbital Elements from the Comet Observation Home Page which are in turn from IAU Circulars. The Hale-Bopp ephemeris uses elements having an epoch in 1997.)

YEAR DATE	RA	DEC	DELTA	R	MAG	ELONG	TAIL	TPA	TLAU	RISE	TRAN	S ET	TWI	MOON	AL1	AL2	AZ1	AZ2	EMS	EMC
1996 NOV 15.05	17 51.2	− 2 19	3.021	2.349	4.4	39.9	1.5	63	0.33	8:18	14:13	20:09	18:07S		22	22	247		50	23
1996 NOV 17.05	17 52.8	− 2 8	3.013	2.324	4.3	38.7	1.5	62	0.34	8:11	14:07	20:03	18:06S		21	0	248		76	46
1996 NOV 19.05	17 54.5	− 1 57	3.004	2.299	4.3	37.5	1.5	60	0.34	8:05	14:01	19:57	18:05S		20	0	249		103	73
1996 NOV 21.04	17 56.3	− 1 46	2.993	2.275	4.3	36.4	1.5	58	0.35	7:58	13:55	19:52	18:04S		20	0	250		129	99
1996 NOV 23.04	17 58.1	− 1 34	2.982	2.250	4.2	35.3	1.5	56	0.35	7:51	13:49	19:46	18:03S		19	0	251		154	125
1996 NOV 25.04	17 59.9	− 1 21	2.970	2.225	4.2	34.3	1.6	53	0.36	7:44	13:43	19:41	18:02S		18	0	252		175	149
1996 NOV 27.04	18 1.8	− 1 8	2.957	2.200	4.1	33.3	1.6	51	0.37	7:38	13:37	19:36	18:01S		17	17	253		157	163
1996 NOV 29.04	18 3.8	− 0 54	2.943	2.175	4.1	32.4	1.6	49	0.37	7:31	13:31	19:31	18:01S		16	16	254		135	153
1996 DEC 1.04	18 5.8	− 0 40	2.928	2.150	4.0	31.5	1.6	46	0.38	7:24	13:25	19:26	18:01S		16	16	256		113	133
1996 DEC 3.04	18 7.9	− 0 25	2.912	2.125	4.0	30.7	1.6	44	0.39	7:18	13:19	19:21	18:01S		15	15	257		92	111
1996 DEC 5.04	18 10.0	− 0 10	2.894	2.100	3.9	30.0	1.6	41	0.40	7:11	13:13	19:16	18:01S		14	14	258		70	89
1996 DEC 7.04	18 12.2	+ 0 7	2.876	2.075	3.9	29.3	1.6	38	0.40	7:04	13:08	19:11	18:01S		13	13	259		47	66
1996 DEC 9.04	18 14.5	+ 0 24	2.857	2.049	3.8	28.8	1.7	35	0.41	6:58	13:02	19:06	18:01S		12	12	260		23	42
1996 DEC 11.04	18 16.8	+ 0 41	2.837	2.024	3.8	28.3	1.7	32	0.42	6:51	12:57	19:02	18:01S		11	11	262		6	22
1996 DEC 13.04	18 19.1	+ 0 60	2.816	1.999	3.7	27.8	1.7	29	0.43	6:45	12:51	18:57	18:02S		10	10	263		31	26
1996 DEC 15.04	18 21.5	+ 1 19	2.794	1.974	3.6	27.5	1.8	26	0.44	6:38	12:46	18:53	18:02S		9	9	264		58	50
1996 DEC 17.04	18 24.0	+ 1 39	2.771	1.948	3.6	27.2	1.8	23	0.45	6:32	12:40	18:49	18:03S		8	0	265		85	76
1996 DEC 19.04	18 26.5	+ 2 0	2.747	1.923	3.5	27.0	1.9	19	0.46	6:25	12:35	18:44	18:04S		7	0	267		111	101
1996 DEC 21.04	18 29.1	+ 2 22	2.722	1.898	3.5	26.9	2.0	16	0.47	6:18	12:29	18:40	18:05S		6	0	268		136	125
1996 DEC 23.05	18 31.8	+ 2 45	2.696	1.872	3.4	26.9	2.0	13	0.48	6:12	12:24	18:36	18:06S		5	0	269		159	147
1996 DEC 25.05	18 34.5	+ 3 9	2.669	1.847	3.3	26.9	2.1	10	0.49	6:05	12:19	18:33	18:07S		4	0	270		174	159
1996 DEC 27.05	18 37.3	+ 3 35	2.641	1.822	3.3	27.1	2.2	6	0.50	5:59	12:14	18:29	18:08S		3	3	272		154	150
1996 DEC 29.05	18 40.1	+ 4 1	2.613	1.796	3.2	27.3	2.3	3	0.51	5:52	12:09	18:25	18:09S		3	3	273		133	132
1996 DEC 31.05	18 43.0	+ 4 28	2.583	1.771	3.1	27.6	2.5	0	0.52	5:46	12:04	18:22	18:11S		2	2	275		112	112
1997 JAN 2.05	18 46.0	+ 4 57	2.553	1.746	3.0	27.9	2.6	358	0.53	5:39	11:59	18:19	5:56R	18:12T	3	1	86	276	90	90
1997 JAN 4.54	18 49.8	+ 5 34	2.514	1.714	3.0	28.4	2.8	354	0.55	5:31	11:53	18:15	5:56R		4	4	86		61	63
1997 JAN 6.54	18 53.0	+ 6 6	2.482	1.689	2.9	28.9	3.0	352	0.56	5:25	11:48	18:12	5:56R		5	5	87		36	40
1997 JAN 8.54	18 56.2	+ 6 39	2.449	1.664	2.8	29.4	3.2	349	0.57	5:18	11:44	18:09	5:56R		7	7	87		9	26
1997 JAN 10.54	18 59.6	+ 7 13	2.415	1.638	2.7	29.9	3.4	347	0.58	5:12	11:39	18:06	5:56R		8	8	87		20	35
1997 JAN 12.54	19 3.0	+ 7 49	2.381	1.613	2.6	30.5	3.6	344	0.60	5:05	11:35	18:04	5:56R		9	9	87		47	57
1997 JAN 14.54	19 6.5	+ 8 27	2.346	1.588	2.6	31.2	3.8	342	0.61	4:59	11:30	18:02	5:55R		10	10	88		73	82
1997 JAN 16.54	19 10.1	+ 9 7	2.310	1.563	2.5	31.8	4.1	340	0.63	4:52	11:26	18:00	5:55R		11	11	88		98	105
1997 JAN 18.54	19 13.9	+ 9 48	2.274	1.538	2.4	32.5	4.4	338	0.64	4:46	11:22	17:58	5:54R		12	12	88		122	126

(continued)

YEAR	DATE	RA	DEC	DELTA	R	MAG	ELONG	TAIL	TPA	TLAU	RISE	TRAN	SET	TWI	MOON	AL1	AL2	AZ1	AZ2	EMS	EMC
1997 JAN	10.54	19 17.8	+10 31	2.237	1.513	2.3	33.2	4.7	337	0.66	4:39	11:18	17:57	5:53R		14	14	88		145	144
1997 JAN	22.54	19 21.7	+11 16	2.199	1.488	2.2	34.0	5.0	335	0.67	4:32	11:14	17:55	5:53R		15	0	87		166	152
1997 JAN	24.54	19 25.9	+12 3	2.162	1.464	2.1	34.7	5.4	334	0.69	4:26	11:10	17:55	5:52R		16	0	87		170	146
1997 JAN	26.53	19 30.2	+12 52	2.123	1.439	2.0	35.4	5.8	332	0.71	4:19	11:07	17:54	5:50R		17	0	87		149	130
1997 JAN	28.53	19 34.6	+13 44	2.085	1.415	1.9	36.2	6.2	331	0.73	4:13	11:03	17:54	5:49R		17	0	87		127	111
1997 JAN	30.53	19 39.2	+14 38	2.046	1.391	1.8	36.9	6.6	330	0.74	4:06	11:00	17:54	5:48R		18	0	86		105	92
1997 FEB	1.53	19 44.0	+15 34	2.007	1.367	1.7	37.7	7.1	329	0.76	4:00	10:57	17:54	5:46R		19	0	86		81	71
1997 FEB	3.53	19 49.0	+16 33	1.968	1.343	1.6	38.4	7.6	328	0.78	3:53	10:54	17:55	5:44R		20	20	85		56	52
1997 FEB	5.53	19 54.3	+17 34	1.929	1.319	1.5	39.1	8.2	327	0.80	3:47	10:51	17:56	5:43R		21	21	84		29	37
1997 FEB	7.53	19 59.8	+18 38	1.890	1.296	1.4	39.8	8.7	327	0.82	3:40	10:49	17:58	5:41R		22	22	83		3	37
1997 FEB	9.53	20 5.5	+19 45	1.851	1.273	1.3	40.5	9.3	326	0.84	3:33	10:47	18:00	5:39R		22	22	82		27	53
1997 FEB	11.53	20 11.6	+20 54	1.812	1.251	1.2	41.1	10.0	326	0.87	3:27	10:45	18:03	5:37R		23	23	81		54	74
1997 FEB	13.52	20 18.0	+22 7	1.774	1.228	1.1	41.8	10.7	326	0.89	3:20	10:44	18:07	5:35R		23	23	80		80	94
1997 FEB	15.52	20 24.8	+23 22	1.736	1.206	1.0	42.3	11.4	326	0.91	3:14	10:42	18:11	5:32R		24	24	79		103	112
1997 FEB	17.52	20 32.0	+24 40	1.699	1.185	0.9	42.9	12.1	326	0.93	3:08	10:42	18:16	5:30R		24	24	77		126	127
1997 FEB	19.52	20 39.6	+26 0	1.663	1.164	0.8	43.4	12.9	326	0.96	3:01	10:41	18:22	5:27R		25	25	76		147	138
1997 FEB	21.52	20 47.7	+27 23	1.627	1.143	0.7	43.9	13.7	326	0.98	2:55	10:42	18:28	5:25R		25	0	74		169	141
1997 FEB	23.52	20 56.2	+28 48	1.593	1.123	0.6	44.3	14.5	327	1.00	2:49	10:43	18:36	5:22R		25	0	73		169	135
1997 FEB	25.51	21 5.6	+30 15	1.560	1.103	0.5	44.7	15.4	328	1.03	2:43	10:44	18:45	5:19R		25	0	71		147	123
1997 FEB	27.51	21 15.5	+31 44	1.529	1.085	0.4	45.0	16.2	329	1.05	2:37	10:46	18:55	5:17R		25	0	69		124	108
1997 MAR	1.51	21 26.2	+33 13	1.499	1.066	0.3	45.3	17.1	330	1.07	2:31	10:49	19:06	5:14R		25	0	67		101	92
1997 MAR	3.09	21 35.2	+34 24	1.476	1.053	0.2	45.4	17.8	331	1.09	2:27	10:51	19:16	5:11R		25	0	65		81	79
1997 MAR	5.51	21 49.9	+36 12	1.444	1.032	0.1	45.6	18.8	333	1.12	2:21	10:57	19:33	5:08R		24	24	62		49	60
1997 MAR	7.50	22 3.1	+37 39	1.420	1.017	0.0	45.7	19.7	336	1.14	2:16	11:02	19:48	5:05R		24	24	60		22	49
1997 MAR	9.50	22 17.3	+39 3	1.399	1.002	−0.1	45.7	20.4	338	1.17	2:12	11:08	20:05	5:02R		23	23	57		7	46
1997 MAR	11.50	22 32.4	+40 22	1.380	0.988	−0.1	45.7	21.2	341	1.19	2:08	11:16	20:23	4:58R		22	22	55		34	53
1997 MAR	13.50	22 48.5	+41 36	1.364	0.975	−0.2	45.6	21.9	344	1.21	2:05	11:24	20:42	4:55R		21	21	53		60	66
1997 MAR	15.49	23 5.6	+42 42	1.351	0.963	−0.3	45.4	22.5	347	1.23	2:04	11:33	21:02	4:52R		20	20	50		84	80
1997 MAR	17.49	23 23.6	+43 39	1.340	0.952	−0.3	45.2	23.0	351	1.24	2:04	11:43	21:23	4:48R		18	18	48		106	95
1997 MAR	19.49	23 42.3	+44 25	1.333	0.943	−0.4	44.9	23.4	355	1.26	2:06	11:54	21:42	4:45R		17	17	45		128	108
1997 MAR	21.49	0 1.7	+44 59	1.329	0.935	−0.4	44.6	23.7	359	1.27	2:10	12:05	22:01	4:42R		15	15	43		150	121
1997 MAR	23.11	0 17.7	+45 17	1.328	0.929	−0.4	44.3	23.8	3	1.28	2:15	12:15	22:15	19:35S		16	0	317		169	131
1997 MAR	25.11	0 37.6	+45 28	1.330	0.923	−0.4	43.9	23.9	8	1.29	2:25	12:27	22:29	19:37S		17	0	316		169	139

(continued)

357

YEAR DATE	RA	DEC	DELTA	R	MAG	ELONG	TAIL	TPA	TLAU	RISE	TRAN	S ET	TWI	MOON	AL1	AL2	AZ1	AZ2	EMS	EMC
1997 MAR 27.11	0 57.4	+45 25	1.335	0.918	−0.5	43.5	23.9	12	1.30	2:38	12:39	22:40	19:39S		19	19	314		146	140
1997 MAR 29.11	1 16.9	+45 9	1.343	0.915	−0.4	43.0	23.7	17	1.31	2:53	12:51	22:48	19:42S		20	20	313		123	132
1997 MAR 31.11	1 35.9	+44 40	1.354	0.913	−0.4	42.4	23.4	21	1.31	3:10	13:02	22:53	19:44S		21	21	312		99	118
1997 APR 2.12	1 54.2	+43 60	1.368	0.913	−0.4	41.9	23.0	26	1.31	3:29	13:12	22:56	19:46S		21	21	310		73	99
1997 APR 4.12	2 11.6	+43 10	1.384	0.914	−0.4	41.3	22.4	30	1.31	3:47	13:22	22:56	19:49S		22	22	309		47	78
1997 APR 6.12	2 28.0	+42 12	1.403	0.917	−0.3	40.7	21.8	35	1.31	4:06	13:30	22:54	19:51S		22	22	307		19	56
1997 APR 8.12	2 43.4	+41 6	1.424	0.920	−0.3	40.0	21.1	39	1.30	4:24	13:38	22:52	19:54S		22	22	306		9	36
1997 APR 10.12	2 57.9	+39 56	1.448	0.926	−0.2	39.4	20.4	43	1.29	4:40	13:44	22:48	19:56S		22	22	304		35	27
1997 APR 12.12	3 11.3	+38 41	1.473	0.932	−0.2	38.7	19.5	46	1.28	4:56	13:50	22:44	19:59S		22	22	303		60	37
1997 APR 14.13	3 23.8	+37 23	1.500	0.940	−0.1	38.0	18.7	50	1.26	5:10	13:54	22:39	20:01S		22	0	301		83	54
1997 APR 16.13	3 35.4	+36 3	1.528	0.949	0.0	37.3	17.8	53	1.25	5:23	13:58	22:33	20:04S		21	0	300		106	74
1997 APR 18.13	3 46.1	+34 42	1.558	0.959	0.0	36.5	16.9	57	1.23	5:34	14:01	22:27	20:07S		21	0	299		128	95
1997 APR 20.13	3 56.1	+33 21	1.588	0.971	0.1	35.8	16.0	60	1.21	5:45	14:03	22:21	20:09S		20	0	298		150	116
1997 APR 22.13	4 5.5	+32 0	1.620	0.983	0.2	35.0	15.2	63	1.19	5:54	14:04	22:15	20:12S		19	0	297		172	138
1997 APR 24.14	4 14.1	+30 40	1.652	0.997	0.3	34.3	14.3	66	1.17	6:02	14:05	22:09	20:15S		18	0	296		164	159
1997 APR 26.14	4 22.3	+29 21	1.685	1.012	0.4	33.5	13.4	69	1.15	6:09	14:05	22:02	20:17S		16	16	295		140	165
1997 APR 28.14	4 29.9	+28 3	1.718	1.027	0.5	32.7	12.6	71	1.13	6:15	14:05	21:55	20:20S		15	15	294		115	145
1997 APR 30.14	4 37.1	+26 47	1.752	1.044	0.6	31.9	11.8	74	1.11	6:21	14:04	21:48	20:23S		14	14	294		89	120
1997 MAY 2.14	4 43.8	+25 33	1.786	1.061	0.7	31.2	11.0	77	1.08	6:25	14:03	21:41	20:26S		12	12	293		63	94
1997 MAY 4.14	4 50.2	+24 20	1.820	1.079	0.7	30.4	10.3	80	1.06	6:29	14:02	21:34	20:28S		11	11	293		37	66
1997 MAY 6.15	4 56.2	+23 9	1.854	1.097	0.8	29.7	9.6	83	1.03	6:32	14:00	21:27	20:31S		9	9	293		11	39
1997 MAY 8.15	5 2.0	+21 60	1.887	1.117	0.9	28.9	9.0	86	1.01	6:35	13:58	21:20	20:34S		7	7	293		17	13
1997 MAY 10.15	5 7.4	+20 52	1.921	1.137	1.0	28.2	8.4	89	0.99	6:37	13:55	21:13	20:37S		6	6	293		41	13
1997 MAY 12.15	5 12.8	+19 47	1.955	1.157	1.1	27.5	7.8	91	0.96	6:39	13:53	21:06	20:40S		4	0	293		64	37
1997 MAY 14.15	5 17.7	+18 43	1.988	1.178	1.2	26.8	7.3	94	0.94	6:41	13:50	20:59	20:42S		2	0	293		86	60
1997 MAY 16.16	5 22.5	+17 40	2.021	1.199	1.3	26.2	6.8	97	0.92	6:41	13:47	20:52	20:45S		1	0	293		108	83
1997 MAY 18.00	5 26.7	+16 44	2.050	1.220	1.4	25.6	6.4	100	0.90	6:42	13:44	20:45			0	0	0		129	105
1997 MAY 20.00	5 31.2	+15 44	2.082	1.242	1.5	25.0	5.9	104	0.87	6:42	13:40	20:38			0	0	0		152	129
1997 MAY 22.00	5 35.5	+14 46	2.114	1.264	1.6	24.5	5.5	107	0.85	6:42	13:37	20:31			0	0	0		174	153
1997 MAY 24.00	5 39.7	+13 49	2.145	1.287	1.7	24.0	5.2	110	0.83	6:42	13:33	20:24			0	0	0		159	175
1997 MAY 26.00	5 43.7	+12 54	2.176	1.310	1.8	23.5	4.8	114	0.81	6:42	13:29	20:17			0	0	0		133	153
1997 MAY 28.00	5 47.7	+11 59	2.206	1.334	1.9	23.1	4.5	117	0.79	6:41	13:25	20:09			0	0	0		108	127
1997 MAY 30.00	5 51.5	+11 6	2.235	1.358	1.9	22.7	4.3	121	0.77	6:40	13:21	20:02			0	0	0		82	100
1997 JUN 1.00	5 55.2	+10 14	2.264	1.381	2.0	22.4	4.0	124	0.75	6:39	13:17	19:55			0	0	0		56	73

GLOSSARY

albedo   The reflectivity of an object.

anti-tail   A comet tail that, because of our viewing angle, appears to point toward the Sun rather than away from it.

aphelion   The outermost point in an orbit around the Sun (the *perihelion* is the innermost point).

apparition   The display that a celestial object puts on during the time it is observed. (Periodic comets return after years of unviewability for additional apparitions.)

ascending node   The place where the plane of one object's orbit cuts northward through the plane of a second object's orbit. The second object is typically Earth, so a comet at its ascending node is passing northward through the plane of Earth's orbit—perhaps far inside or outside of Earth's orbit itself. (The *descending node* is the place where the plane of the comet's orbit passes southward through the plane of Earth's orbit.)

asteroids (also known as **minor planets**)   Predominantly rocky bodies smaller than the major planets and mostly found between the orbits of Mars and Jupiter.

asteroid belt   A broad region between the orbits of Mars and Jupiter in which a majority of the asteroids are found.

astrobleme   A giant impact crater.

astronomical unit or **a.u.**   A unit of distance equal to the average distance of Earth from the Sun. One astronomical unit is about 92,955,807 miles, or 149,597,870 kilometers.

averted vision   Technique of looking just a bit away from a faint object so as to bring its image to the most light-sensitive area of the retina.

Bobrovnikoff method   One of several methods for estimating the brightness of a comet. The observer defocuses the images of stars near the comet until they are similar in size to the comet's image and then compares the brightness of the comet to that of stars of known brightness.

celestial pole   Point in the heavens directly over the north (or south) pole of Earth (the star Polaris is located quite near the north celestial pole, but in our era of history no bright star is near the south celestial pole).

central condensation   The more concentrated area in the midst of a comet's coma. (In a very diffuse coma, no central condensation is seen; in some comets a smaller and even more intense *false nucleus* appears.)

circumpolar   Close enough to a celestial pole so as never to rise or set but instead to complete an entire circling of the pole above the horizon. (Which constellations and comets appear circumpolar depends on one's latitude: The Big Dipper scrapes just above the northern horizon at its low point as seen from about 40° N latitude.)

clathrate hydrate   A substance in which a crystalline structure contains pockets of other chemical molecules and atoms. The ice of a comet nucleus is an example.

coma   The cloud of gas and dust that forms around the solid *nucleus* of a comet when it is heated by the Sun (or, in some cases, when it is worked on by other forces, such as meteorite impact or cosmic ray bombardment). The nucleus and coma together form the *head* of a comet.

comets   Predominantly icy bodies, tremendously less massive than planets, that produce clouds of gas and dust when sufficiently heated by the Sun (or, occasionally, by other forces).

comet seeker   Old-fashioned name for a telescope especially suited for use in the observation and discovery of comets.

comet shower   A hypothetical barrage of perhaps thousands of comets at a time passing through the inner solar system every year for maybe a million years.

contact surface   In a comet, the boundary separating the region closer to the nucleus that is pure cometary plasma from the region outside where the solar wind mixes with the cometary plasma.

dark adaptation   The process in which the rod cells in our retinas become chemically more sensitive to dim light over the passage of time (largely the first 20 or 30 minutes) in a low-light environment.

daughter molecules   See *parent molecules*.

DC   Abbreviation for *degree of condensation*.

DE   Abbreviation for *disconnection event*.

declination   The coordinate for the starry heavens (the "celestial sphere") that corresponds to latitude on the Earth. Thus the north celestial pole corresponds to the north pole on Earth, though instead of being said to lie at 90° N like Earth's north pole, it is said to lie at +90°. (Declinations south of the celestial equator are preceded by a minus sign.) See also *right ascension*.

deep-sky objects   Astronomical objects—such as galaxies, nebulae, and star clusters—that are beyond our solar system.

degree of condensation   The degree to which the light of a comet's coma is concentrated toward its center. This *DC* ranges from 0 (completely diffuse, no brightening toward center) to 9 (entire coma star-like).

degrees (and minutes and seconds) of arc   Angular measure in which a circle around the entire heavens above and below the horizon can be divided into 360 degrees of arc, in which 1 degree can be divided into 60 minutes (') of arc and in which 1 minute can be divided into 60 seconds (") of arc.

descending node   See *ascending node* for definition.

"dirty snowball" model   Fred Whipple's model of the comet nucleus as a solid body coposed of ice (mainly water ice) with embedded dust (mostly silicate).

disconnection event   A spectacular detachment and flying off of a comet's current gas tail (quickly replaced) that may occur because the comet passes rapidly through a boundary between sectors where the interplanetary magnetic field is oriented in different directions. Sometimes called a *DE* for short.

dust tail   The broad, curving fan of dust that solar radiation pressure pushes away from a comet's coma. (Also called a *Type II tail.*)

eccentricity   An *orbital element* that defines how elongated or even how open-ended the orbit of a comet (or other astronomical object) is. A circle has an eccentricity of 0, an ellipse of between 0 and 1, a parabola of 1, and a hyperbola of greater than 1.

ecliptic   The line that marks the Sun's apparent path around the starry heavens during the year. It is really Earth that is doing the moving around the Sun, of course, so the *ecliptic plane* is the plane of Earth's orbit around the Sun.

ecliptic plane   See *ecliptic.*

elliptical orbit   An orbit that is a closed curve like an elongated circle, with the Sun (or other primary body around which the orbiting occurs) located at one of the two foci to either side of its center.

elongation   The angular separation (generally given in degrees) of one celestial body from another, usually that of a solar system body from the Sun in our sky.

envelopes   Shells of dust or gas expanding out from a comet's nucleus. (Also called *halos.*)

ephemeris   A table of information listed by date. In the case of comets, typically a table giving a comet's position, its predicted brightness, and sometimes its distance from the Sun and Earth and other information, on various dates.

extended radius vector   The line outward from the Sun to a comet (radius vector) extended onward beyond the comet's head (the gas tail usually lies almost along the extended radius vector).

false nucleus   An especially concentrated region in the coma of some comets, representing a densest cloud of inner coma rather than the much smaller true nucleus. (Also called *apparent nucleus* and *pseudonucleus.*)

fluorescence   Process in which a high-energy photon is absorbed by an atom or molecule and re-emitted as several photons of lower energy.

"flying sandbank" model   The model of the comet nucleus as a swarm of ice-coated dust, abandoned in favor of Whipple's *"dirty snowball" model*.

gas tail   The ionized gas component of a comet's tail, driven nearly straight away from the Sun by the solar wind. (Also called *ion tail*, *plasma tail*, and *Type I tail*.)

halos   See *envelopes*.

head   The *nucleus* and surrounding *coma* of a comet.

hydrogen coma (or **H coma**)   The cometary cloud of hydrogen, detectable in ultraviolet light, that is immensely bigger than even the huge visible coma it surrounds. It is produced by the dissociation of water into hydrogen and oxygen and by other processes set into motion by solar radiation and the solar wind.

hyperbolic orbit   An orbit that is an open curve whose ends get wider apart at any rate between that of an ellipse and a straight line. Some comets' orbits become hyperbolic through the gravitational influence of a planet the comet passes near.

"icy conglomerate" model   More formal name for Whipple's *"dirty snowball" model* of the comet's nucleus.

impact winter   The enormous drop in temperature and the related effects of the shrouding of Earth with soot and dust particles after the planet is struck by a sizable comet or asteroid. Such a phenomenon is believed to have killed off the dinosaurs. (Less severe but similar are the *volcanic winter* that follows certain of the greatest sulfur dioxide–producing volcanic eruptions and the deadly *nuclear winter* which could be produced by even a relatively small number of nuclear explosions in a short period of time.)

inclination   An *orbital element* that defines the angle between the plane of an object's orbit and the plane of Earth's orbit (the *ecliptic plane*).

ionization   The process by which an atom loses or gains electrons and ends up with an electric charge.

ion rays   The thin glowing streamers in a comet's *ion tail*.

ion tail   See *gas tail*.

jets   Fountain-like formations of gas and/or dust gushing out from compact regions on a comet's nulceus.

Kreutz sungrazer   Member of a family of comets (believed to have all once been a single large comet nucleus) that includes most of the comets known to pass grazingly close to the surface of the Sun and become extremely bright.

K-T event   The Cretaceous–Tertiary boundary event that destroyed the dinosaurs and a majority of other species on Earth. (This event is now believed to have been the impact of a 6-mile-wide asteroid or, perhaps more likely, a 10-mile-wide comet nucleus and its aftereffects, including a severe *impact winter*).

Kuiper Belt   A cloud of comets just at and beyond the orbits of the outermost known planets. The cloud is accurately called a belt because it is flattened to its greatest density of comets in the plane of the planets' orbits. The Kuiper Belt is believed to be the direct source of short-period (periodic) comets, the *Oort Cloud* the direct source of long-period comets.

libration Various rocking motions of objects or their orbits in astronomy. A *libration cycle* would be the period of the rocking from one extreme to the other. The orbit of Halley's Comet librates, the distance from the Sun of its ascending node and descending node changing.

light pollution Excessive or misdirected outdoor artificial lighting. It brightens the sky and has a particularly damaging effect on the observation of comets.

long-period comets Comets with orbital periods of more than 200 years. Contrast with *short-period comets* (*periodic comets*).

magnitude A measure of brightness in astronomy in which a difference of five magnitudes is 100 times, and the lower a magnitude, the brighter the object. Thus a star with a magnitude of 1.5 (brilliant to the naked eye) is five magnitudes (100 times) brighter than a star of magnitude 6.5 (on the border line of naked-eye visibility under excellent conditions).

Messier objects Over a hundred relatively bright *deep-sky objects* that Charles Messier catalogued—at least in part to prevent himself from mistaking them for comets.

meteor The streak of light caused when a *meteoroid* enters Earth's atmosphere and becomes incandescent, mostly from friction with the air at high speed.

meteorite The piece of rock or iron found on the surface of Earth when a *meteoroid* survives its trip through Earth's atmosphere as a meteor.

meteoroid A piece of silicate and/or metallic matter in space ranging in size from tiniest grains up to the width of the smallest asteroids. (There is no clear-cut dividing line between the size of the smallest asteroids and that of the largest meteoroids, but an object larger than a few hundred meters across would usually be called an asteroid.)

meteoroid stream The meteoroids distributed all along an orbit (presumably always an active or extinct comet's orbit) and diffused somewhat around it.

meteoroid swarm A relatively dense collection of meteoroids at certain spots along some meteoroid streams.

meteor shower An increased number of meteors, all appearing to diverge from the direction of a single area among the constellations. Meteor showers occur annually on the same dates, the dates when Earth crosses through a meteoroid stream.

meteor storm An extremely intense meteor shower, in which hundreds or even many thousands of meteors per hour may be observed.

noctilucent clouds The highest clouds in Earth's atmosphere, visible only during twilight, and consisting of micrometeoroidal particles (often the dust from past comet tails) coated with ice from water lofted up from the lower atmosphere.

node The crossing point of two things—in astronomy, of two orbits. (See *ascending node*.)

"nongravitational forces" The forces of *jets* from a comet's nucleus that can cause a rocket-like effect and alter a comet's direction of motion slightly.

nucleus   The solid, more permanent part of a comet; the source of the coma.

Oort Cloud   The vast cloud of comets between about 50,000 and 150,000 a.u. out from the Sun that is the source from which the supply of long-period comets entering the inner solar system is replenished.

Oort sense, new in the   Said of a comet making its first visit from the Oort Cloud (having been perturbed from its original orbit by passing stars).

orbital elements   A set of quantities that specifies the size and shape of an orbit and the times when the object on that orbit reaches key positions. The six orbital elements that are required to predict a comet's location on various dates are the eccentricty, inclination, longitude of the ascending node, argument of the perihelion, longitude of the ascending node, and perihelion date. The period (the amount of time the object takes to complete an orbit) is an orbital element, but it is not required for calculating where an object will be because it is directly related to the size of the orbit, which is already given by the other elements. For a full definition and discussion of these orbital elements, see Appendix E.

outburst   A fairly brief period of unusually strong gas and/or dust production from a comet nucleus.

outer shock front   The boundary separating the region where there is only solar wind from the region where the solar wind mixes with the comet's plasma.

PA   Abbreviation for *position angle*.

parabolic orbit   An orbit that has the shape of the open-ended curve called a parabola, the limiting case between an *elliptical orbit* (*eccentricity* less than 1) and a *hyperbolic orbit* (eccentricity ≤ 1).

parallax   The apparent shift of a nearby object's position in relation to more distant ones when the nearby object is observed from different viewing angles.

parent molecules   The molecules initially produced when a comet nucleus sublimates, soon changed to different *daughter molecules* because of the effects of solar radiation.

perigee   The nearest point to Earth in an orbit.

perihelion   The nearest point to the Sun in an orbit or, to put it another way, the innermost point in an orbit around the Sun. (The *aphelion* is the outermost point in such an orbit.) Perihelion is the most important position in most comets' orbits—the place where the comet is most affected by the Sun.

period   The amount of time a comet or other astronomical body takes to complete an orbit.

periodic comets   Comets with periods of less than 200 years. (Also called *short-period comets*.)

perturbation   Gravitational effect of a third body that causes an alteration in the orbit of a body going around the Sun.

planetesimals   The smaller bodies in the history of the solar system that in some cases clumped together to form major planets but in other cases were ejected from the solar system or remained relatively small and became asteroids (rocky planetesimals in the inner solar system) and comets (predominantly icy planetesimals in the outer solar system).

plasma   A gas composed of electrons and positive ions.

plasma tail   See *gas tail.*

position angle   The direction of a secondary body or feature (such as the dimmer member of a double star system or a comet tail) from a primary (such as the brighter member of a double star system or the central condensation of a comet), measured in the system of 0° = north, 90° = east, 180° = south, and so on to 359° and back to 0°. Abbreviated as *PA.*

precession   An ongoing, regularly recurring change in the orientation of a rotation or an orbital revolution.

radiant   The point or region among the constellations from which meteors in a meteor shower appear to diverge.

right ascension or **RA**   The coordinate for the starry heavens (the "celestial sphere") that corresponds to longitude on Earth, though it is measured somewhat differently, in 24 "hours" of RA running all the way around the sky from the "0^h" of RA that goes through the vernal equinox point in the heavens. (See Appendix D and *declination.*)

semimajor axis   The *major axis* of an orbit is the line that passes through the Sun and extends from perihelion to aphelion, bisecting the orbit; the semimajor axis is therefore half the major axis and is the mean radius of the orbit.

"shadow of the nucleus"   A dark (or, rather, less bright) lane that appears behind the coma in some comets. It is not the shadow of the true nucleus but sometimes may be a region of the near-tail that lies behind the densest part of the inner coma and therefore receives less sunlight.

short-period comets   See *periodic comets.*

skyglow   The brightening of the sky above a city (or other source of light pollution) caused by the *light pollution* scattering off dust, humidity, and air.

solar nebula   The cloud of interstellar gas and dust from which the Sun and the rest of the solar system initially formed.

solar radiation pressure   The force exerted by photons of visible light and other wavelengths of electromagnetic radiation coming from the Sun. (Solar radiation pressure is what pushes a comet's dust outward to form a dust tail.)

solar wind   The *plasma* constantly traveling outward from the Sun.

spectrogram   A photograph of an astronomical object's *spectrum.*

spectrum   A display of the distribution by wavelength of the light (or other electromagnetic radiation) of a radiant object.

spine   A very narrow line of light extending back from the coma into the tail of some comets.

sporadic meteors   *Meteors* not associated with any known *meteor shower.*

striae   Secondary *synchrones* that originate at a certain point out in some comets' dust tail, a point where for some reason the dust particles have fragmented.

sublimation   The process whereby a substance goes from a solid directly to a gaseous form.

sungrazer   A comet that passes extremely close to the Sun's surface, in some cases close enough almost literally to skim the surface.

Swan bands   Three prominent bands in the spectrum of comets, caused by diatomic carbon (C_2).

synchrone   A line connecting particles in a comet tail that left the comet at the same time.

syndyne or **syndyname**   A line connecting particles in a comet tail that left the comet at the same rate.

tail   A formation of gas and/or dust that streams away from the coma of many comets under the influence of *solar radiation pressure* and the *solar wind*.

Type I tail   The *gas tail* of a comet.

Type II tail   The *dust tail* of a comet.

Type III tail   Seemingly a third major type of comet tail seen in some comets but really just an additional, short, stubby section of the dust tail that trails farthest behind the comet's motion because it is composed of the largest silicate particles.

zenith   The point directly overhead.

zodiac   The band of constellations through which the *ecliptic* runs and in which the Sun, Moon, and planets—but not necessarily comets—are found.

SOURCES OF INFORMATION

An asterisk (*) indicates the entry contains some technical material which the novice may find challenging.

BOOKS

Chapman, Robert D., and John C. Brandt. *The Comet Book*. Boston: Jones and Bartlett, 1984.

Freitag, Ruth S. *Halley's Comet: A Bibliography*. Washington, D.C.: Library of Congress, 1984.

Hartmann, William K., and Ron Miller. *The History of Earth*. New York: Workman Publishing, 1991. For discussion of the K-T event and a few other impact- and comet-related matters.

Kronk, Gary W. *Cometography: Volume 1*. Due from Cambridge University Press in late April 1997. The remaining three volumes of this enormous descriptive catalog of every comet in history are scheduled to be published in the following few years. (A few sample entries on famous comets are available from Kronk's comet home page; see online sources listing below.)

Kronk, Gary W. *Comets: A Descriptive Catalog*. Hillside, NJ: Enslow Publishers, Inc., 1984. (This is the original work which Kronk has now greatly expanded to make the volumes of *Cometography*.)

Levy, David H. *The Quest for Comets*. New York: Avon Books, 1995.

Liller, William. *The Cambridge Guide to Astronomical Discovery*. New York: Cambridge University Press, 1992.

Marsden, Brian G. *Catalogue of Cometary Orbits,* * 6th ed. Hillside, NJ: Enslow Publishers, Inc., 1995.

Olson, Roberta J. M. *Fire and Ice: A History of Comets in Art*. New York: Walker and Company, for the National Air and Space Museum, Smithsonian Institution, 1985.

Ottewell, Guy, and Fred Schaaf. *Mankind's Comet: Halley's Comet in the Past, the Future, and Especially the Present*. Greenville, SC: Astronomical Workshop, 1985. Lots of information on comets in general, too. Can presently (1996) be ordered for $11 (half of original price), plus $3 shipping and handling from: Universal Workshop, Furman University, Greenville, SC, 29613.

Peltier, Leslie C. *Starlight Nights: The Adventures of a Star Gazer*. Cambridge, MA: Sky Publishing Corporation, 1980.

Sagan, Carl. *Pale Blue Dot*. New York: Random House, 1994. For discussion of the dangers of comet and asteroid impacts on Earth and what we should do about the threat.

Sagan, Carl, and Ann Druyan. *Comet*. New York: Random House, 1985.

Sagan, Carl, and Richard Turco. *A Path Where No Man Thought: Nuclear Winter and the End of the Arms Race*. New York: Random House, 1990. For insight into impact winter, too.

Schaaf, Fred. *The Starry Room*. New York: John Wiley & Sons, 1988. Two chapters on Halley's Comet (and three on meteor showers, fireball meteors, and meteorites).

Seargent, David A. *Comets: Vagabonds of Space*. New York: Doubleday, 1982.

Wilkening, Laurel L., with assistance of Mildred Shapley Matthews. *Comets.* * Tucson, AZ: Univ. of Arizona Press, 1982.

Yeomans, Donald K. *Comets: A Chronological History of Observation, Science, Myth, and Folklore*. New York: John Wiley & Sons, 1991.

PERIODICALS

General

Astronomy Magazine. Kalmbach Publishing, P.O. Box 1612, Waukesha, WI 53187.

Sky & Telescope. P.O. Box 9111, Belmont, MA 02178. Subscription orders by phone: 800-253-0245. Articles on important comets, plus semi-regular "Comet Digest" column by John Bortle.

Specific

International Comet Quarterly.* See online resources below.

Comet Rapid Announcement Service (CRAS). The notices of this service (conducted by an amateur astronomer, Stephen Smith) are sent out "promptly" after the discovery of any comet expected to be magnitude 11.0 or brighter and include detailed finder charts, ephemerides, and other information. Subscription to CRAS includes "The Shallow Sky Bulletin," all for $16/year. CRAS, P.O. Box 110282, Cleveland, OH 44111.

ANNUALS

Astronomical Calendar. By Guy Ottewell at Universal Workshop (for address, see book *Mankind's Comet* above). Includes each year about six atlas-sized pages of giant sky charts and orbital diagrams plus many thousands of words of commentary on all recently past, current, and soon-upcoming comets.

PHONE HOTLINE

"Skyline": 617-497-4168. This service of *Sky & Telescope* is a recorded phone message about current news and sights in astronomy which announces comet discoveries and provides information, including sky positions, for well-placed comets visible in fairly small telescopes. The message is updated each Friday, or sometimes more frequently as circumstances dictate. (A version of the "Skyline" text, usually with some illustrations or additional information, is presented online as "S&T News Bulletin"—available in America OnLine's Astronomy section or from the Internet in SKY Online, listed below).

ONLINE SERVICES

General

SKY Online. (http://www.skypub.com) Home page of *Sky & Telescope* magazine and Sky Publishing Corporation. Has a "Comets" page.

Specific

BAA Comet Section Home Page. (http://www.ast.cam.ac.uk:80/~jds/)

Chiron Information and Documentation. (http://www.vub.ac.be/STER/www.astro/chihp.htm)

Comet Hale-Bopp Home Page. (http://www.hale.bopp.com/)

Comet Hyakutake Home Page (JPL). (http://www.jpl.nasa.gov./comet/hyakutake/)

Comet Observation Home Page (JPL—Charles Morris). (http://encke.jpl.nasa.gov/index.html) The best all-around site for comet observers!

Comets and Meteor Showers (Gary Kronk). (http://medicine.wust.edu/~kronkg/index.html) Includes excerpts on several historic comets from Kronk's *Cometography*.

Comets Online. (http://fly.hiwaay.net/~cwbol/astron/comet.html)

Comet Shoemaker-Levy 9 Home Page (JPL). (http//:www.jpl.nasa.gov/sl9/)

IAU: Central Bureau for Astronomical Telegrams.* (http://cta-www.harvard.edu/cfa/ps/cbat.html)

International Comet Quarterly (ICQ). (http://cfa-www.harvard.edu/cfa/ps/icq.html)

Kuiper Belt Home Page. (http://www.ifa.hawaii.edu/~jewitt/kb.html)

STARDUST Mission Home Page. (http://pdcsrva.jpl.nasa.gov/stardust/home.html)

AUDIOVISUAL MATERIALS AND SOFTWARE

Astronomical Society of the Pacific. 390 Ashton Ave., San Francisco, CA 94112. (415)337-1100. This century-old educational organization offers a catalog with a wide variety of excellent books, posters, slide sets, computer programs, and CD-ROMs.

Sky Publishing Corporation. P.O. Box 9111, Belmont, MA 02178. Phone: (800)253-0245. E-mail: orders@skypub.com. The publishers of *Sky & Telescope* offer a catalog with books, star atlases, computer programs, CD-ROMs, slide sets, posters, and even globes (including ones of other planets).

CREDITS

Text

Illustrations on the following pages were created specially for this book by Guy Ottewell: 7, 8, 10, 13, 16, 151, 155, 216, 217, 224, 226, 233, 234, 236, 242, 243, 244, 245, 249, 250, 253, 257, 258, 292, 313, 318, 324.

Illustrations on the following pages are reproduced with permission from Guy Ottewell and Fred Schaaf, *Mankind's Comet: Halley's Comet in the Past, the Future, and Especially the Present,* © 1985 by Guy Ottewell, Greenville, SC: Astronomical Workshop (Furman University).

Illustrations on the following pages are reproduced with the kind permission of Ruth Freitag and the Library of Congress: 31, 35, 64, 87, 97, 104, 168, 183, 214, 219, 220, 223 (*top*), 239, 246.

p. 48: From Zdenek Sekanina, "Properties of Dust Particles in Comet Halley from Observations Made in 1910 . . .," NASA/ESA working group proceedings, 1981, Fig. 5.

p. 58: Guy Ottewell, *Astronomical Calendar 1995,* © 1995 by Guy Ottewell, Greenville, SC: Astronomical Workshop (Furman University).

p. 70: From George F. Chambers, *A Handbook of Descriptive and Practical Astronomy,* fourth edition, Oxford: The Clarendon Press, 1889.

p. 148: Drawings by Angelo Secchi with a 9 1/2-inch refractor at Rome on July 11 and 31, 1862, respectively. Reproduced from *Sky and Telescope,* April 1971, p. 214.

p. 182: Courtesy of Yerkes Observatory

Color Insert

1: (*top*) Dennis Milon. 20-second exposure on High-Speed Ektachrome with a 50mm f/1.9 Miranda camera. (*bottom*) Ray Maher. 90-second exposure on Kodak Gold 200.   2: (*top*) Dennis and Betty Milon. Fujica camera with 45mm f/1.8 lens. (*bottom left*) Courtesy of Space Telescope Science Institute (*bottom right*) © Gordon Garradd. Photographed at 18:32 UT from Loomberah NSW, Australia. North is down, the field approximately 1 degree, 16-minute exposure on Hypered Fuji Super HGV400 with 25cm f/4.1 telescope.   3: (*top left*) Courtesy of Ruth Freitag and the Library of Congress. (*top right*) Phil Hudson, 28mm f/3.5 lens, Konica 3200 film. (*center*) Richard A. Keen, 12-inch f/4 telescope. (*bottom*) © Paul Ostwald, 50mm f/2 lens, 1600 film.   4: (*top left*) © Paul Ostwald, 50mm f/2 lens, 1600 film. (*top right*) Dennis di Cicco, 2-minute exposure on Fuji R-100 film with 8-inch f/1.5 Schmidt camera. (*center*) Johnny Horne, 4-minute exposure on Fujicolor HGV-400 with an 8-inch Celestron Schmidt camera. (*bottom left*) Ray Maher, 20-second exposure on Kodak 1000 film. (*bottom right*) William Liller.

Black-and-White Insert

1: Courtesy of Ruth Freitag and Library of Congress.   2: (*top*) Courtesy of Ruth Freitag and Library of Congress.   3: Courtesy of Ruth Freitag and Library of Congress.   4: (*top*) Courtesy of Ruth Freitag and Library of Congress.   6: (*top*) Dennis Milon, 25-second exposure. (*bottom*) Helen and Richard Lines, 3-minute exposure on Tri-X, f/8 Newtonian focus of homemade 16-inch reflector.   7: (*bottom*) Dennis di Cicco, 30-second exposure on Tri-X with a 58mm f/1.4 lens.   8: (*top*) Bob Ross, 60-second exposure with 180mm lens using a SBig ST-6 CCD camera. (*bottom*) Dennis di Cicco, 2-minute exposure on Tri-X with a 58mm f/2 lens.   9: (*top*) Dennis di Cicco. (*bottom*) Johnny Horne, 5-minute exposure on Kodak Technical Pan with an 8-inch Schmidt astronomical camera.   10: (*top*) Glen D. Gould. (*bottom*) © Gordon Garradd. Photographed at 11:03 UT on July 24, 1995, from Loomberah NSW, Australia, with HI-SIS 22 CCD and 25cm f/4.1 telescope.   11: (*top*) © Paul Ostwald, 30-second exposure with 135mm f/4 lens on 1600 film.   12: (*top*) Max Wolf, 15-minute exposure with a 28-inch telescope. Drawing by Max Wolf. Reproduced from *Atlas of Cometary Forms: Structures Near the Nucleus*, 1969. (*bottom*) Courtesy of Space Telescope Science Institute and NASA.   13: Courtesy of Freeman Miller, University of Michigan.   14: (*top*) Courtesy of Freeman Miller, University of Michigan. (*bottom*) William Liller, photographed from Easter Island.   15: William Liller.   16: William Liller.

GENERAL INDEX

Entries for individual comets appear in the separate Comet Index following this one.

COMET INDEX

This index lists individual comets mentioned in this book. The listing is given by proper name if the comet has one, with additional designation supplied only when it is needed to distinguish the entry from other comets in the index. Comets known only by their year or year and Roman numeral in order of perihelion passage are listed first, chronologically, below.

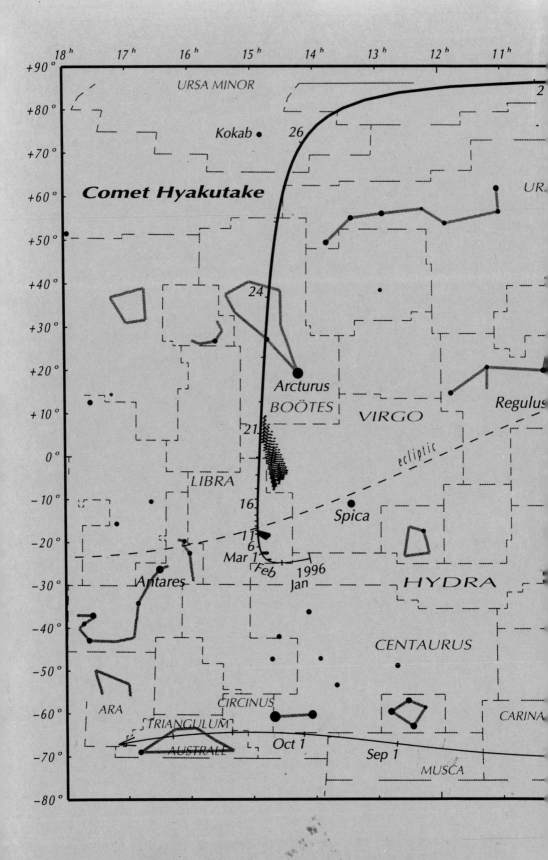